WITHDRAWN
C0-AZG-971

3 1205 00056 5679
DANVILLE PUBLIC LIBRARY

OK

AN INTRODUCTION TO PALAEONTOLOGY

FRONTISPIECE

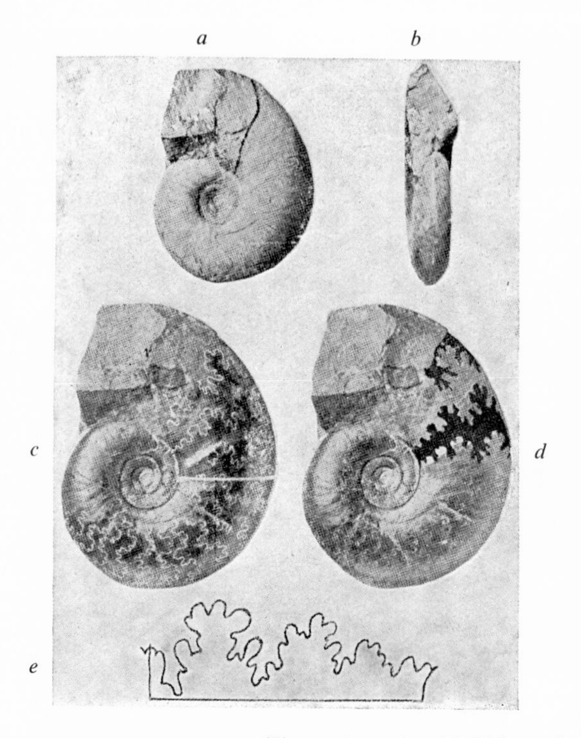

AMAUROCERAS FERRUGINEUM (Simpson), an Amaltheid from the Upper Pliensbachian of Whitby

Photographs by the late J. W. Tutcher of the holotype in Whitby Museum, reproduced from "Type Ammonites," by S. S. Buckman, vol. iii (1919).

a, Side view ($\times \frac{3}{2}$), radial line painted black; *b*, Peripheral (apertural) view ($\times \frac{3}{2}$); *c*, Side view ($\times$ 2), suture-lines strengthened on negative, guide-line in white; *d*, Side view ($\times$ 2), intervals between certain sutures painted on print; *e*, Right half of suture-line, straightened out ($\times$ 4).

A. MORLEY DAVIES

D.SC., A.R.C.S., F.G.S.

Honorary F.R.G.S., Corresponding Member, Geological Society of Belgium: Sometime Reader in Palaeontology, University of London (Imperial College of Science and Technology)

AN INTRODUCTION TO PALAEONTOLOGY

THIRD EDITION

Revised and partly re-written

by

C. J. STUBBLEFIELD

D.SC., A.R.C.S., F.G.S., F.R.S.

Director, Geological Survey of Great Britain (formerly Chief Palaeontologist)

London

THOMAS MURBY & CO.

First Published in 1920
Revised Second Edition 1947
Second Impression (Second Edition) 1949
Third Impression 1951
Fourth Impression 1956
Revised Third Edition (Sixth Impression) 1961
Seventh Impression (with minor emendations) 1962

George Allen and Unwin Ltd.
Ruskin House, Museum Street, London, W.C.1
are now the proprietors of Thomas Murby and Co.

Printed in Great Britain in 10 pt. Times Roman by Unwin Brothers Ltd., Woking and London

PREFACE TO THIRD EDITION

In the present revision, the original plan of my former teacher's book has been retained. Some reviewers and others have criticized this plan but I see adequate reasons for its retention. There is much in favour of starting a course in palaeontology with macroscopic rather than microscopic fossils. Furthermore, the long and wide outcrop of the richly fossiliferous Jurassic System in Britain, thanks to William Smith, gave impetus to the serious study of fossils. The late Dr. Davies presumably singled out the brachiopods as the first group for study because of their local abundance in the Jurassic and additionally they are plentiful in the Lower Carboniferous strata which have well-developed outcrops in the United Kingdom. Having dealt with brachiopods it is helpful for the reader to consider the other bivalved group, the lamellibranchs, whose shells are especially familiar objects on our sea-shores and in Cainozoic deposits.

The order of description of fossil animals in a supposedly biological evolutionary sequence would be subjective after dealing with the Protozoa, Porifera and Coelenterata. Some authors follow Coelenterata with Polyzoa, others with Annelida; some treat with Echinoderma before Mollusca and there is much variation in the choice of group selected to precede Chordata.

Revision has been effected in most chapters, especially in those dealing with Brachiopoda, Trilobita and other Arthropoda, Echinoderma and Coelenterata. Information new to the book has also been incorporated into the chapters on the collection and preservation of fossils and on the rules of nomenclature; concerning the latter I am grateful to Mr. R. V. Melville for advice. In the stratigraphical tables I am indebted to Dr. R. Casey for guidance concerning the Cretaceous, to Dr. M. R. House for the Devonian and to Professor O. M. B. Bulman for the graptolite zones. In these tables and elsewhere in the book I have drawn freely on the recently published volumes of the *Treatise on Invertebrate Paleontology*.

Acknowledgement is made to the Geological Survey of Great Britain and to Her Majesty's Stationery Office for the use of a few illustrations in the chapters on Brachiopoda and Trilobita from the British Regional Geology Handbooks; additionally four illustrations are acknowledged from the *Treatise* volume dealing with Trilobitomorpha.

C. J. STUBBLEFIELD

30*th December*, 1959

PREFACE TO SECOND EDITION

THE first edition of this book, published over a quarter-century ago, had two novel features: the detailed description of selected fossil species as preliminaries to the more general treatment of the groups of which they were representative; and the order in which those groups were taken—an order which, in my opinion, avoids the difficulties associated with the more usual methods of beginning with either the simplest or the most advanced organisms.

I have not altered this plan, but the advances in palaeontological science during these 26 years have necessitated considerable revision. Some portions have been cut out as unsuitable or out of date, others have been more or less completely rewritten, and there are many improvements in detail. In this I have been greatly helped by friends more familiar with recent advances than I am. In particular I must thank Dr. W. J. Arkell for suggestions on the general plan and on the tables of Jurassic stratigraphy; to Dr. O. M. B. Bulman, F.R.S., for advice on Graptolites and on methods of research; to Dr. L. F. Spath, F.R.S., for information on Cephalopoda and Mesozoic stratigraphy; Dr. C. J. Stubblefield, F.R.S., on Brachiopoda, Trilobita and Palaeozoic stratigraphy; and Dr. H. Dighton Thomas, on Corals. Responsibility for inaccuracies is my own. I am also indebted to Dr. Alan Wood for facilities for examining and drawing specimens in the Imperial College collections.

Eighteen new illustrations have been added and a number of alterations made in old ones. The new ones are partly taken from my larger work, *Tertiary Faunas*, partly drawn by myself from the fossils, partly copied from standard works. I must particularly mention the four drawings of restored extinct land-vertebrates, permission to reproduce which was readily given both by the artist, Mr. J. F. Horrabin, and by the author of the work for which they were drawn, *The Outline of History*. That author was my old friend and fellow-student, Dr. H. G. Wells, whose death as I was writing this preface deprives me, among greater things, of the satisfaction of showing him the finished work, in the preparation of which, as of my other books, he took a kindly interest.

A. MORLEY DAVIES

17*th August*, 1946

CONTENTS

ILLUSTRATIONS

I
THE BRACHIOPODA

On a large-scale geological map of England there is seen a narrow, sinuous band of colour traceable almost continuously from the Dorset coast to that of Yorkshire, explained on the index as "Cornbrash." This corresponds to the outcrop on the actual ground of beds of rubbly limestone, in which are many small quarries or pits very attractive to the fossil collector. Various kinds of fossils may be picked up, but among the most noticeable, and often the most beautifully preserved, are the forms represented in Figs. 1 and 2.

It is seen at once that these two forms, which are between 3 or 4 centimetres in length, though differing in details, have many features in common which distinguish them from their companions in the Cornbrash. Each one shows perfect symmetry about a single plane; each has at one end a rounded projection perforated by a circular opening; and the surface of each, examined under a lens, shows, if in a good state of preservation, a regular pattern of minute pits (Fig. 1*c*).

Although, as a rule, each fossil appears at first sight to be a single solid body, examination soon shows that it is really a hollow calcite (calcium carbonate) shell composed of two main portions in contact along their edges. These are called the **valves** of the shell, and the valve which carries the circular opening is called (for reasons to be seen presently) either the **ventral valve** or the **pedicle valve**; the other is called the **dorsal** or **brachial** valve. The rounded protuberance of the ventral valve is called its **beak** or **umbo.** The dorsal valve has a similar, but much less prominent umbo, without any opening. The end of the shell at which lie these **umbones** is regarded as the hind or **posterior** end, the opposite end being the front or **anterior**; the **length** of the shell is from one to the other. The **breadth** of the shell is measured at right angles to the length, in the plane separating the two valves, while the **thickness** is from the surface of one valve to that of the other where they are farthest apart.

The surface is nearly smooth, but at irregular intervals it is marked by fine lines nearly parallel to the valve margins, but becoming crowded together as they approach the umbo. Evidently these lines, like the rings of a tree or those of a ram's horn, mark stages of growth (**growth-lines**); they must, in fact, represent the margins of the valves at successive intervals, and it is evident that the earliest formed shell must be that in the region of the umbo.

The region of the umbones may be obscured by rock substance, which it is tedious to remove, but when this is done some interesting features are revealed. The anterior side of the circular opening (**foramen**) is seen to be bounded differently from the remaining portion of its circumference. It is not bounded by the ventral valve proper, but by the oblique ends of two small plates of triangular shape: these are called the **deltidial plates**, and the complete gap in the shell (of which

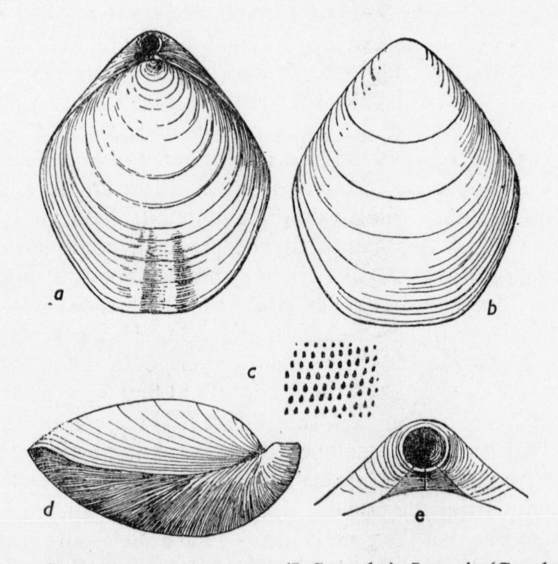

FIG. 1.—CERERITHYRIS INTERMEDIA (J. Sowerby), Jurassic (Cornbrash)
a, Dorsal view; *b*, ventral view; *c*, punctation; *d*, side view; *e*, umbo, foramen, and deltidial plates of ventral valve. *a*, *b*, *d*, Natural size; *e*, x 2; *c*, greatly enlarged (Original).

the foramen is only a part) which would be exposed by their removal is called the **delthyrium.** On either side of these plates the surface of the ventral valve shows a somewhat crescent-shaped, concave area, having the appearance of being carved from the shell: this is the **interarea.** Along this interarea the ventral valve margin is seen to be pressed close against and slightly overlapped by that of the dorsal valve: it is easy to see that here the valves are hinged together, while elsewhere they may have been free to gape apart, the gaping being greatest at the anterior end. Indeed, although the great majority of these fossils are found with the valves tightly closed, one is occasionally found in which they yawn more or less.

All the description so far given will apply equally well not only to the two fossil species from the Cornbrash, but to many others found in older or younger strata. All these were long united under the common name *Terebratula*, and were said to constitute a **genus** of animals. At the present time the name *Terebratula* is used in a more restricted sense, but we can still speak of all these forms as Terebratuloids. To distinguish these two **species** from one another James Sowerby, over a century ago, named one *Terebratula intermedia* and

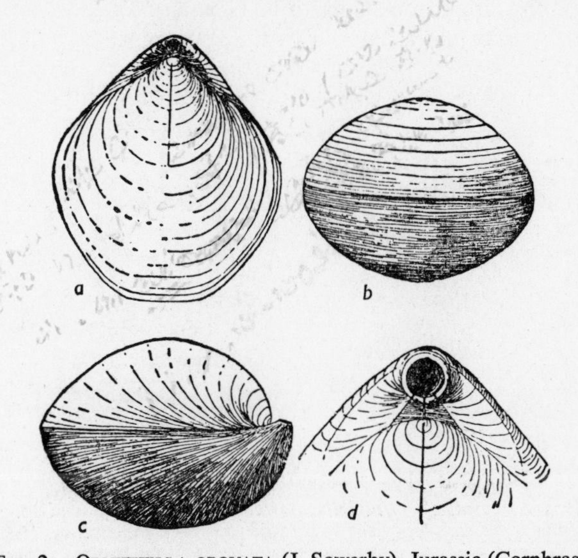

FIG. 2.—ORNITHELLA OBOVATA (J. Sowerby), Jurassic (Cornbrash)
a, Dorsal view; *b*, anterior view; *c*, side view; all natural size: *d*, umbonal region of both valves, x $\frac{5}{2}$, showing interarea, foramen, and deltidial plates in ventral valve, umbo and dark streak indicating internal septum in dorsal valve (Original).

the other *Terebratula obovata*. Let us now see some of the differences that underlie their resemblances.

The two Cornbrash species obviously differ in proportions. The ratio of length, breadth, thickness is in *T. intermedia* roughly 12, 10, 6, in *T. obovata* 11, 10, 8, the latter being thus broader and thicker. In the latter also the boundary between the interarea and the general surface of the ventral valve is definitely angular, while in the former it is rounded off so as to be somewhat indefinite. The foramen of *T. intermedia* is much larger in proportion to the size of the whole shell, and the outer edges of the deltidial plates, if continued, would form an obtuse angle; while in *T. obovata* the foramen is small, and the corresponding edges would form a right or slightly acute angle.

If these two shells are broken open, they will be seen either to contain crystals or some rock matrix or they may be hollow, and projecting into the cavity from near the posterior end there is what looks like a looped ribbon covered with little crystals of calcite. In *T. obovata* this **loop** extends forward nearly to the front end of the cavity; in *T. intermedia* it is less than half this length (Fig. 3). In the former also there runs along the middle line of the dorsal valve an internal ridge, forming an incomplete partition (**septum**). The presence of this can be detected from the exterior by the appearance of a dark streak

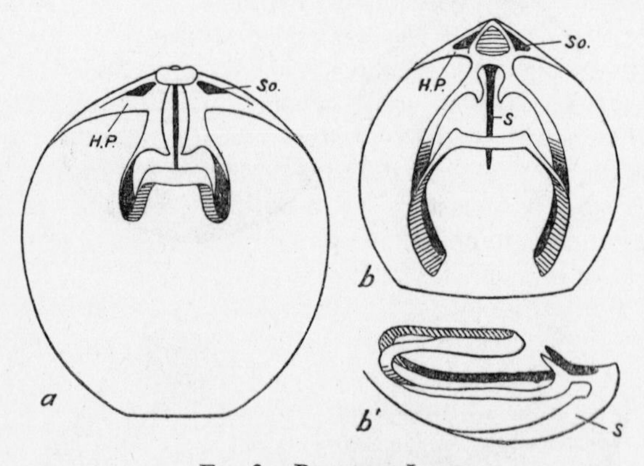

FIG. 3.—BRACHIAL LOOPS

a, *Cererithyris intermedia*; *b*, *Ornithella obovata*; *a,b*, interior views of dorsal valve; *b′*, side view, remainder of shell removed to expose the septum; *H.P.*, hinge-plate; *S.*, septum; *So.*, dental sockets. The cardinal process is seen in the uppermost portion of fig. *a* (Natural size; after Davidson).

(Fig. 2): little or no trace of such can be seen externally in *T. intermedia*. It has been found that the presence of such a dark streak, indicating a strong dorsal septum, is constantly associated with the presence of a long internal loop, and its absence with a short loop, in all Terebratuloids. We may therefore divide these into long-looped and short-looped groups.

This difference in the length and form of loop is regarded by palaeontologists, for reasons that will appear later, as a more fundamental difference than any of the others that have been given. It is considered too important to be expressed by a mere difference of specific name: the generic (and even the family) names must also be different. Neither species being really referable to the true genus *Terebratula* (which occurs much later in time), they are nowadays called *Cererithyris intermedia* and *Ornithella obovata*.*

Shells resembling these are not only found fossil in many different strata but they are also dredged, with the animal still living of which they form part, from the ocean in most parts of the world. Altogether over 100 species of Terebratuloids have been found in the exploration of the sea, as well as 58 other species which, though not actual Terebratuloids, are so near them in structure that they are united with them in zoological classification under the name **Brachiopoda** (or **brachiopods**). They are purely marine in habitat, no species having ever been found in any lake or river; the largest number of species live on the sea-bottom between the depths of 30 and 100 fathoms; many species (and perhaps the greatest number of individuals) live in water shallower than this, even in a few cases above low-water mark; only a few species live in water beyond the 100-fathom line. The coasts of Japan furnish the largest number of species for any one area: those on the east coast differing almost entirely from those on the west (20 species in eastern, 11 in western waters, of which only 2 species are common to both). The geographical range of some species is wide, that of others restricted.

A living Terebratuloid, such as *Magellania flavescens* of the Southern Seas, lives attached to the sea-bottom (usually many individuals in a cluster) by a short muscular stalk (**pedicle**) which extrudes through the foramen in the ventral valve. The animal feeds entirely upon microscopic organisms, both animals and plants (but probably in the main the minute plants known as diatoms) which drift about helplessly in the currents in the water. So enormously abundant are such organisms in the ocean, and so quickly do they grow and reproduce themselves, that they form the food, at first, second, or third hand, of nearly all the animals in the sea. It has well been said that "all fish is diatom" in the sense that "all flesh is grass." The larger sea-weeds are not used as food by many of the marine animals.

Animals feeding upon such helpless microscopic prey can do without many of the organs that are essential to the familiar animals of the dry land—eyes and ears to see and hear their prey, arms and jaws to seize them, swimming or running organs to pursue them are all needless. Shut up within its shell or **exoskeleton** a brachiopod is comparatively safe from enemies; fixed by its pedicle, only very violent storms can throw it up to die on the shore. All it needs to assist feeding and the normal life processes are (1) some means of creating a constant

* For further explanation of the terms species, genus, family, etc., see Chap. XIII.

water current into and out of its shell-cavity, to bring in food and oxygen and carry out waste products; (2) a shell so arranged that it can be closed tightly when necessary for safety, and opened at will to allow ingress and egress of the water-currents. These necessities control and determine the structure both of the soft parts of the animal's body and of the shell which it secretes.

The actual "body" of the Terebratuloid, containing the digestive tube, central nervous system and other fundamental organs, occupies a surprisingly small portion of the interior of the shell. From it there arise a pair of muscular sheets which form a lining to the two valves, and by which the valves are in fact secreted: these are called the **mantle-lobes,** and the cavity enclosed by them is called the **mantle-chamber.** Hanging freely in this mantle-chamber is the **lophophore,** composed of a pair of spiral ribbons (so-called "arms," loops, or **brachia***), fringed with tentacles along their whole length and covered with microscopic vibratile hairs (**cilia**). The constant movement of these creates the water-current which is so vitally necessary, and the lophophore acts as the breathing organ for the exchange of gases between blood and sea-water. The shelly "loop" described above as occurring in the fossil forms serves as a supporting internal skeleton of the lophophore. In *Magellania* the loop is long and doubled back on itself, the two halves of this doubled-back portion being distinguished as "ascending lamellae" from the longer "descending lamellae" (Fig. 11, p. 30).

If the interior of the valves of a modern Terebratuloid is examined certain details of structure are seen more easily than on a fossil specimen; though afterwards we shall find we can recognize the same details in the fossils. The two valves are hinged together by means of a pair of projections from the ventral valve (**hinge-teeth**) which fit into two hollows (**dental sockets**) in a horizontal shelf (**hinge-plate**) in front of the umbo of the dorsal valve. To give leverage for the opening of the valves there is a short knob-like projection (**cardinal process,** Fig. 3*a*) from the dorsal valve into the umbonal cavity of the ventral valve. The line along which the margins of the two valves meet in the neighbourhood of the hinge-teeth is called the **hinge-line.** To effect the opening of the valves there are muscles fixed at the umbonal end of the dorsal valve to the cardinal process and at the other end to the inner surface of the ventral valve (**divaricator** muscles); to close the valves there are other muscles running across from valve to valve (**adductor** muscles). (See Figs. 6*b*, *d*, and Figs. 7*d*, *d'*.) All muscles act by contraction which pulls their two attachments towards one

* Hence the name Brachiopoda from two Greek words—Brachion, arm, and pous, podos, foot.

another. The dorsal attachments of these two sets of muscles being on opposite sides of the fulcrum (the **hinge-line**), they work against one another. The muscles of the pedicle are attached to the inner surface of the ventral valve.

When the muscles decay after death, the areas of the shell to which they were attached are left as slight depressions devoid of the smoothness of the rest of the internal surface. Thus it is possible to tell in a fossil brachiopod the presence and arrangement of these muscles. Although the details differ greatly among the many forms of brachiopods, the adductor impressions in the ventral valve are always close together, and the divaricator impressions lie more or less to the right and left of them.

The body proper of a brachiopod being so small, it is not surprising that the reproductive organs should extend into the mantle which lines and secretes the two valves; impressions of the ovaries are in many species to be seen on the inner surface of the shell; as also are the branching impressions of vessels belonging to a circulatory system (**vascular markings**). In forms like the Terebratuloids which have a punctate shell, each punctation contains a minute outgrowth of the mantle, and in at least one recent species this has been shown to have the structure of a simple sense-organ, though the nature of the sense is unknown.

The larval brachiopod escapes from the egg as a microscopic body covered with vibratile cilia, by means of which and the action of currents it gets the only chance in its life to migrate away from the fixed home of its parents. Commonly this free-swimming larval stage lasts but 10 or 12 days thus it is not likely that any one generation travels very far, yet some modern species are world-wide, thanks to the cumulative effect of small journeys through thousands of generations. The length of life of individual brachiopods in general is not known, but in one species there is evidence of five years' life.

Very soon the larva fixes itself by its hind portion which forms a rudimentary pedicle and it begins to secrete two valves. In all observed cases this first shell, or **protegulum,** has a similar form, which resembles that of one of the primitive brachiopod shells in the Cambrian rocks (Fig. 7*a*). This resemblance favours the view that the development of the individual (**ontogeny**) repeats, in an abbreviated and more or less imperfect manner, the ancestral history of the race (**phylogeny**). Although biologists have of late thrown doubt upon anything approaching the universal operation of this supposed principle of **recapitulation** (or **palingenesis**) it may still be applied with caution in the case of those animals whose shell-growth is of the kind found in brachiopods, where the early stages of the shell are preserved and added to, to form the adult (growth by **accretion**). This question will be discussed again in Chap. IV (pp. 101, 118).

The protegulum of a brachiopod forms the apex of each valve. Too often, owing to the exposed position of the umbo, especially in the ventral valve, it may be rubbed away in later life; in some forms it is absorbed during the enlargement of the pedicle-opening; but in many cases it persists through life. Growth consists in adding to it both on its inner surface and on its margins—to the greatest extent at the anterior margins. During this growth the shape of the shell may change, perhaps more than once; it may acquire a surface pattern or ornament, and this, in later formed parts of the shell, may alter or not be formed; but (apart from wear and tear) every stage is retained in the adult shell, and if the characters of an early stage closely resemble those of the adult shell of a brachiopod of earlier date, it is believed by some authors that the earlier species may be ancestral to the one under consideration.

When *O. obovata* has attained its full length, it may still continue to grow in thickness, the anterior ends of the valves no longer meeting in an acute edge, but as a flattened surface. This is a sign of old age—a senile or **gerontic** character.

Growth by accretion is not possible for all parts of the skeleton: the loop, once formed, could not grow and retain its shape. As the valves increase in size, repeated resorption and fresh secretion of the shelly substance of the loop must take place. The difference in development between the valves and the loop is like the difference between that of a ram's horn (which grows continuously) and that of a stag's antler (shed and renewed annually). Hence the adult loop retains no evidence of the stages through which it has passed. These can only be determined by detailed observation of the development in living brachiopods, and by the fortunate preservation of immature as well as mature specimens among fossils. The former method has yielded results which afford interesting examples of palingenesis, for it has been claimed by Beecher and Thiele that the loops of the modern Terebratuloids *Magellania* and *Macandrevia*, though much alike in the adult stage, pass through two quite different series of metamorphoses, and in each case the successive stages correspond to the adult loop in a number of other genera, some still living, others extinct (Fig. 11).

It will be convenient here, before returning to the fossil forms, to consider some of the other brachiopods found living in the sea today.

Hemithyris psittacea (Fig. 4) is a form living in the cold northern waters: its shells are often cast up by storms on the coast of Labrador.

It agrees with the Terebratuloids in having dissimilar symmetrical valves, the larger being perforated for a pedicle, and in having the delthyrium partly closed in by a pair of deltidial plates. But these plates, instead of bounding the foramen on its anterior side only, bound it laterally and even tend to meet on the posterior side just under the umbo (Fig. 10, p. 28). Looking at the anterior end (Fig. 4*c*) we see that the junction (or **commissure**) of the valves is not rectilinear (as in Fig. 2*b*) but **sinuate,** arching up dorsalwards in the centre. This is a feature found in many brachiopods of quite different groups, and is one means of increasing the efficiency of the water-circulation, by keeping the lateral inflowing currents at a different level from the median outflowing current. In *Hemithyris* this marginal sinuation has no effect on the rotundity of the outline, but in many brachiopods it results in a pinching-up or **median fold** on the dorsal valve, and a

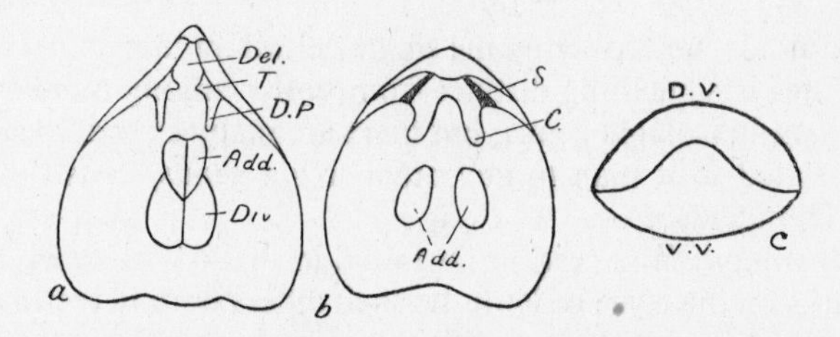

FIG. 4.—HEMITHYRIS PSITTACEA (Chemnitz) Recent
a, Ventral; *b*, dorsal valve (interior views); *c*, anterior view; *Add*., adductor muscle-impressions; *C*, crus; *Del*., delthyrium, the deltidial plates removed; *Div*., divaricator impressions; *D.P*., dental plates; *D.V.* dorsal valve; *S*, socket; *T*, hinge-tooth; *V.V*., ventral valve (Slightly enlarged; original).

depression or **sinus** on the ventral. (See, for instance, Figs. 7*l* and 9*b*′). On examining the surface with a lens, instead of the pattern of punctations so characteristic of the Terebratuloids a silky, fibrous appearance is observed.

Internally, still more striking differences are found. It is true that the cardinal process, hinge-plate and sockets, hinge-teeth and the several pairs of muscle-impressions are all present with only slight differences of arrangement; but although spiral brachia exist there is no loop made of calcium carbonate. From the hinge-plate two short processes project into the cavity, and correspond to the beginning of a loop: they are termed the **crura** (plural of **crus**). Again, in the umbonal cavity of the ventral valve there are a pair of vertical partitions which, as they extend into and strengthen the teeth, are called

dental plates*: these, however, are also found in some Terebratuloids.

Forms more or less similar to *Hemithyris psittacea* are commonly found fossil, in the Cornbrash among other strata. In the works of J. and J. de C. Sowerby they were included under the broad name of *Terebratula*, but afterwards they were distinguished by the separate name of *Rhynchonella* by Fischer von Waldheim. This name in turn was restricted to a portion only of these forms, but we can still conveniently speak of them as Rhynchonelloids.

Lacazella mediterranea is a small living brachiopod with a very thick shell. It has no pedicle, but is found cemented to other shells by its ventral umbo. The long and narrow delthyrium occupies the middle of a large interarea, and is completely closed, not by a pair of plates, but by a single one (**deltidium**). Internally there are no crura even, but the thick dorsal valve has deep furrows with ridges between, corresponding to the course of the spiral brachia. The other structures do not differ essentially from those of Rhynchonelloids and Terebratuloids, but evidently the single deltidium in place of two deltidial plates is something distinct from both. Among modern forms *Lacazella* is an isolated genus; but there are many extinct forms more nearly related to it than to any other living genus, constituting the family Thecideidae.

Discinisca lamellosa (Fig. 7*c*) is found in shallow water or between tide-marks on the coast of Peru. Its shell differs both in texture and in shape from those hitherto considered, for it has a horny appearance; a quarter of its composition consists of organic matter with the remaining three quarters being largely calcium phosphate with but a small percentage of calcium carbonate. Each valve is nearly circular in outline, with the umbo near the centre. The ventral valve resembles a low cone of which one portion of the surface (on the posterior side) has been crushed in. In the middle of this depressed area is a narrow slit through which the pedicle passes obliquely. The dorsal valve is a still flatter cone. Internally, there are no hinge-teeth, no cardinal process or hinge-plates, no crura or loop; in fact, the only internal features of the shell are the muscle-impressions and a small median septum in the ventral valve anterior to the delthyrial slit. Here is a brachiopod differing greatly from all the forms previously discussed, especially in not having its valves hinged together, but only attached by muscles and other soft tissues. There are eight species of *Discinisca* living and one species each of the allied genera *Discina* and *Pelagodiscus*: they may be spoken of as Discinids. Many fossil Discinids are known, but owing to the inarticulate (unhinged) character of the

* It has been pointed out that these would be more accurately termed **delthyrial supporting-plates**, but the shorter term is not likely to be dropped.

shell it is commoner to find isolated valves than complete shells—the reverse is the case in Terebratuloids and Rhynchonelloids, and indeed most brachiopods.

Lingula unguis (Fig. 7*b*) is a brachiopod which burrows umbones downwards in the sea-bottom in shallow water; it has a very long pedicle which it pushes into the sediment and which it uses to withdraw itself from the surface into its burrow when necessary. Like *Discinisca* it has a horny shell but this is about one-half phosphatic in composition and the organic matter and phosphate are here in alternating layers and not uniformly admixed as in *Discinisca.* In shape it is somewhat oblong, the two valves being very nearly alike, each having the umbo at the extreme posterior end. There is no delthyrium, the valves diverging at the posterior end for the passage of the pedicle. Internally the valves show no hinge-structures or brachial skeleton, though the spiral brachia are present as usual; but in the dorsal valve there is just below the umbo a reflected portion of the shell resembling an interarea. The muscular impressions are faint, but are more complex than in previous cases, the absence of articulation making it possible to have muscles for sliding the valves sideways as well as for opening and closing the shell. Fifteen species of living Lingulids are known, and a very great number of fossil species.

The resemblance of many fossil brachiopod shells to those of living forms compels us to believe that the former were also once parts of living organisms which lived in much the same manner. Where therefore we find brachiopod shells in a stratum we must conclude that it was originally a deposit on the sea-bottom, and we may even venture in certain cases to estimate the depth below the surface at which it was deposited from a consideration of the abundance and generic identity of the brachiopods. But the fossils also serve another very useful purpose.

As long ago as 1688, the famous Dr. Robert Hooke foresaw the possible utility of fossils as time-markers. "However trivial a thing a rotten shell may appear to some," he wrote, "yet these monuments of nature are more certain tokens of antiquity than coins or medals . . . and though it must be granted that it is very difficult to read them and to raise a chronology out of them . . . yet it is not impossible." It was William Smith who, at the end of the eighteenth century, was first able to "raise a chronology out of them," by showing that in the series of strata that lie one upon the other in the Bath district (and elsewhere) each division is distinguished by particular species of fossils.

Thus William Smith found that underneath the Cornbrash was a series of beds of hard flaggy limestone, which he termed the "Forest

Marble"; below these again the Bradford Clay; below this the thick mass of oolitic limestone for which Bath is famous, and so on. In Fig. 5 are shown species of brachiopods that are characteristic of two of these formations. The value of such fossils for identifying the stratified rocks of the earth in systematic order will be realized when we remember that though in a hilly district with many quarries and other openings into the rocks like that around Bath, or along a cliffy coast, it may be easy to see clearly how one group of strata overlies another, yet when efforts are made to follow the strata from place to place and especially into regions where exposures of the rocks are few and poor, many difficulties are met. A stratum may change its character (thus the Forest Marble gradually changes from limestone to clay in the Midlands; or it may "thin out" and disappear altogether; or a "fault" may displace the outcrop greatly. Amid these difficulties, the

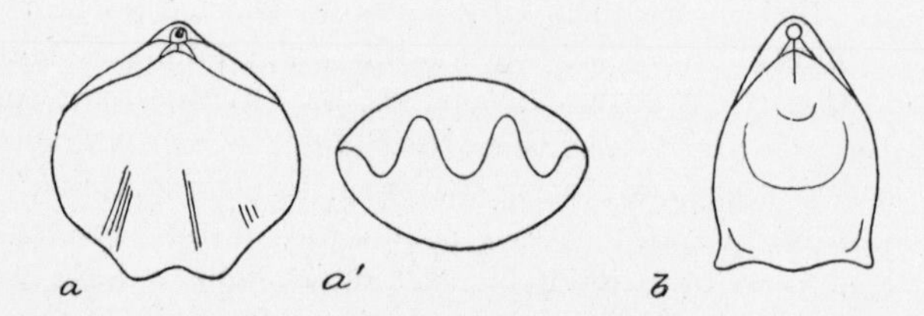

FIG. 5.—BATHONIAN TEREBRATULOIDEA
a, a'. Epithyris bathonica S. Buckman, Bath Oolite, × $\frac{1}{2}$; *b, Ornithella digona* (J. Sowerby), Bradford Clay, × $\frac{3}{4}$ (After Davidson).

finding of distinctive fossils will on occasions save the geological surveyor from a mistake which might perhaps lead someone to sink a mine for coal or to bore for water in a wrong place.

The whole group of animal species found together in the same beds are spoken of as the **fauna** of those beds, just as the collection of species now living in any given area is the fauna of that area. In past times, as at present, a fauna had a definite **habitat** or distribution in space; so that the nature of a fossil fauna in Australia being different from one in England does not prove that they were not contemporaneous. This may at first sight seem to make it impossible to apply William Smith's principle over great distances; but how it can be done will be explained later. At present we must consider another point.

Several thousand species of fossil brachiopods have been described, and of these only about 25 are identical with living species, while about 150 living forms are unknown as fossils. These thousands of species represent the members of a great number of successive faunas that have succeeded one another throughout geological time. If species followed one another without any guiding principle of succession, the human memory would be unable to grapple with the geological sequence. Fortunately this is not the case. While species are short-lived, geologically speaking, genera are longer lived, but yet most of them are restricted to a limited portion of geological time; and the abundance of certain genera will characterize large groups of strata, just as the abundance of certain species marks a smaller group. The smallest thickness of rock that can be distinguished by means of its fossils over a wide geographical area is called a **zone**: it is named after its dominant species, but locally it may be divided into **subzones,** characterized by shorter-lived species. The actual thickness of a zone may vary in different places from an inch or even less to several hundred feet, according to the net rate at which sediment was accumulating. The time during which the strata constituting a zone were deposited is called a **hemera.** We are unable at present to bring this unit of geological time into relation with any of our astronomical time divisions.

A number of consecutive hemerae (varying in different cases from three to ten) are grouped as an **age,** corresponding to the development of a genus or a family: the corresponding thickness of strata is called a **stage.** Ages are again grouped into **epochs,** these into **periods,** and these into **eras,** the bounds of these being determined by great and rapid changes of fauna. These terms and their equivalents may be tabulated thus:

TIME-DIVISIONS	ROCK-DIVISIONS
Era	Group
Period	System
Epoch	Series
Age	Stage
Hemera	Zone

Cererithyris intermedia dates one of the subzones of the Lower Cornbrash, and the brachiopods shown in Fig. 5 mark other subzones of the Bathonian Stage, taking its name from Bath, around which city these zones are well displayed. This stage is part of the Middle Jurassic Series of the Jurassic System, so named from the Jura Mountains. This in turn is part of the Mesozoic Group.*

The sequence of fossils in successive divisions of geological time may appear to be casual and meaningless, but that is probably only owing to the imperfection of our knowledge. In a number of cases a

* See Appendix for tabulated statements of time-divisions. Reference to these tables will frequently be necessary during the reading of what follows.

succession can be traced which has the appearance of being an actual genealogical tree, and as we pass back into the earlier geological times we find a convergence of the lines of descent towards a few very simple forms. At the beginning of the palaeontological record we seem to be very near the beginning of the Brachiopoda.

An ideal classification should be based on structure and blood-relationship. A full justification for the following classification (which is based on that formulated by Beecher, 1893), modified by Cooper, R. C. Moore, Muir-Wood and others cannot be given here but it is offered as the best of many attempts to express the inter-relationships of Brachiopoda.

The Brachiopoda are so well defined and sharply marked off from all other animals that they have been accorded the rank of a **phylum** or primary branch of the Animal Kingdom*. Brachiopoda are marine animals fixed to the sea-floor or to objects lying on it by a pedicle or otherwise, feeding on microscopic floating organisms by means of spiral "brachia" and secreting a shell with dorsal and ventral dissimilar symmetrical valves.

In the opinion of most students of the phylum, the most fundamental plane of cleavage is that between forms like *Lingula* and *Discinisca*, in which the valves are kept attached to each other only by muscles and are not articulated by teeth and sockets, and those like *Terebratula* in which they are articulated together. These two divisions constitute the **class** Inarticulata and the **class** Articulata respectively. The characters of the delthyrium afford the next means of subdivision. Inarticulata such as *Lingula* which have no delthyrium constitute the **order** Atremata; those with a delthyrium enclosed by the ventral valve (at least during early life), the Neotremata; Articulata with a median triangular plate (**deltidium**) were formerly considered to form the order Protremata and those with a pair of deltidial plates, the Telotremata. However it has lately been demonstrated that deltidial plates occur in some Protremata and that a deltidium occurs in some Telotremata so the use of the terms Protremata and Telotremata has been abandoned. The Articulata and each of the Orders of the Inarticulata are divided into **suborders** (or **superfamilies**) and these into **families**, made up in turn of one or more **genera**, each with one or more **species**.

The beginner in Palaeontology is often troubled by the importance attached in classification (or taxonomy) to the internal characters of

* Some zoologists unite the Brachiopoda with the Bryozoa in a phylum Molluscoidea, of which they form two classes. The differences between them appear to the present writers too profound to be expressed as merely class differences.

fossils or other features commonly invisible in ordinary specimens, and cannot understand how identification is possible in such cases. It may be pointed out that if one single specimen shows all the characters that are necessary to determine its systematic position, other specimens can be recognized as identical with it by their minor features. Hence the apparent paradox that it is often easier to determine the species of a fossil than its genus, which is quite contrary to the ordinary conception of classification. For the specific character is commonly some feature of external shape or surface-pattern that is easily recognized but might be found in any one of a number of genera, while the generic character may be some feature which only an exceptionally well-preserved specimen will show. One must never forget that classifications are not usually framed as means for the easy naming of specimens, but are intended to indicate, as well as possible, the natural relationships of species.

In the particular case of the internal characters of brachiopods, however, these are not so inaccessible as a beginner may think. It often happens that the interior of a shell may be filled, after death, with foreign material such as mud or sand, and that later the shell may be dissolved, leaving a space bounded by consolidated mud or sand forming the **internal** and **external moulds** of the shell. On the internal mould all the structures on the inner surface can be recognized—hollows in the shell being represented by elevations, and projections by depressions. It is even advisable, when the infilling material is suitable and specimens are abundant, to make moulds artificially by burning away or otherwise removing the shell. The external mould preserves the features of the outer shell surface.

CLASS INARTICULATA

Valves inarticulate, lacking teeth and sockets, usually with several groups of muscles for opening and closing the shell and for sideways movements of valves. Mantle lobes, separate, developed directly, not rotated. Pedicle in recent forms attached by muscles to one valve only. In general, delthyrium or notothyrium not developed. Lophophore without calcite spicules likewise mantle except in Craniacea. Shell chitino-phosphatic or chitinous, exceptionally calcareous. Shell layers may be punctate. Lophophores unsupported by brachiophores or crura. Lower Cambrian–Recent.

ORDER I: ATREMATA

Inarticulate Brachiopoda with horny and calcium phosphate shells except in the Trimerellacea where the shell is composed of calcium carbonate; the pedicle emerging between the divergent valves, which show little or no tendency to enclose it in a delthyrium.

This is the most primitive and conservative order, especially characteristic of the Older Palaeozoic era, only a single family, the Lingulidae, surviving through later eras to the present day. It is divided into three superfamilies.

1. **Obolacea.**—Rounded and lenticular in form, with short pedicle and thick shell; probably fixed to floating seaweed. These lived in the Older Palaeozoic era, and were formerly believed to include the supposed brachiopods, *Fermoria* and *Protobolella* of the Vindhyan of India. *Rustella* of the Cambrian has been claimed to be near the ancestral form of all later brachiopods, through the Obolidae to the highly specialized Trimerellacea of the Ordovician and Silurian. In *Obolus* (Figs. 7*a*, *a′*), each valve has a well-marked interarea, that of the ventral valve with a longitudinal pedicle-groove, and its umbo being more prominent. Lower Cambrian–Ordovician.

2. **Lingulacea.**—Elongated in form, biconvex, burrowing, with thin shell and long pedicle. Lower Cambrian–Recent.

These begin in the Cambrian with *Lingulella*, which retains the Obolid characters of an interarea, with pedicle-groove in the dorsal valve only, and of a more prominent ventral umbo. It is followed in the Ordovician by *Lingula* (Figs. 7*b*, *b′*), in which the valves are almost equal, and the pedicle groove is wanting. This genus survives to the present time, and so is one of the longest-lived of all genera, not only of Brachiopoda, but of all but some microscopic animals. In the later Palaeozoic and younger deposits, the Lingulidae tend to occur in poorly aerated brackish water deposits commonly to the exclusion of other fossils. *Lingula* (Ord.–Rec.) usually less than 1 cm. long but may reach 5 cm. in some species.

3. **Trimerellacea.**—Biconvex shells of circular or pear-shaped outline with a length up to 8 cm.; impunctate; composed of calcium carbonate. Muscles attached to platforms raised above the floors of both valves and supported by a median septum. The internal moulds of Trimerellidae are thus complex. Ord.–Sil. *Trimerella* (Sil.) up to 6 cm. long.

ORDER II: NEOTREMATA

Inarticulate Brachiopoda, in which one or other valve acquires a more conical form, and the pedicle becomes surrounded by the ventral valve and comes to emerge close to the umbo and at a distance from the margin. In all superfamilies except the Craniacea the shell is horny and phosphatic.

1. **Paterinacea.**—Ventral valve with high interarea. "Delthyrium" closed by a **homeodeltidium,** apparently no pedicle opening. Confined to Lower and Middle Cambrian. Nowadays considered specialized and not as formerly held ancestral to Articulata. *Micromitra,* commonly with straight hinge line; usually less than 8 mm. wide.

2. **Acrotretacea.**—Ventral valve generally conical with very high interarea with a median furrow or trough. Pedicle foramen apical or just behind apex. Dorsal valve weakly convex and may have a median septum. L. Cam.–U. Ord. *Acrotreta,* very small.

3. **Siphonotretacea.**—Biconvex elongated valves, both with obtuse-angled interareas. Pedicle foramen apical or anteriorly subapical. L. Cam.–M. Ord. *Siphonotreta.*

4. **Discinacea.**—Ventral valve flat or concave with longitudinal pedicle slit, dorsal valve conical with eccentric umbo. Sub-circular. Mid. Ord.—Recent. *Trematis* (Ord.), pedicle slit extends from posterior margin to umbo; shell surface punctate or pitted. *Orbiculoidea* (Ord.–Cret.), pedicle slit restricted by a plate called a **listrium** growing between apex and slit as growth proceeds; pedicle passes obliquely through remaining portion of slit emerging as a small tube on interior surface near posterior margin. *Discinisca* (Tert.–Rec., Figs. 7*c*, *c′*), *Discina* (Rec.), *Schizocrania* (Ord.–L. Dev.).

5. **Craniacea.**—Aberrant forms in which shell is subcircular, punctate and of calcium carbonate. Fixation by cementation of the ventral valve to a foreign body; adults lack a pedicle foramen. Dorsal valve more or less conical; in one genus (*Petrocrania,* Ord.–Perm.) reflects in markings those of host shell. Ord.–Rec. *Craniops* (Sil.), very small. *Crania* (Ord.–Rec.).

CLASS ARTICULATA

Valves articulated by hinges, generally bearing teeth and sockets. Mantle lobes revolved from a posterior to an anterior direction during larval stage. Pedicle, cartilaginous not elastic, attached by muscles to both valves. Delthyrium and in some forms notothyrium developed. Shell of calcium carbonate in two layers; the inner may be fibrous (prismatic), the outer fibrous or lamellar with external chitinous periostracum. Shell layers punctate, impunctate or pseudo-punctate. Lophophore attached to brachiophores or crura or supported by calcareous loops or spiralia. Calcite spicules developed in mantle lobes and in some forms also in lophophores. Lower Cambrian–Recent.

Although the Orthoidea are the oldest of the Articulate groups with *Kutorgina* in the Lower Cambrian as an early member, in this book it is convenient first to consider the Strophomenoidea which group was probably derived from the Orthoidea, by early loss of the functional pedicle.

SUBORDER 1. **Strophomenoidea.**—The calcium carbonate shell is impunctate but the inner or fibrous layer contains rod-like calcite bodies, the ends of which project into the body cavity; on weathering these rods dissolve more rapidly than the fibrous material, leaving cavities simulating punctations. The shell is thus said to be **pseudo-punctate.** Generally the hinge line is the widest part of the shell. One valve is characteristically convex and the other concave. Surface radially costate or spinose or both. L. Ord.–Rec. but most important from Ord. to Perm.

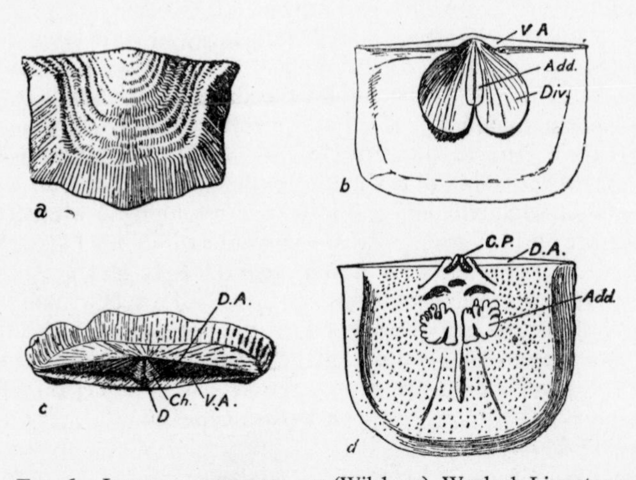

FIG. 6.—LEPTAENA RHOMBOIDALIS (Wilckens). Wenlock Limestone
a, Ventral view; *b*, interior of ventral valve; *c*, posterior view; *d*, interior of dorsal valve; *Add.*, adductor impression; *Ch.*, chilidium (showing median pedicle-groove); *C.P.*, cardinal process; *D.*, delthyrium; *D.A.*, interarea of dorsal valve; *Div.*, divaricator impression; *V.A.*, interarea of ventral valve (Natural size; original except for *a* which is after Geological Survey).

Strophomenacea. With well-defined interareas in each valve and lacking spines. L. Ord.–Rec.

Leptaena rhomboidalis (Fig. 6) of the Wenlock Limestone is a broad shell (length: breadth is about 2:3) with a quadrate or semi-elliptical outline, the straight hinge-line forming the greatest breadth of the shell. The early (posterior) part of the shell is flat, but after growing to a length of about two centimetres a sudden change takes place and the surface of the ventral valve becomes bent at about right angles (**reflected**), making that valve in the adult highly convex externally, while the dorsal valve is correspondingly concave, so that the change does not result in great increase of thickness. (Contrast the effect in the biconvex shell of *Ornithella* in old age, p. 8.) Concavo-convex shells are common among Palaeozoic brachiopods: when, as in *Leptaena*, the dorsal valve is concave, they are said to be **normally concavo-convex;** when the ventral is concave they are **reversed concavo-convex** (or convexi-concave).

The surface of the shell has radiating ridges or **costae,** crossed by much finer concentric lines. In addition to these there appear at a little distance from the umbo coarser and irregular concentric corrugations or **rugae,** which rapidly increase in size up to the line of growth-change (reflexion), beyond which they cease, as though they were premature attempts to make the change in growth-direction.

Each valve has a wide and low interarea, extending all along the hinge-line. The delthyrium notches the interareas of the ventral valve and the **notothyrium** notches that of the dorsal valve equally so that the joint cavity is rhomboidal in shape. In the ventral valve there is a small deltidium close to the umbo; in the dorsal a much larger convex plate (**chilidium**), with a median groove, covers the whole notothyrium. The pedicle-opening is small, between umbo and deltidium (encroaching on the former in old age); the pedicle probably rested on the groove of the chilidium.

In the interior of the ventral valve, well-marked muscle-impressions and teeth are seen. The raised rims of the divaricator-areas join on to the teeth and delthyrial margin like rudimentary dental plates, and there is a slight median septum. In the dorsal valve, the cardinal process is short and bifid and fits closely under the chilidium. Dental sockets are conspicuous.

Genera allied to *Leptaena* (Ord.–L. Carb.) are an Ordovician genus *Rafinesquina* with fine radial costae alternately larger and smaller, and with ridged muscle-areas and *Stophodonta* [*Stropheodonta*] (Sil.–Dev.) with fine denticulations along the hinge-line, and an inconspicuous chilidium. Neither of these has the concentric corrugations of *Leptaena*. *Davidsonia* of the Devonian fixed itself by cementation on to other shells, and its brachial valve shows internal spiral markings which are interesting as evidence that this animal probably had a spirally shaped lophophore.

Of genera with reversed concavity, *Strophomena* (M.–U. Ord.) might be described as a reversed *Rafinesquina*, and *Strophonella* (Sil.–L. Dev.) as a reversed *Strophodonta.*

The family Orthotetidae (Figs. 7*f–i*) is characterized, in general, by well-marked dental plates, which may extend far forward in the pedicle-valve—sometimes parallel to one another as in *Meekella*

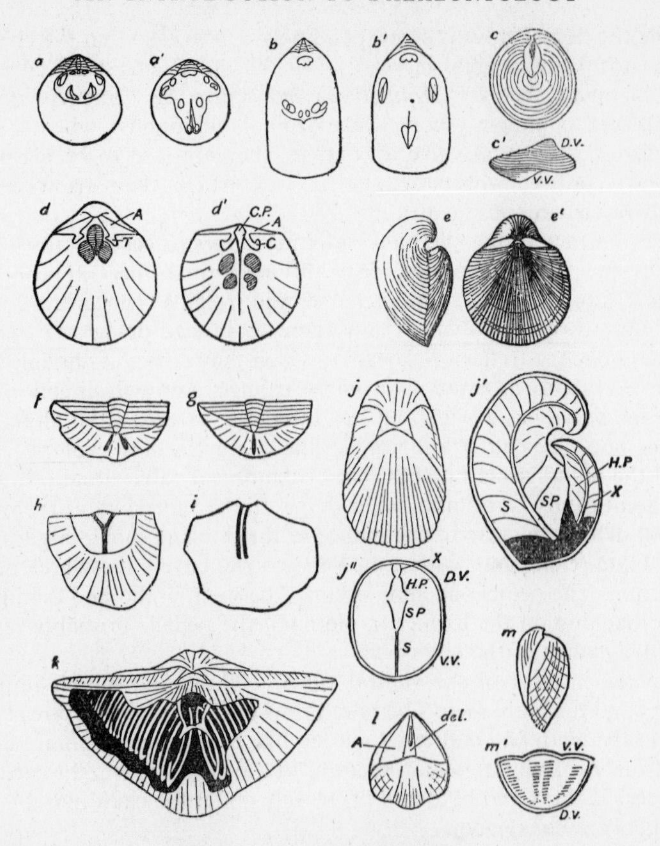

FIG. 7.—VARIOUS BRACHIOPODS

a, *a′*, *Obolus appolinis* Eichwald, Cambrian (Tremadoc), ventral and dorsal valves; below the umbo in each is the interarea (shaded), that of the ventral valve having a median pedicle-groove; muscle-impressions in outline (natural size); *b*, *b′*, *Lingula unguis* (Linné), Recent, ventral and dorsal valves: parts as in *a*, *a′* (× $\frac{1}{3}$); *c*, *c′*, *Discinisca lamellosa* (Broderip), Recent, ventral and side views; foramen black (natural size); *d*, *d′*, *Resserella* [*Orthis*] *elegantula* (Dalman), Silurian, ventral, and dorsal valves: muscle-impressions shaded (× $\frac{3}{2}$); *e*, *e′*, dorsal and side views (× 1); *f*, *g*, Umbonal views of ventral valves of *f*, *Schellwienella* (Lower Carboniferous); *g*, *Derbyia* (Carboniferous); areas shaded, dental plates (black) showing through the test; *h*, Umbonal region of *Orthotetes* (Coal Measures), ground down to show dental plates and septum (black); *i*, Internal mould of ventral valve of *Meekella*, dental plates black; *j*, *Conchidium knighti* (J. Sowerby), Silurian, dorsal view; *j′*, naturally split section; *j″*, cross-section (× $\frac{1}{3}$); *k*, *Spirifer striatus* (Martin), Lower Carboniferous, dorsal view, with test partly removed, showing spiralia (× $\frac{1}{2}$); *l*, *Cyrtia exporrecta* (Wahlenberg), Silurian, dorsal views (× $\frac{1}{2}$); *m*, *m′*, *Atrypa reticularis* (Linné), Silurian, side view and cross-section showing spiralia (× $\frac{2}{3}$); *A*, Interarea; *C*, crus; *C.P.*, cardinal process; *del.*, deltidium; *D.V.*, dorsal valve; *H.P.*, crural plates (cruralium); *S*, septum;

(Carb.–Perm.), or diverging as in *Schellwienella* (Carb.) or converging ventrally to form a median septum as in *Orthotetes* (Carb.–Perm.). In *Derbyia* (Carb.–Perm), however, the dental plates are greatly reduced, and the well-developed septum extends to the apex of the umbo. In *Schuchertella* (L. Dev.–Perm.) dental plates are wanting, while the dorsal valve has very short, stout, curved crural lamellae. In *Streptorhynchus* (Carb.) dental plates are vestigial, the ventral umbo is irregularly twisted and cemented to some foreign body. Externally all the genera of this family are radially ribbed; the costae are generally coarse in *Meekella* with finer ones superimposed; in other genera the costae are generally fine. Certain species belonging to different genera are so closely similar externally, that it is impossible to discriminate between them without examining the interior, for which the grinding of sections may be necessary: such species are said to be **homoeomorphs** of one another. The failure to discriminate between homoeomorphs has frequently led to mistakes in the correlation of strata.

Productacea. Concavo-convex shells. Interareas though present in some families are not conspicuous in any. Spines along posterior margin or on other parts of shell. U. Ord.–Trias.

Productus productus (commonly, but wrongly, termed *Productus martini*, see p. 270) is one of many species of *Productus* that abound in the Lower Carboniferous limestones and shales. In general form it is not unlike *Leptaena*, but the convex ventral valve shows a stronger and more uniform curvature, and its umbonal region is much larger and more rounded; there is no interarea and no delthyrium. The dorsal valve is, for the first 2 cm. from the umbo, almost flat, very slightly concave; it also has no interarea and only the cardinal process projects beyond the hinge-line which is long and straight, but less than the greatest breadth of the shell. The profile of the ventral valve first forms an arc which is almost a semicircle, of which the dorsal valve forms the chord, and at 2 cm. from the umbo the valves meet as though the shell were complete. From this distance on, however, the ventral valve is continued with very slight curvature for a considerable length as a **trail,** while the dorsal valve bends abruptly and a short partition grows transversely towards the ventral valve forming a **diaphragm,** the two valves continue in close contact with only a very narrow space between them when the shell is closed (Figs. 8*a*, *b*). This trail is very easily broken off during extraction of the

---

FIG. 7.—VARIOUS BRACHIOPODS (*continued*).
SP., spondylium (dental plates); *T*, hinge-tooth; *V.V.*, ventral valve; *X*, septal plate. *a*, *b*, after Walcott; *e*, after Geological Survey; *f–i*, after Thomas; the rest after Davidson.

fossil from the rock, and what remains has all the appearance of a perfect shell, 2 cm. long and about 2·5 cm. broad.

The surface is marked with close-set radiating costae, rounded in section, with a few concentric corrugations near the umbones. Here and there, especially near the sides, may be seen round circlets, proved by comparison with better-preserved specimens to be the

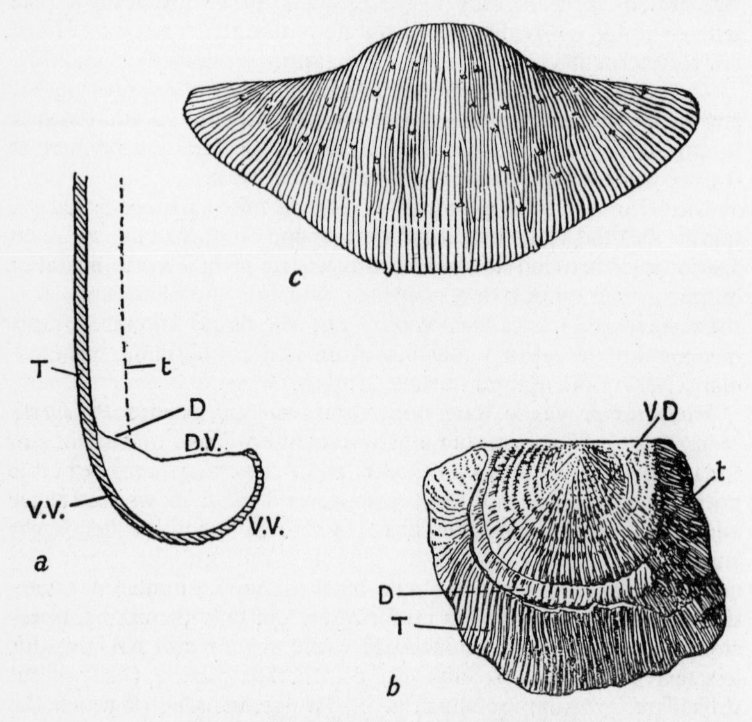

FIG. 8.—SOME PRODUCTID BRACHIOPODS

a, *Productus productus* (Martin), Lower Carboniferous, diagrammatic longitudinal section; *b*, dorsal valve and part of trail of ventral valve; *D*., diaphragm; *DV*., dorsal valve; *T*., trail of ventral valve; *t*, trail of dorsal valve; *VD*., visceral disc of dorsal valve; *VV*., ventral valve (× 1; after Muir-Wood); *c*, *Productus* (*Gigantoproductus*) *latissimus* J. Sowerby, Lower Carboniferous (× 1; after Geological Survey).

stumps of long curved spines. These spines are commonly developed in the family Productidae, and they served perhaps to anchor or to support the brachiopod on the sea-bottom mud, thus enabling it to dispense with a pedicle. In *Etheridgina* (known from the Carboniferous of Scotland) the spines grip a crinoid stem tightly and the ventral valve is cemented to it. In the Permian *Strophalosia* also, the spinose ventral valve is fixed to foreign bodies. The Permian genus *Prorichthofenia* has the ventral valve developed as a cone with anchoring spines and the dorsal valve is lid-like.

Internally, *Productus* does not differ greatly from *Leptaena*; the divaricator-impressions in the ventral valve are more widely separated, and in front of each one there is a spiral thickening of the interior of the valve like that of *Davidsonia*; in the dorsal valve there is a pair of somewhat similar, reniform (kidney-shaped) elevations.

The Productid genus *Gigantoproductus* (Carb., Fig. 8*c*) provides the largest known brachiopod, in one species reaching 30 cm. between its cardinal extremities. The Productidae are especially prominent in the Carboniferous, but there are Permian genera.

The Chonetidae superficially resemble the Strophomenacean family Plectambonitae of the Ordovician and Silurian in their shallow concavo-convex shells but they carry spines on the posterior margin of the ventral valves directed outwards and backwards from the centre. *Chonetes* (Sil.–Perm.) has a ventral interarea more developed than is the dorsal, also teeth and delthyrium. *Lissochonetes* (Carb.) has a smooth surface. *Cadomella* is a Liassic descendant.

Thecideacea. These have been regarded as degenerate Strophomenoidea, small in size and always cemented. They are represented today by only one genus, *Lacazella* (p. 10) which has survived from the Tertiary. Many fossil forms are known, though often overlooked because of their small size; all are said to have a dense irregularly punctate shell. Rhaetic–Rec.

SUBORDER 2. **Orthoidea.**—The shells have a circular or semi-elliptical outline; they may be biconvex, concavo-convex or exceptionally, convexo-concave. The hinge-line may or may not form the greatest breadth. Each valve has an interarea; crura occur in the dorsal and teeth in the ventral valves. Two groups are recognized; the **Orthacea** with impunctate shells is known from the Lower Cambrian to the Middle Devonian and is thought by some to provide the root stock of all Articulata; the **Dalmanellacea,** with punctate shells, ranges from Middle Ordovician to Permian. An example of the Dalmanellacea is *Resserella* [*Dalmanella*] *elegantula* of the Wenlock Limestone, a small shell, nearly circular in outline, little over a centimetre long and about as broad (Figs. 7*d*, *e*). The hinge-line is less than the greatest breadth of the shell. The ventral valve is very convex, with a well-defined slightly concave interarea, a triangular delthyrium without any deltidium, and teeth supported by dental plates. The dorsal valve is nearly flat, with a very narrow interarea, and a very small, bifid cardinal process occupying the centre of a small notothyrium sited at the end of a slight median septum; the inner side of each dental socket is produced into what at first sight looks like a tooth, but as these do not fit into sockets in the ventral valve they cannot be teeth but must be crura like those of the Rhynchonelloids. The muscle-scars in the ventral valve are very short, only occupying about one-third the length of the valve. The surface is marked by rather fine radial costae, which here and there increase in number by bifurcation or intercalation. The genus *Resserella* ranges from Ordovician to Silurian.

Other genera of the Dalmanellacea are—*Rhipidomella* (Sil.–Perm., not above Coal Meas. in Great Britain), with very short hinge-line, both valves convex, slight dorsal sinus, finely costate, ventral muscle-scars longer than in most Orthoidea, divaricators completely surrounding adductors; *Schizophoria* (Sil.–Perm.), with rather long hinge-line, dorsal valve more convex than ventral, dorsal fold and ventral sinus, surface with extremely thin radial costae, ventral muscle-scars rather long; *Dicoelosia* [*Bilobites*] (U. Ord.–L. Dev.), each valve with a median constriction, making the valves bilobate. Among the impunctate group (**Orthacea**), *Billingsella* occurs in the Middle Cambrian; it has a broad ventral interarea and a delthyrium partly closed by a deltidium. Other genera include *Orthis* (L.–M. Ord.), with long hinge-line, coarse radial costae not increasing in number; *Dinorthis* (M.–U. Ord.) with convexi-concave shell with rounded outline and strong costae; *Nicolella* (M.–U. Ord.), hinge-line provides greatest width of shell, dorsal valve flat or slightly concave, strong costae duplicated by later intercalation.

SUBORDER 3. **Pentameroidea.**—The uppermost Ordovician and lowest Silurian beds in parts of England and Wales both contain calcareous sandstones crowded with fossils which are largely in the form of internal moulds. Many of these fossils differ by characters too slight to be detected in the field; but the geological surveyor knows at once that he has Silurian and not Ordovician rocks before him when he sees brachiopod moulds of rounded outline cut almost in half by deep fissures. These fissures in the mould correspond to large internal plates in the shell, such as only occur in the Pentameroids, a group of brachiopods which migrated in force into the British area at the opening of the Silurian period.

The suborder ranges from Ordovician to Devonian.

One of the most familiar Pentameroids is *Conchidium* [*Pentamerus*] *knighti* (Fig. 7*j*) of the Aymestry Limestone. This is a large shell, attaining a length of 7·5 cm., and then about 4 cm. broad and 5·5 cm. thick. Both valves are very convex, with the umbo of each greatly curved over; there is no interarea; the delthyrium is broad and bears a concave deltidium (often lost); the hinge-line is short and curved. The surface is marked by coarse rounded costae which increase in number sometimes by bifurcation, sometimes by intercalation of a new costa. There are no regular concentric markings, but occasional stoppages in growth are indicated by growth-lines.

It is not easy to extract a perfect specimen, because the fossil splits so readily along the large internal plates; and it is not difficult to get an almost median section. By combining what we see in such a section with the evidence of a transverse section (Figs. 7*j*′, *j*′′) we can realize that the mantle-cavity in the umbonal region and for some distance forwards was divided into three chambers, the middle one containing the body proper, the lateral chambers probably containing the spiral brachia. Towards the anterior end the three chambers united. In the ventral valve there is first a large median plate, which on closer inspection is seen to be two septa in contact with each other, a great development of the little septum seen in *Leptaena*. At its inner edge this is continuous with a pair of diverging plates which on being traced to the hinge can be recognized as greatly-developed dental plates. We have already seen that in *Leptaena* the rims of the muscle-areas are continuous with the dental plates, and it is evident that in *Conchidium* the development of these plates must lift the muscles well away from the inner surface of the shell into the median chamber. Such a pair of muscle-bearing dental plates is called a **duplex spondylium** as opposed to a **simplex spondylium** where the base rests on a single septum.

In the dorsal valve instead of a median septum there is a pair of **septal plates** a little on either side of the middle line; from these diverge at a low angle another pair of **crural plates** the free edges of which are in contact with those of the dental plates for some distance from the hinge, thus bounding the central chamber.

Although *Conchidium knighti* is often loosely referred to as *Pentamerus*, that genus only contains smooth-shelled species (e.g., *P. oblongus* of the Upper Valentian or Llandovery Series). Other genera closely allied are *Stricklandia* (e.g., *S. lens*, of the Lower Valentian or Llandovery), with a straight hinge-line and without the greatly-curved umbones, and *Gypidula* (e.g., *G. galeata*, Wenlock Limestone) in which there is a median elevation (**fold**) on the ventral valve, and corresponding depression (**sinus**) on the dorsal, a reversal of the more usual arrangement.

The Pentameroids all have biconvex impunctate shells with small obtuse-angled interareas in both valves. The delthyrium and notothyrium when present have no plates. The spondylium is characteristic of the group and some genera have a similar structure in the dorsal valve developed from the crura. In the **Pentameracea** elongated shells are more general than are transverse; this superfamily was probably derived from the **Syntrophiacea** which appeared in the Middle Cambrian and is abundant in the Ordovician and to the Middle Devonian. The Pentameracea are specially characteristic of the Silurian and Devonian. A third superfamily the **Clitambonitacea,** probably derived from Billingsellid Orthacea, is prominent in the Ordovician of Russia with the genera *Clitambonites* and *Kullervo*.

SUBORDER 4. **Triplesioidea.**—A small group formerly placed with the Strophomenoidea but with impunctate biconvex smooth or costate shells with a ventral sinus and dorsal fold. The hinge-line is short and normally each valve has an interarea. The cardinal process is strong and bifurcated. Ord.–Sil.

SUBORDER 5. **Rhynchonelloidea.**—Shell fibrous, impunctate, except for Rhynchoporacea (Carb.–Perm.) which are punctate. Pedicle functional, deltidial plates more or less embracing foramen, which is small, often elliptical, and does not encroach on the umbo (**hypothyrid**). Typical form short and stout (length, breadth, and thickness not very unequal) with short curved hinge-line, usually with a fold on the dorsal valve and sinus on the ventral; surface smooth or with fine radial lines or strong radial costae, sometimes spinose. Teeth supported by dental plates. Cardinal process usually wanting. Crura usually present. Ord.–Recent.

The Rhynchonelloids can generally be recognized by their form. The presence of a sinus in the ventral valve and fold in the dorsal, with consequent **V** or **W** shaped plication of the valve margin, is generally characteristic of this suborder; it is one method of separating the inflowing and outflowing water-currents.

In some cases they approach Terebratuloids in shape, but generally they can be distinguished by the impunctate shell, the surface of which, under a lens, commonly shows a silky fibrous appearance. At one time all the species of the suborder were included in the one genus *Rhynchonella*.

Among Palaeozoic genera are *Camarotoechia* (Sil.–Carb.; Fig. 9*c*) with a septum in the dorsal valve and prominent dental lamellae; *Sphaerirhynchia* (formerly called *Wilsonia*) (Sil.; Fig. 9*a*) similar, but it has a more globose shape and finer costation, bifid anteriorly in *S. wilsoni*; *Rhynchotreta* (Sil.) with acuminate beak on which the foramen encroaches, conspicuous deltidial plates, long dental plates, a thick median dorsal septum and a cardinal process; *Pugnax* (Dev.–Carb.) has no median septum in the dorsal valve but has a very strong fold and sinus, and with only a few strong costae or even smooth, as in the familiar Carboniferous Limestone species, *P. acuminatus* (Fig. 9*b*).

Among Mesozoic genera the true *Rhynchonella* (*R. loxia*) is rare, being only known from the Upper Jurassic of Russia: it is like *Pugnax acuminatus* in form, the sinus producing an acute anterior projection of the ventral valve. Other genera with a somewhat similar tetrahedral shape are *Homoeorhynchia* (mainly Inferior Oolite or Bajocian), and *Tetrarhynchia* (Middle Lias) but in the latter the

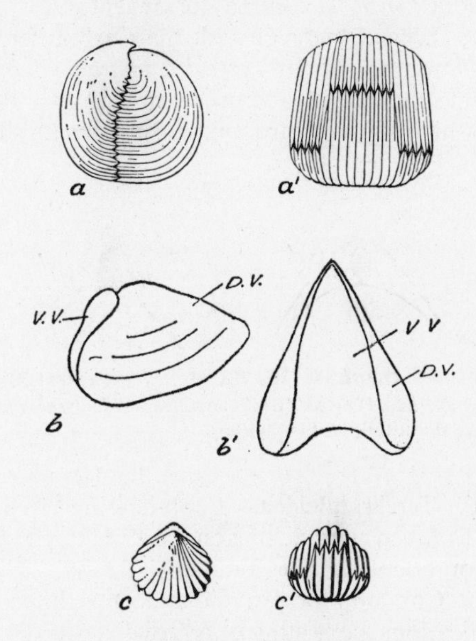

FIG. 9.—PALAEOZOIC RHYNCHONELLOIDS

a, a', *Sphaerirhynchia wilsoni* (J. Sowerby): Silurian (Wenlock Limestone), × 1; *b, b'*, *Pugnax acuminatus* (Martin); Lower Carboniferous × $\frac{1}{2}$. *a, b*, side-views; *a', b', c'*, anterior views; *c*, dorsal view; *c, c'*, *Camarotoechia nucula* (J. de C. Sowerby); Silurian (Upper Ludlow); × 1.

anterior projection is broad and deep, not pointed. *Rhynchonelloidea* is more globose than these, and with finer costation; in *Kallirhynchia* (Fig. 10*b*), *Burmirhynchia* (Fig. 10*a*) and *Rhactorhynchia* the fold and sinus are less pronounced, the former genus being finely, and the two latter more coarsely costate. These four genera are mainly Bathonian, with a varying range below and above. The fundamental distinctions between all these Jurassic genera are not, however, external, but internal—mainly seen in the muscle-scar patterns. *Furcirhynchia* is a Pliensbachian genus in which the costation is at first very fine, and suddenly changes to coarse. *Acanthothyris* (Aalenian–Bajocian) and *Acanthorhynchia* (Upper Jurassic) have thin costae, from which delicate spines arise.

The commonest Cretaceous species belong to the genus *Cyclothyris*: they are rather broad forms, with thin and close-set costae, distinct fold and sinus, and deltidial plates protruding more or less as a tube around the pedicle. Such are *C. latissima* of the Lower Greensand, *C. grasiana* of the Upper Greensand, *Cretirhynchia plicatilis* of the Chalk by contrast has a median dorsal septum.

The aberrant **Stenoscismatacea** with the Carboniferous and Permian genus *Stenoscisma* (formerly called *Camarophoria*) has a prominent spondylium simplex with a low median septum and in the dorsal valve there is a high septum surmounted by a trough-like plate and crura.

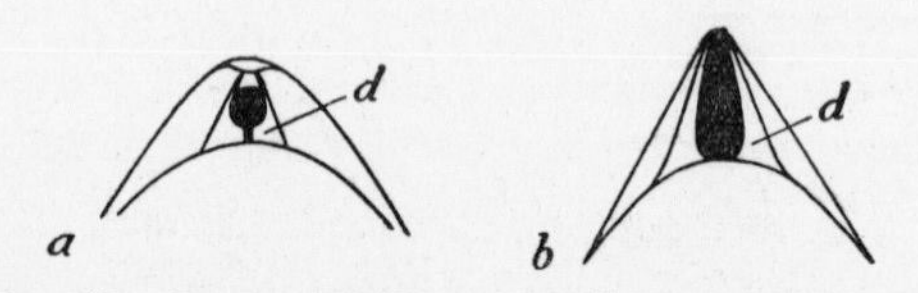

FIG. 10.—DELTIDIAL PLATES OF RHYNCHONELLOIDS
a, *Burmirhynchia obsoleta* (Davidson); *b*, *Kallirhynchia concinna* (J. Sowerby), young × 2; *d*, deltidial plate (Original).

SUBORDER 6. **Terebratuloidea.**—Loop-bearing Articulata, with punctate shell, foramen encroaching on the umbo (**epithyrid**), and deltidial plates not bounding the foramen laterally. Dental lamellae normally absent. Common form, ovoid and smooth; sometimes with a few coarse plications, consisting of a radial elevation (**fold**) on one valve, and a corresponding depression (**sulcus** or **sinus**) on the other; very rarely with numerous costae. Ventral umbo and foramen usually much larger than in Rhynchonelloids.

The Terebratuloids first appear late in the Silurian period. One of the Devonian genera has so many peculiarities that it is best described apart from the bulk of the suborder.

Stringocephalus burtini of the Middle Devonian is a large shell, nearly circular but for its straight hinge-line and high and pointed (rostrate) ventral umbo. Unlike other Terebratuloids it is hypothyrid, the deltidial plates arching over and meeting above the foramen. The shell does not show the typical Terebratuloid pattern of punctation. The loop is long and wide, parallel to the margin of the dorsal valve; and there is a median septum in each valve. The cardinal process is so long that it bifurcates to avoid the ventral septum.

The remainder of the Terebratuloids are roughly divided into **short-looped** forms with obsolete dorsal septum (**Terebratulacea**) and rarely with dental plates, and **long-looped** forms (**Terebratellacea**) with strong dorsal septum, and in most cases dental plates.

The short-looped forms begin with the Devonian *Centronella*, smooth, with a very simple loop, not doubled back like that of later genera. *Dielasma* (Dev.–Perm.) has a doubled-back loop, large crural lamellae running towards one another from the top of the descending lamellae, and strong dental plates. Similar forms without dental plates are common in the Jurassic, such as *Lobothyris* (Sinemurian–Kimmeridgian) a smooth form, longer than broad, without any fold; *Cererithyris* (Bathonian; Figs. 1 and 3), broader, with large foramen and slight double fold; *Epithyris* (Bajocian–Bathonian; Fig. 5), broad, with strong double or quadruple fold; *Sphaeroidothyris* (Pliensbachian–Callovian), globose, beak short, foramen small, with wavy anterior margin, but no actual plication; *Plectothyris* (Bajocian only), which abruptly develops coarse ribbing after a smooth youth; *Pseudoglossothyris* (Bajocian only) with large foramen, the dorsal valve much flattened, and unusually long loop; and *Dictyothyris* (Bathonian), very broad, with two strong ventral and one slighter dorsal folds, with very fine costae which become crossed by ridges producing a delicate cross-hatched pattern. As with the contemporary Rhynchonelloids, the muscle-scars and other internal features are essential to generic identification, and there are homoeomorphs of misleading similarity in outward form.

Terebratulina (Jur.–Recent) is finely costate and has a ring-like (annular) loop; *Gryphus* (Cret.–Rec.) has a number of radial grooves in the inner surface of the ventral valve (well seen in internal moulds).

Although some Devonian forms have a long loop, the typical series of long-looped forms (Terebratellacea) begins in the Lower Jurassic, with the simple and minute *Zellania*, in which the loop is incomplete and the dorsal septum anterior in position. This is followed by *Megathyris*, a broad form with a loop not unlike that of *Stringocephalus*, but supported by two lateral septa as well as the median septum. Several allied genera are known in the sea today, survivors of this primitive group. The more recent genera fall into families, one in northern, the other in southern seas, and the latest in each case (*Dallina*, northern; *Magellania*, southern) has a very similar adult loop, but they attain it respectively through two quite distinct series of metamorphoses, and the temporary loop-forms of these genera correspond to adult loop-forms of other genera, mostly extinct. These facts are illustrated in Fig. 11. It will be noticed that three of the intermediate stages are represented by recent, not fossil, genera (*Platidia*, *Megerlina* and *Magasella*). This can be accounted for by the imperfection of the geological record, for the genera in question are small forms, easily overlooked by collectors.

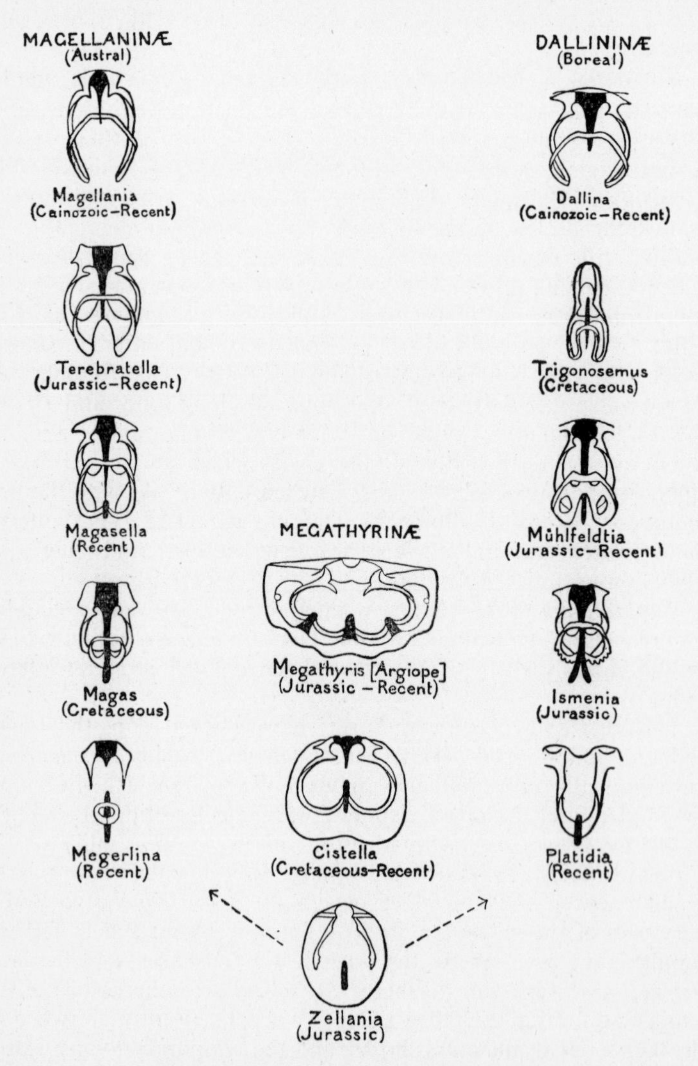

FIG. 11.—EVOLUTION OF THE LOOP IN LONG-LOOPED TEREBRATULOIDEA
Loop in outline, septum black (After Beecher).

In addition to the genera illustrated in Fig. 11, the following are noteworthy:

Ornithella (Jurassic; Figs. 2, 5*b*), smooth, with straight anterior margin; *Rugitela* (Jur.), similar, with longer dorsal septum; *Aulacothyris* (Jur.), smooth, with ventral fold and dorsal sinus; *Terebrirostra* and *Trigonosemus*, Cretaceous, plicate forms, with narrow, high interareas.

SUBORDER 7. **Spiriferoidea.**—Articulata, with calcereous **spiralia** or spirals; deltidial plates in some cases fused into a single plate. Ord.–L. Jurassic.

Spirifer striatus (Fig. 7*k*) of the Carboniferous Limestone is a very broad shell, being just about twice as broad as long. The greatest breadth is at the hinge-line, which is straight. Both valves are strongly convex, but the ventral has a deep median sinus, the dorsal a corresponding fold. The surface is patterned with numerous radiating costae. The umbo of the ventral valve is prominent, but not large, and below it is a large interarea showing growth-striations at right-angles to the hinge-line; it includes a triangular delthyrium, in which a large deltidium partly closes the pedicle-foramen. In the dorsal valve there is a much narrower interarea.

The internal structure can only exceptionally be made out. The chief feature is the pair of spirals. Starting from the crura the ribbons first approach one another, and then, as the coiling begins, diverge from one another, so that the apices of the spiral cones are near the outer ends of the hinge-line. From this it would seem that, unlike most other brachiopods, *Spirifer* had a median inhalent current and a pair of lateral exhalent currents. The first turn of one spiral is joined to that of the other, on the dorsal side, by a simple bar, the **jugum.** In the ventral valve there are short dental plates.

Various other species of *Spirifer* answer this same description closely, e.g., *S. bisulcatus* of the Carboniferous Limestone, or *S. verneuili* of the Devonian. The genus in its widest sense ranges from Devonian to Permian, but there is an allied form, *Eospirifer* with fine filiform costae in the Silurian, e.g., *E. radiatus* of the Wenlock Limestone; the genus ranges locally into the Lower Devonian.

Other genera closely allied to *Spirifer* are—*Martinia* (Carb.–Perm., without dental plates and with a tendency to smoothness of surface); *Syringothyris* (U. Dev.–L. Carb.) with a very large ventral interarea situated at right-angles to the plane separating the valves, instead of parallel to it as usual; there is in the delthyrium a peculiar "split-tube" or syrinx penetrating a transverse plate lying between the dental plates; and *Cyrtia* (Sil.–Dev.; Fig. 7*l*), with a high ventral interarea, and narrow delthyrium with the deltidium perforated centrally for the pedicle. Not quite so near are *Cyrtina* (Dev.) and *Davidsonina* (L. Carb.) with a punctate shell and a ventral median septum supporting a spondylium; in *Cyrtina* the spondylium is simplex, in *Davidsonina* it is duplex: otherwise they resemble in form *Cyrtia* and *Spirifer* respectively. *Spiriferina* (Carb.–Jur.) also has a punctate shell and radial costae, but the ventral median septum is not attached to the dental plates. All these, and various other genera, are usually united into one superfamily **Spiriferacea.**

In the superfamily **Atrypacea** we find the Ordovician spire-bearers, such as *Zygospira*, in which not only are the spirals short and simple, but their apices are directed towards one another, so that they must have had a median exhalent current as in normal brachiopods. Thus in every respect they are the most primitive spire-bearers. The later (Sil.–Dev.) and more familiar *Atrypa* (Fig. 7*m*) has the spiral cones parallel, with apices turned dorsalward. In this genus the dorsal valve is more inflated than the ventral; both valves are costate and may have concrescent lamellar outgrowths. In the smooth-shelled *Dayia* (Sil.–Dev.) the jugum is anteriorly situated and drawn out posteriorly into a simple short stout process from which arose laterally directed spiral cones; ventral valve has two chevron-shaped muscle-scars arranged on divergent projections from the inner-shell surface.

Lastly, the superfamily **Rostrospiracea,** which includes the family Athyridae with a most complex brachial skeleton. The spirals point away from one another as in *Spirifer*, but they start by a sharp bend back from the crura, and the jugum is never a simple bar, as in the two other superfamilies: its least complex form is a Y, but it may be scissors-shaped or have still more elaborate forms. Genera in this superfamily usually have a rounded form and short hinge-line. *Meristina* is a Silurian genus, with a median dorsal septum, and paired ventral septa (the converse of the contemporary Pentameroids); *Athyris* (Dev.–Carb.) and *Composita* [*Seminula*] (Carb.–Perm.) are both smooth-shelled forms, the latter mimicking the contemporary *Dielasma*, but distinguishable from it not only by its internal spirals, but also by its non-punctate shell. *Cleiothyridina* (Carb.–Perm.) is like *Athyris* and has lamellose concentric shell outgrowths but they are in the form of projections.

The Spiriferoidea appear in the Ordovician, where only primitive Atrypacea and Rostrospiracea are found; the Spiriferacea appear in the Silurian and all continue to the Carboniferous, when the Atrypacea die out. The other two superfamilies survive to the earlier part of the Jurassic period, some species of *Spiriferina* being common locally in certain zones of the Lias.

The Upper Lias seems to mark the time of the final extinction of all Palaeozoic types of brachiopods—not only the Spiriferacea (*Spiriferina*) and Rostrospiracea (*Koninckella*), but also the Chonetacea (*Cadomella*) making their last appearance.

Short Bibliography

This and succeeding short bibliographies are intended to give the student some idea of the original works from which he may obtain more detailed information than can be given in any textbook, and of which he must study those which are relevant if he is engaged in research—even the simple research involved in naming accurately the fossils he collects. In selecting works for these lists, consideration has been given to accessibility (to English students) as well as to importance. Reference should also be made to Chapter XIII and the bibliography there given. In these short bibliographies, where reference is made to a paper printed in a scientific periodical, the number of the volume is given in Roman numerals (e.g., xxvi) and the first page of the paper quoted in Arabic (e.g., 26), the abbreviations "vol." and "p." being usually omitted. An Arabic numeral, thus (3), preceding volume no., denotes Series.

ARBER, Muriel A.—Papers on Structures in Strophomenoids, *Geol. Mag.*, lxxvi, 82; lxxvii, 161; lxxix, 179 (1939–42).

BUCKMAN, S. S.—"Brachiopoda of Namyau beds," *Palaeont. Indica* (n.s.), iii, mem. 2 (1917).

COBBOLD, E. S.—"The Cambrian Horizons of Comley (Shropshire) and their Brachiopoda . . . ," *Quart. Journ. Geol. Soc.*, lxxvi for 1920, 325 (1921).

COOPER, G. A.—"Chazyan and related Brachiopods," *Smithson. Misc. Coll.*, cxxvii (2 pts.) (1956) 269 pls.

DAVIDSON, T.—"British Fossil Brachiopoda," *Palaeontographical Society*, 6 vols. (1851–86). The last volume is a full bibliography of Brachiopoda up to date of publication.

GEORGE, T. N.—Papers on Carboniferous Brachiopods, *in Geol. Mag.*, lxiv, 106, 193 (1927); lxvii, 554 (1930); and *Quart. Journ. Geol. Soc.*, lxxxvii, 30; and lxxxviii, 516 (1931–2).

HALL, J., and CLARKE, J. M.—"Paleozoic Brachiopoda," *Paleont. of New York*, vol. viii (1892–4). A fundamental work on classification, summarized in (2) "Introduction to the Study of the Brachiopoda," *Rep. New York State Geologist*, pts. 1 and 2 (1894–5).

JONES, O. T.—"*Plectambonites* and some allied genera," *Mem. Geol. Surv. Gt. Britain: Palaeont.*, i, pt. 5 (1928).

KOZLOWSKI, R.—"Les brachiopodes gothlandiens de la Podolie polonaise," *Palaeont. polonica*, i (1929).

MUIR-WOOD, Helen M.—(1) "British Carboniferous Producti," *Mem. Geol. Surv. Gt. Britain: Palaeont.*, iii, pt. 1 (1928); (2) "Internal Structure of some Mesozoic Brachiopoda," *Phil. Trans. Roy. Soc.* (B), ccxxiii, 511 (1934); (3) "Brachiopoda of the British Great Oolite Series, Pt. 1. (Fuller's Earth)," *Palaeont. Soc.* (1936); (4) *A history of the classification of the phylum Brachiopoda*. Brit. Mus. (Nat. Hist.) publ. (1955).

MUIR-WOOD, Helen M., and COOPER, G. A.—"Morphology, Classification and Life Habits of the Productoidea (Brachiopoda)," *Mem. Geol. Soc. Amer.*, 81, 447 pp. (1960).

SCHUCHERT, C., and COOPER, G. A.—"Brachiopod Genera of the sub-orders Orthoidea and Pentameroidea," *Mem. Peabody Mus. Yale*, iv, pt. 1 (1932).

THOMAS, I.—(1) "British Carboniferous Orthotetinae," and (2) "British Carboniferous Producti," *Mem. Geol. Surv. Gt. Britain: Palaeont.*, pts. 2 (1910) and 4 (1914).

THOMSON, J. A.—"Brachiopod Morphology and Genera (Recent and Tertiary)," *N.Z. Board of Science and Art Manual*, no. 7, Wellington, N.Z. (1927).

VAUGHAN, A.—(1) "The Carboniferous Limestone Series (Avonian) of the Avon Gorge," *Proc. Bristol Nat. Soc.* (4), i (1906); (2) Various papers in *Quart. Journ. Geol. Soc.* (1905–15).

WAAGEN, W.—"Salt Range Fossils: Brachiopoda," *Palaeontologia Indica* (13), vol. i, pt. 4 (1882–85).

WALCOTT, C. D.—Cambrian Brachiopoda, *Monograph U.S. Geol. Surv.*, 51 (1912).

WILLIAMS, A.—"North American and European Stropheodontids: their Morphology and Systematics," *Mem. Geol. Soc. Amer.*, no. 56 (1953).

II

THE LAMELLIBRANCHIA

ASSOCIATED with the Terebratuloids in the Cornbrash, though for the most part less well preserved, are other bivalved calcareous shells clearly differing from brachiopods. In many of them the two valves are known from the physiology of living lamellibranchs to be right and left, not dorsal and ventral; they are like mirror-images of one another, and though in some the two valves are unequal, they do not show the perfect symmetry of brachiopod-valves. They never have a pedicle-perforation, and though there is a hinge-line and in many cases a cardinal area, the internal muscle-impressions are quite unlike those of brachiopods, and a brachial skeleton is never present.

These bivalves belong to a phylum distinct from the Brachiopoda —the great phylum Mollusca, of which they constitute a class, the Lamellibranchia or Pelecypoda. The names respectively refer to the plate-like gills or hatchet-shaped foot of the living animal. The resemblances between the shells or exoskeletons of lamellibranchs and brachiopods is due to the animals leading a similar life and having the same needs; the differences are due to their different ancestry; that is, a different initial structure was adapted to the same needs.

As in brachiopods, each valve has an umbo and near it the hinge-line, but the anatomy of the animal shows that these structures mark the **dorsal** region, not the posterior as in brachiopods. The region where the valves separate most widely when the shell opens is therefore **ventral** (instead of anterior). The measurement from the dorsal to the ventral edge is the **height** of the shell, the measurement at right angles to this (in the plane separating the valves) is the **length,** being from the anterior to the posterior end; and the measurement across both valves, at right angles to both height and length, is the **thickness.**

The shell is chiefly composed of calcium carbonate in the form of calcite or aragonite or both with a small amount of organic matter, **conchiolin,** which forms the outermost layer or **periostracum.** Typically the carbonate part of the shell consists of two layers, an outer prismatic calcite layer and an inner lamellar layer. In some genera very thin lamellae of aragonite produce an inner **nacreous** or pearly layer. The prismatic layer may be absent in some families, e.g., Chamidae.

Lamellibranchia are not found in the strata older than Tremadoc; they were rarer than Brachiopoda throughout the Palaeozoic era; as the Brachiopoda diminished in numbers, the Lamellibranchia increased, until in the Cainozoic era and at the present time they far outnumber the Brachiopoda. They are less restricted in their habitat, a few genera being found in fresh waters, though the great majority are **marine.**

Though the mode of life is similar to that of brachiopods, they are more free to move about. A few fix themselves after the larval stage by cementation (as the oysters) or by silky threads (**byssus**), as the mussels, but none is fixed by a muscular pedicle; many burrow in sand, or mud, a few bore into harder materials, the majority move sluggishly about the bottom, and a very few (as the scallops) swim by a series of jumps.

In the majority of cases, the animal is bilaterally symmetrical, and the right and left valves are as nearly counterparts as is possible, except that internally the hinge-teeth of the one must come opposite the sockets of the other (**equivalve** shells). But in some families, especially those with a fixed habit, the valve on which the animal lies is the larger (**inequivalve shells**). This may be the left valve, as in the oyster, or the right as in scallops.

1. **Glycymeris** [*Pectunculus**]. As a first example of a lamellibranch, this genus is chosen as it is one very common at the present day and in the Cainozoic era. Familiar species are *G. glycymeris* of the Red Crag (Figs. 12, 13), and *G. deletus* of the Barton Beds. The shell is circular in outline and lenticular in shape, the valves are symmetrical to one another and very nearly symmetrical in themselves, the distinction between anterior and posterior ends being very slight. In fact the only external distinction in most species is that the umbo is very slightly nearer to the anterior end and faces towards the posterior end, and these facts would not help us much in deciding which end was which, for although in most lamellibranchs the umbo is nearer the anterior end, yet there are exceptions; and on the other hand the umbo usually points towards the anterior end. The only absolute test of the anterior end is the soft anatomy of the animal, but in the majority of lamellibranchs there are adequate clues in the shape of the shell.

The external surface is marked by radial and concentric lines: the

* A name in square brackets, as in this case, is an obsolete name which the student is likely to find used in many works in place of the correct name. See Chap. XIII.

latter are always very slightly marked, the former vary in different species from very faint striae to coarse costae.

Beneath the umbo and above the hinge-line there is an obtusely triangular **cardinal area,** which in a fossil shell is seen to be bounded above by a slightly raised margin, and to bear a number of ridges arranged like a set of chevrons or inverted **V**'s, one within the other (Fig. 13). In a living animal this area is concealed by a brown leathery mass uniting the two valves, known as the **elastic ligament,** composed of an organic substance, conchiolin. This elastic ligament is unknown in brachiopods, and it is connected with a fundamental difference in the method of opening the shell in the two groups. The lamellibranch has no divaricator muscles: when its adductor muscles contract so as to close the shell tightly, the elastic ligament is subjected to tension;

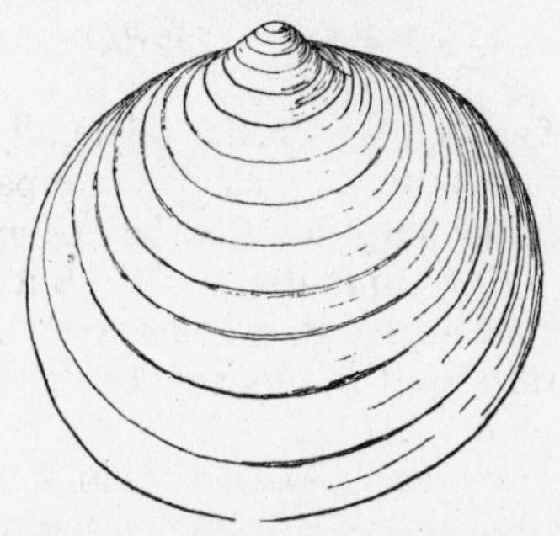

FIG. 12.—GLYCYMERIS GLYCYMERIS (Linné), Red Crag (Pleistocene) Exterior of right valve (Natural size, not full grown; original).

when they relax, the elasticity of the ligament draws the two cardinal areas towards one another and the valves open. Throughout life, however, the adductors are never completely relaxed, and the ligament is always more or less stretched, so that it is not until after death that the ligament has full play and the valves then yawn apart more than ever during life.

Fossil lamellibranch shells are most commonly found in one of three conditions, according to the rapidity of their burial after death. (1) When quickly buried (or when the animal lived and died in a burrow), the valves remain tightly closed, the weight of the sediment counteracting the tendency of the ligament to open them: in this case the interior is filled with calcite or other material deposited from solution (except in those genera whose valve-margins do not meet for their whole length). (2) If the muscles decay away before burial, the valves yawn open, and sediment drifting in prevents their closing, but they remain united. This case is well illustrated by the specimens of *Glycymeris brevirostris* that abound in the Bognor Rock, a band of hard sandstone in the London Clay of the Sussex coast. In some cases, as with *Dunbarella papyracea* from a marine band in the Coal Measures, flattish valves remain united when opened out to 180°. (3) If burial is long delayed, the valves are drifted about until not only the muscles but the ligament also is decayed; then the valves become completely separated. This is the case with the abundant valves of *G. glycymeris* in the Red Crag of Suffolk. Each of these three conditions may be represented by internal and external moulds, the material of the internal mould in the two latter cases being rock-

FIG. 13.—GLYCYMERIS GLYCYMERIS (Linné), Red Crag
Interior of left valve. From above down are seen in order—umbo, cardinal area curved row of teeth, edge of hinge-plate, adductor muscle-impressions, pallial line, crenulate ventral margin (Natural size; original).

matrix, in the first case usually calcite or other materials deposited from solution.

On examining the interior of a valve of *Glycymeris*, below the cardinal area a curved hinge-plate is seen bearing a row of **teeth,** all practically alike, usually slightly **V**-shaped and about twenty or more in number. Comparison of young and old shells of the same species shows that as growth proceeds new teeth are added at each end of the row, while the cardinal area, as it increases in size, encroaches on the central teeth until they become obliterated.

Around the margin of the ventral half of the valve is a series of radial grooves and ridges, by which one valve locks into the other when tightly closed. These simulate a continuation of the row of hinge-teeth, but are separated from it at either end by a short smooth area. Margins provided with such ridges and furrows are said to be **crenulate.**

The **adductor** muscle-impressions are two in number, anterior and posterior in position. They differ slightly in shape, concavity and size, more in some species than in others. From the inner margin a distinct line runs up to the umbo, marking off the path of shifting of each muscle as the growth of the shell proceeded.

From one adductor to the other there runs, parallel to the ventral margin, a distinct thin groove or line, the **pallial line.** This takes its name from the soft mantle or pallium, the two halves of which line the interior of the two valves, much as in brachiopods. The edges of these mantle-folds are thickened, both because this is the region by which new shell is secreted, and because there is here an important muscle (**orbicular** muscle) by which the lips of the mantle can be pressed together even when the valves are not tightly closed. The pallial line marks the inner margin of this thickening of the mantle. The mantle-edges separate only at certain points—(1) antero-ventrally, where the foot may be protruded; (2) posteriorly, where they enclose two openings like a figure 8; the lower opening being that by which water is sucked in (**inhalent aperture**), the upper that by which it is expelled (**exhalent**). These two apertures are in some lamellibranchs extended into long projecting siphons composed of soft tissues (not preserved fossil), and in that case the necessary local increase in the orbicular muscle, to form a **retractor** for the siphons, causes the pallial line to be more or less indented as it approaches the posterior adductor. Such an indentation is termed a **pallial sinus**: it is not present in *Glycymeris.*

It is necessary now to point out which of the characters seen in the shell of *Glycymeris* are not characteristic of all lamellibranchs, and for these technical terms must be given.

The shell of *Glycymeris*, having the umbo facing the hind end, is **opisthogyral.** The elastic ligament, as it lies entirely dorsal to the hinge-line, is an **external ligament** or **tension-ligament.** As it extends without interruption fore and aft of the umbo, it is **amphidetic:** the same adjective may be applied to the cardinal area. The hinge, being set with numerous teeth, generally similar except for size, is **taxodont.** As there are two adductor muscles, the shell is **dimyarian;** and, as these two are approximately equal in size, it is **isomyarian.** As there is no pallial sinus, it is **integripalliate.**

2. **Nucula** is a genus represented by several hundred species ranging

from Cretaceous to Recent, and allied forms are found as far back as the Silurian. The following description applied specially to the Pliocene and Recent species, *N. nucleus*, but in all essentials it will apply to any other (Fig. 14).

The shell outline is ovate tending to triangular, equivalve, the surface (as preserved in the fossil state) nearly smooth with very delicate concrescent striae. The umbones are opisthogyral, and situated much nearer the posterior end, so that without a knowledge of the living animal the posterior end would be taken for anterior. There is nothing like the amphidetic cardinal area of *Glycymeris*, but the region just behind the umbo is slightly flattened or concave, and in some species forms a large depressed area called an **escutcheon.** Internally, the hinge is taxodont, there being a narrow hinge-plate bearing a long

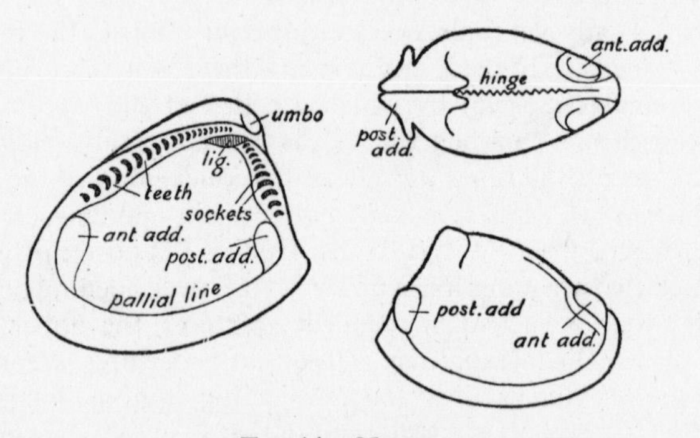

FIG. 14.—NUCULA

Left-hand figure, internal view of right valve of *N. margaritacea* Lamarck, Eocene (× 2; after Deshayes). Right-hand figures, dorsal view (above) and right side view of internal mould of *N. pectinata* J. Sowerby, Albian (Gault) (Natural size; after Woods).

row of simple, slightly curved teeth. Just below the umbo this row is bent and interrupted, and the hinge-plate extends downwards in a projection that bears a deep hollow facing its fellow in the other valve. These hollows are **ligament-pits,** the elastic ligament (rarely found fossil) running across the median plane from one to the other. Such a ligament, lying below the hinge-line, is described as an **internal ligament,** or better as a **resilium,** for when the valves are closed it is under compression not under tension like the external ligament of *Glycymeris*, and it is its resiliency (or tendency to recover its shape when the pressure is removed) that causes the valves to open. *Nucula* is isomyarian and integripalliate; its valve-margins are slightly crenulate. An important difference from *Glycymeris* is that the inner shell-layer is nacreous (pearly), that is to say it is composed of thin oblique carbonate laminae bound together by organic matter (**conchiolin**), and the reflection of light from the surfaces of such thin laminae results in interference and consequent iridescent colours.

3. **Trigonia** is a genus very abundantly represented by well-preserved specimens in the Jurassic system, to a less extent in the Cretaceous, and very rare in later formations, though still surviving

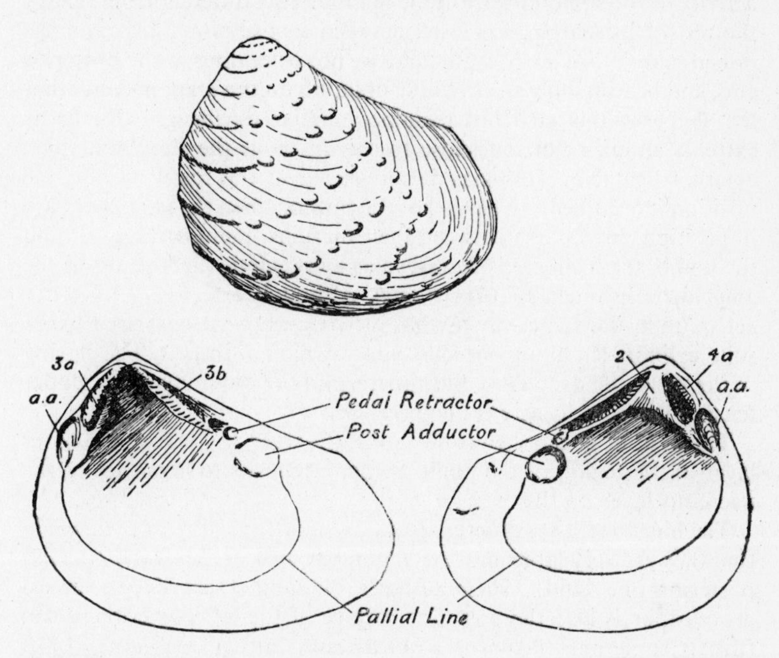

FIG. 15.—TRIGONIA BRONNI L. Agassiz, Upper Jurassic (Corallian), Normandy
Upper figure, left valve, exterior; left lower figure, right valve, interior; right lower figure, left valve, interior. *a, a.*, Anterior adductor. Numbers denote hinge-teeth (Natural size; original).

in Australian seas. The Upper Jurassic species *Trigonia bronni* (Fig. 15) is chosen for description, though most that is said, apart from details of shape and outer shell markings, will apply equally to other species.

The shell is equivalve, the outline between oblong and triangular, the posterior end truncated, the height about two-thirds the length; the umbo is near the anterior end and opisthogyral. The greater part of the surface is patterned with tubercles, arranged in rows curved obliquely to the margin and tending to run together anteriorly; fine growth-lines can be traced over the surface of the tubercles where these are not too worn; but from the umbo to the truncated posterior end there extends an area in which the markings are different, the concentric striae being more prominent, the tubercles much smaller and in three radial rows. As it is the edge of the mantle which secretes the shell, this difference in shell markings must depend on the secreting activity of the siphonal part of the mantle being different from that of the rest of its margin. Enclosed between this siphonal area and the dorsal margin is a narrow escutcheon, not extending to the posterior end, and bearing only striae whose obliquity to the margin shows that the shape of this area has remained similar during growth. In its extreme anterior portion, close to the umbo, is the short but thick external ligament. This may be preserved in the fossil; if not, the position it occupied can easily be recognized, when the two valves are in position, by the gap between their margins. Being entirely behind the umbo, the ligament is described as **opisthodetic** (in contrast to the **amphidetic** ligament of *Glycymeris*). Although external, it does not act quite in the same way as that of *Glycymeris*; it is shaped like a solid cylinder slit along one side, and yawning at the slit. The closing of the valves tends to close this fissure, and the outer layers are under tension while the centre is compressed.

The interior of the shell is nacreous; the wear and tear of the outer layer in the region of the umbo commonly leads to exposure of the nacreous layer on the exterior.

The hinge-teeth are different from those of *Glycymeris* or *Nucula* in that they are very large and few in number, two in each valve (the left posterior one bifid). Their surfaces of contact are deeply cross-grooved, as is also the posterior surface of the left posterior socket (this is sometimes taken as a third tooth, but it does not project beyond the median plane as a tooth must do). In the right valve the anterior tooth and in the left valve the anterior and part of the bifid tooth are carried on a kind of buttress (**myophoric lamina**) which also supports the anterior adductor muscle. The posterior adductor is rather larger, and is situated nearly half-way between the posterior end and the umbo, just behind the posterior hinge-teeth. Dorsal to it is a much smaller pit, marking the attachment of a **pedal** muscle, which retracts the foot. The pallial line has no sinus. The margins are not crenulate.

Thus *Trigonia*, *Glycymeris*, and *Nucula* are all three equivalve, opisthogyral, and integripalliate, and *Trigonia* and *Nucula* agree further in having a nacreous interior, and in being very **inequilateral** (though in *Trigonia* it is the anterior region, in *Nucula* the posterior, which is the shorter). The opisthodetic ligament and hinge-teeth are features in which *Trigonia* differs strikingly from both the other genera. This type of dentition is described as **schizodont.** According to the method of notation to be explained immediately, its formula is

$$\frac{3a,\ 3b}{4a,\ 2}.$$

4. **Cyrena semistriata** is an abundant fossil in the Oligocene beds of the Isle of Wight. The outline is sub-trigonal, the umbo being about one-third the total length back from the anterior end; the anterior border is rounded, the posterior part of the dorsal border sloping obliquely back to a narrow rounded posterior end. The outer surface has fine but well-marked concentric ribbing, with a tendency towards the margin to the intercalation of several finer ribs between each of

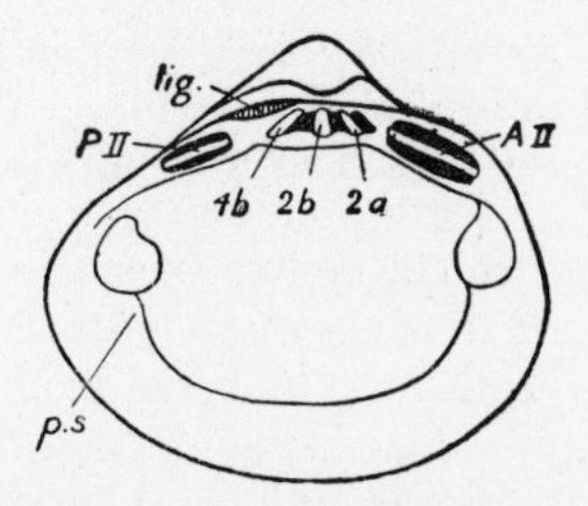

FIG. 16.—CYRENA SP. Recent
Interior of left valve. Teeth white, sockets black. *p.s.*, Pallial sinus; *lig.*, ligament (Natural size; original).

the main ribs. (The rest of the description will apply in general to any other species of *Cyrena*, such as that shown in Fig. 16.) The shell is equivalve. The outer surface is smooth. The interior is not nacreous, but **porcellanous,** having a smooth polished surface. The ligament is opisthodetic and short. Internally, the shell is isomyarian, and there is a very slight pallial sinus, so that it may be described as feebly **sinupalliate.** The hinge-teeth are carried on a well-defined hinge-plate and fall into three distinct sets (**heterodont**): just under the umbo are three short, more or less vertical teeth, radiating from the umbo—the **cardinal** teeth; in the anterior and posterior regions of the hinge-plates there are long teeth, more or less horizontal, parallel to the shell-margin—these are called (inaccurately) the **lateral** teeth. The anterior-lateral teeth come close up to the cardinals, but the posterior-laterals are separated from them by a space. The posterior-laterals are always posterior to the ligament—this fact is the essential distinction between them and the cardinals, the most posterior of which may in many heterodont shells be long and nearly horizontal, so that it might be mistaken for a posterior-lateral if its position relative to the ligament is not noticed. To come to details—in the right valve there are two anterior-laterals, the inner of which is on the inner margin of the hinge-plate, while the outer one is not quite on its outer margin; there are three cardinals, of which the middle one is bifid; and two posterior-laterals, occupying similar positions on the hinge-plate to the two anterior-laterals. In the left valve, there is a well-marked anterior-lateral tooth, not quite on the inner margin of the hinge-plate (since it has to interlock between the two anterior-laterals of the right valve); the outer margin of the hinge-plate is very slightly raised into a vestigial tooth. There are three cardinal teeth, each fitting in behind the corresponding tooth of the right valve. The posterior-laterals are similar to the anterior-laterals.

The general arrangement of the teeth is always the same in all Heterodonts: the lateral teeth of the left valve are always on the outer side of (i.e., above) the corresponding teeth of the right valve. This makes possible a convenient system of symbols, for which we are indebted to the French palaeontologists Munier-Chalmas and Bernard. The lateral teeth are numbered from within outwards, so that those of the right valves have odd numbers, those of the left even: thus the right anterior laterals are AI and AIII, the left AII and AIV; the posterior-laterals PI and PIII, PII, and PIV. The symbols for the cardinals are based upon the facts of early development, which show them to be the detached hook-like ends of laminae of which the main portion forms the anterior-laterals: thus the right middle cardinal is derived from lamina I, and so is symbolized by the arabic numeral 1; the other two are derived from lamina III, and are called 3*a* (the anterior) and 3*b* (the posterior), Similarly in the left valve the anterior and middle cardinals are both derived from lamina II, so they are called 2*a* (anterior) and 2*b* (middle), while the posterior comes from lamina IV, and so is called 4*b*. These facts are expressed diagrammatically in Fig. 17 (upper figure). It will be noticed that the alternation of the cardinal teeth follows that of the anterior-laterals from which they are derived—those numbered 2 fit in between those numbered 1 and 3; those numbered 3 between those numbered 2 and 4.

Heterodont hinges like that described above are known as the Cyrenoid type; there is another type with fewer teeth, the Lucinoid type, shown diagrammatically in the lower figure. In this, the central tooth of the hinge is in the left valve—it is 2, instead of 1. The complete dental formula for each may be written thus:

Lucinoid type ... { AI, III, 3*a*, 3*b*, PI, III.
AII, IV, 2, 4*b*, PII, IV.

Cyrenoid type ... { AI, III, 3*a*, 1, 3*b*, PI, III.
AII, IV, 2*a*, 2*b*, 4*b*, PII, IV.

A shorter formula, omitting the laterals, gives 2, 4*b*/3*a*, b for the Lucinoid and 2*a*, *b*, 4*b*/1, 3*a*, *b* for the Cyrenoid type.

A dentition rarely contains the full set of teeth, but it can still be classed as Lucinoid or Cyrenoid, though sometimes only by comparison with allied genera in which the dentition is complete. The laterals

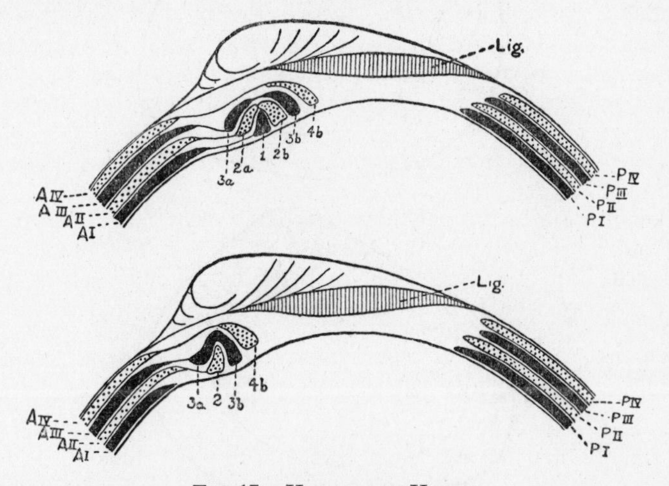

FIG. 17.—HETERODONT HINGES

Diagrammatic representation of Cyrenoid type (above) and Lucinoid type (below). Interior views of right valves, all possible teeth represented, and those originating from the same lamina shown as connected. Teeth of right valve, black; sockets corresponding to teeth of left valve, dotted.

are commonly reduced in number, and when the umbo is far forward the anterior-laterals may be crowded out altogether. The posterior-laterals, again, are a late development: in early forms with a long ligament they are wanting. Rarely, cardinals to be numbered as 4*a* or 5*b* may be developed.

Cyrena is a freshwater lamellibranch, ranging from Jurassic to Recent. Modern species live only in sub-tropical and tropical streams and mangrove-swamps, but the genus (if we include the subgenus

Corbicula, in which the lateral teeth are cross-striated like the teeth of *Trigonia*, but less deeply) survived in Britain until well on in Pleistocene time.

5. **Crassatella sulcata** is a common fossil in the Barton Clay (Upper Eocene) of the Hampshire Basin. In shape it is much like *Trigonia*, but the umbones are forwardly directed **(prosogyral)**. The surface is marked with strong concrescent ridges, but many other species of the

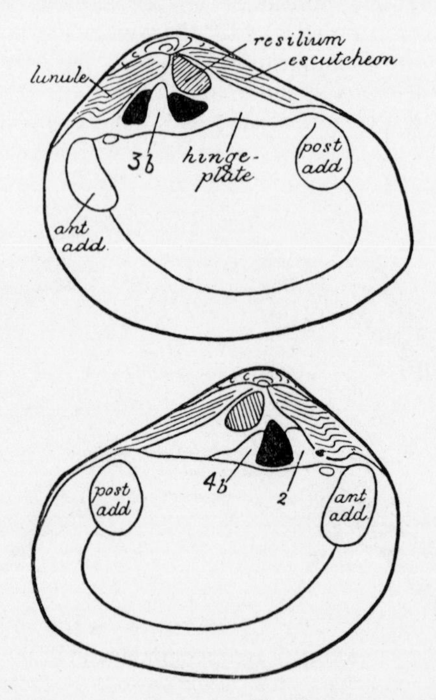

FIG. 18.—CRASSATELLA PONDEROSA (Gmelin), Eocene, Paris Basin
Right valve above, left valve below. Teeth white, sockets black (× ½; after Deshayes).

genus are nearly smooth (e.g., the large *C. ponderosa* of the Paris Basin, Fig. 18). As the hinge-structure is much easier to determine in this latter species, the following description is based mainly upon it. There is an escutcheon, and in front of the umbo there is a similar, but smaller, depressed area, the **lunule.** Internally the shell is porcellanous; it is isomyarian and integripalliate. Behind each anterior-adductor scar is a small pedal muscle-scar (not situated as in *Trigonia*). The hinge-plate is narrow in front and behind, but deep in the umbonal region, where it bears a **resilium-pit,** much larger in proportion than that of *Nucula.* In front of this are two nearly-vertical cardinal teeth: those of the left valve (2, 4*b*) are both large, those of the right valve fit in front of those of the left, but only the posterior tooth (3*b*) is large, 3*a* being hardly recognizable. The articulating surfaces of the teeth are slightly cross-grooved like those of *Trigonia.* Both anterior and posterior laterals are present in *C. sulcata,* one of each in each valve; but in *C. ponderosa* they are **obsolete**—i.e., so reduced as to be scarcely recognizable. As the right anterior-lateral fits under the left, while the right posterior fits over the left, the teeth are probably AI and AII, PIII and PII.

Crassatella, at the present day, is confined to the Indo-Pacific province and the tropical parts of the Atlantic. It is known fossil from Lower Cretaceous times onwards, and was less restricted in its

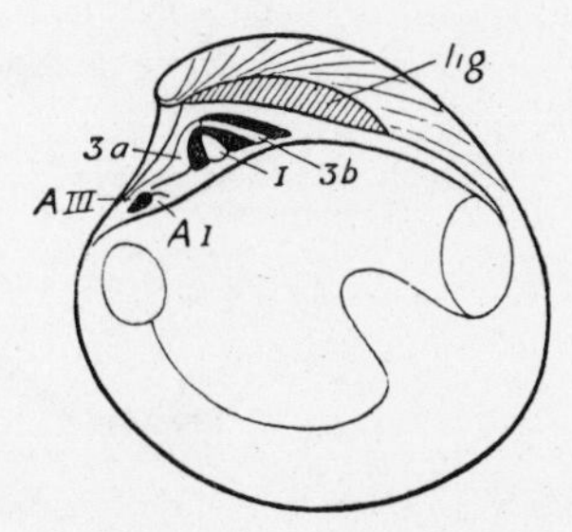

FIG. 19.—CORDIOPSIS SUBORBICULARIS (Goldfuss), Oligocene, Isle of Wight Interior of right valve (Natural size; original).

geographical range until after the Eocene, as is shown by its presence in the Barton Clay and (much more abundantly) in the Eocene of the Paris Basin.

6. **Cordiopsis suborbicularis** (Fig. 19) formerly known as *C. incrassata* is an abundant fossil in the "*Venus* Bed" of the Lower Oligocene of the Isle of Wight and Hampshire—that bed taking its name from the genus in which this species was at one time included.

The shell is rounded and sub-triangular, prosogyral, with fairly well-marked lunule, without escutcheon, but with a fairly long opisthodetic ligament. The surface is marked by concentric striae. The hinge-plate is large and thick, and bears three cardinal teeth in each valve, with two unequal anterior-lateral teeth (AI, AIII) in the right valve, and one (AII) in the left, the hinge being of Cyrenoid type. The absence of posterior-laterals is correlated with the length of the opisthodetic ligament: cnly in genera in which the ligament is shorter do posterior-laterals appear, to strengthen the hinder region, which no longer has a ligamentary connexion. The shell is isomyarian and **sinu-palliate,** there being a well-marked pallial sinus caused by the large size of the retractor muscles of the long siphons. The valve-margins are sharp and not crenulate.

Thus *Cordiopsis* is heterodont and sinu-palliate, and therefore representative of what is regarded as the most morphologically advanced type of lamellibranch—certainly one of the most modern types. The family to which it belongs (Veneridae) has a world-wide distribution today, but is not known earlier than the Jurassic period.

7. **Meleagrinella [Pseudomonotis] echinata** (Fig. 20) is the best-preserved lamellibranch found in the Cornbrash. It is small (length about 14 mm., height 16 mm., thickness 7 mm.), rounded, and

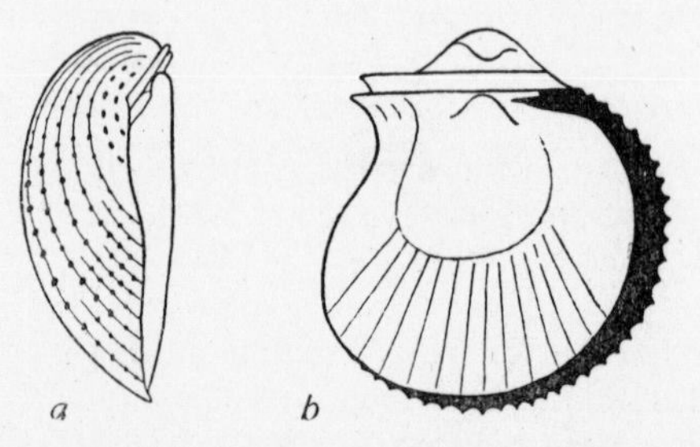

FIG. 20.—MELEAGRINELLA ECHINATA (J. de C. Sowerby), Bathonian
a, Posterior view; *b*, right side view: the black portion is the interior of the left valve where it overlaps the right. At the upper end of this is seen the very small right anterior ear, with the byssal notch below it; between it and the elevated umbo of the left valve is the amphidetic area of the left valve (× 2; original).

strongly inequivalve; the left valve being convex, with prominent umbo, and with spiny radial costae; the right valve being nearly flat, with an inconspicuous umbo, and nearly smooth, with only slight radial striae. The outline of each valve is somewhat gibbous—oval with the axis curved in a posterior direction ventrally; but as the outline is traced round to the umbo posteriorly it is seen to be deflected outwards, forming a flat triangular projection which has the effect of lengthening the hinge-line. Such projections are termed **ears** (or **wings**), and a shell possessing them is said to be **auriculate** (or **alate**). The posterior ears of *Meleagrinella* are fairly large, the left anterior ear is barely recognizable, but the right anterior ear, though small, is quite distinct and has a well-marked notch below it. Such a notch, from knowledge of recent forms, served for the passage from the interior of the shell of the **byssus** (a bunch of silky threads of conchiolin, secreted by the foot) by which the mollusc attached itself to a foreign object, temporarily or permanently. The inequivalve character is here, as in other lamellibranchs, associated with a fixed habit of life.

Between the umbo and the long straight hinge-line there is in each valve a long and low cardinal area, not separated into lumule and escutcheon. As in *Glycymeris*, this is occupied by an amphidetic ligament.

It is not easy to clean matrix from the interior of specimens of *M. echinata*, but from allied forms we may infer that hinge-teeth are feebly developed or obsolete, and that the shell is **aniso-** or **heteromyarian,** there being a large posterior adductor impression from which a pallial line extends indistinctly forward and ends in a very small anterior adductor impression. The edge of the left valve overlaps that of the right, and the small portion of the interior thus exposed is seen to be nacreous.

The genus *Meleagrinella* is found principally in the Jurassic rocks, but allied genera range from the Palaeozoic to Recent times. Among its Recent allies is the pearl "oyster" of the Pacific Ocean, *Pinctada margaritifera*, pearls being pathological secretions of nacre around parasites or other irritating bodies, and mother-of-pearl being the normal nacreous layer of the shell.

8. The genus **Ostrea** includes the common oyster which may live in brackish or in normally saline water. *O. ventilabrum* of the Oligocene (Fig. 21) is a convenient species to describe, but (except for shape and shell-markings many other species will conform to the description. The oysters are fixed (sessile) forms, and consequently very inequivalve. The left valve is fixed by cementation to some solid object—commonly some other shell—and becomes thick and convex; the right valve is nearly flat and forms a lid to the left. The outline (of *O. ventilabrum*) is nearly semicircular, the posterior margin from the umbo back being nearly straight. The height (in the morphological sense) is greater than the length (length 6 cm., height 7 cm., thickness 3 cm.). The shell is opisthogyral. In the left valve, a variable area of the umbonal region is adherent to some foreign body and shows only the impress of that instead of its own ornamentation. Beyond this area the markings consist of coarse diverging costae, subangular to rounded in section, increasing in number by bifurcation or by intercalation of new costae. The whole surface has a scaly look, due to the highly laminated structure of the shell. The right valve is nearly smooth, except for concentric corrugations due to growth irregularities, but also shows this laminated texture. In the umbonal region

there is an area of more irregular character corresponding exactly to the area of attachment of the left valve. That the surface of the foreign body should thus impress itself into the outer structure of a pair of thick valves is at first sight puzzling, especially as no trace of it is seen on the interior of either valve. The explanation is that the oyster fixed itself when very young, both valves of the very thin shell, as well as the animal's body between, being moulded up and down over the irregularities of the surface to which it was attached. As growth proceeded the shell not only extended beyond the area of attachment, but also increased in thickness by the addition of internal layers, which gradually smoothed over the irregularities of the inner surfaces but left those of the outer surfaces unchanged.

Internally, there are no hinge-teeth; there is a large triangular elastic ligament, partly external and partly internal, extending from

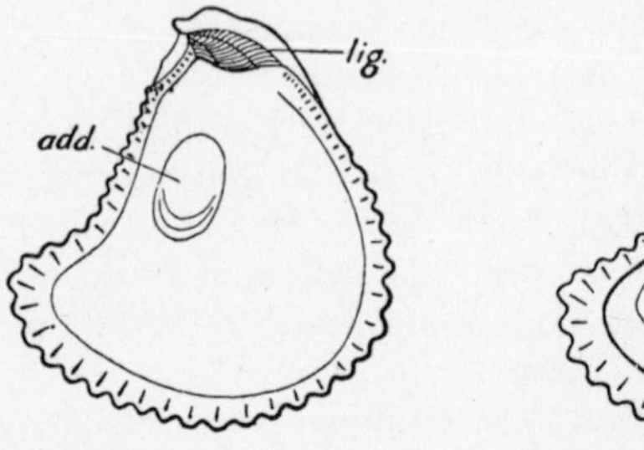

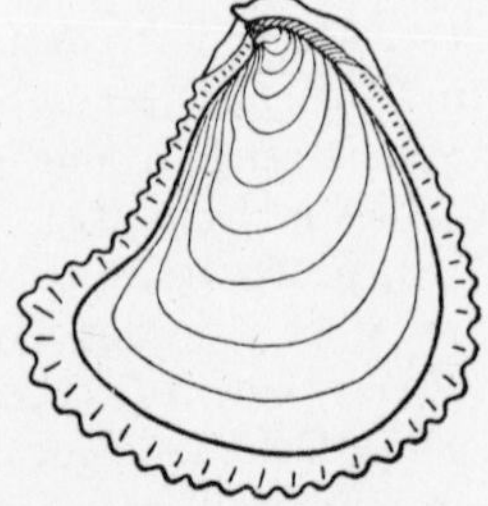

FIG. 21.—OSTREA VENTILABRUM Goldfuss, Oligocene

Right-hand figure shows both valves seen from the right side; in the left-hand figure the right valve is removed. *add.* Adductor impression; *lig.*, ligament (× ½; original).

the umbo to the hinge-line. When the valves are closed, the outer part of the ligament is probably under tension and the inner part under compression. The central part is thicker than the anterior and posterior, so that in the left valve the ligament-area appears divided into three parts, the central concave, the others flat; the whole ligament area is marked by fine horizontal striations.

There is only one adductor muscle-impression, posterior in position, the anterior adductor being entirely aborted in the adult: the shell is therefore **monomyarian.** The interior is **subnacreous,** the iridescence being faint.

The external costae give the valve-margin of the left valve, which projects beyond that of the right, a coarse crenulation. As this is traced towards the hinge it becomes supplanted by a slightly deeper-set series of crenulations, counterparts to which appear in the right valve, together forming an approach to a taxodont dentition. These "teeth" do not interlock tightly, however; they are only found in a few species of *Ostrea*; they are certainly not the relics of the hinge-teeth of the ancestors of the Ostreidae.

The oysters (including allied genera as well as *Ostrea* itself) have the strongest shells of any lamellibranchs, in the sense of best withstanding wear and tear: they may be washed out of their original deposit, rolled about by waves or streams, and retain a recognizable form when most of their associated fossils are pulverized. Hence they

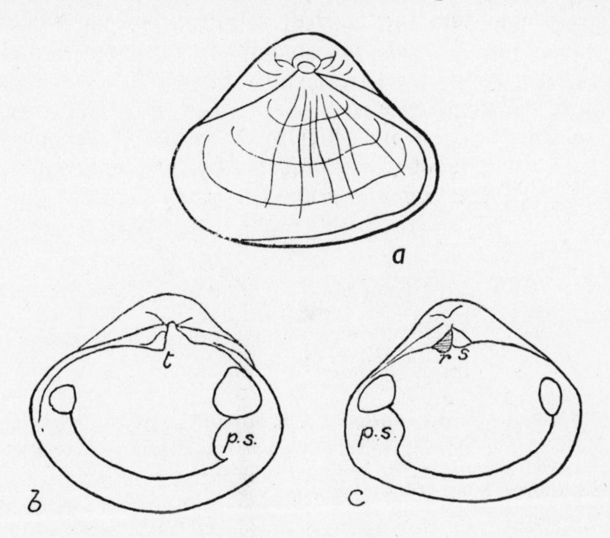

FIG. 22.—CORBULA GALLICA Lamarck, Eocene, Paris Basin
a, complete shell from left side; *b*, interior of right valve; *c*, interior of left valve; *p.s.*, Pallial sinus; *r*, resilifer; *s*, dental socket; *t*, hinge-tooth. *a*, after Deshayes; *b*, *c*, original (All natural size).

are commonly found as **derived fossils** in strata much younger than those to which they belong; Jurassic oysters, for instance, are very commonly found in the glacial gravels of Pleistocene times.

9. **Corbula** is a genus represented by various species with small shells in the Jurassic, Eocene and Oligocene strata of Britian, such as *C. pisum* and *C. revoluta* of the Barton Clay. The shell is inequivalve, the right valve being in every way larger than the left though the two scarcely differ in form; both being oval and more or less drawn out posteriorly (**rostrate**), with no outer shell-markings other than lines of growth. The umbones are nearly central in position and opisthogyral. The hinge-structure is best made out on large-shelled species such as

Corbula gallica of the Eocene of the Paris Basin (Fig. 22). In the right valve there is a sharp anterior-cardinal tooth projecting to an unusual extent towards the left valve and slightly upwards. Behind this is a space with a resilium-pit which instead of being vertical is horizontal, facing downwards, and placed within the umbo; behind this there is a long vestigial posterior-lateral tooth. In the left valve, a deep conical socket corresponds to the right cardinal tooth; behind this is a conspicuously-projecting horizontal plate (**resilifer**), which carried the resilium and underlies the resilium-pit of the right valve. The interiors are not nacreous; the shell is isomyarian and has a short pallial sinus. Fossil *Corbula*, presumably like those now living led a recumbent life, lying on its right valve; it was neither fixed to the substratum, nor a burrower.

10. **Pholadomya** is a genus which made its first appearance in the Lower Lias, was abundantly represented by species throughout the Jurassic, became less common in the Cretaceous, is still represented

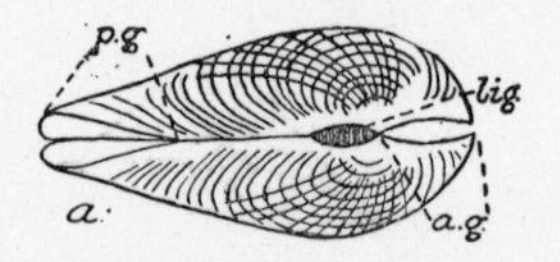

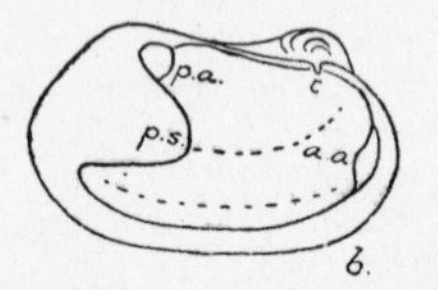

FIG. 23.—PHOLADOMYA CANDIDA (G. B. Sowerby), Recent, West Indies
a, dorsal view; *b*, interior of R. valve; *a.a.*, anterior adductor; *a.g.*, anterior gape; *lig.*, external ligament; *p.a.*, posterior adductor; *p.g.*, posterior gape; *p.s.*, pallial sinus; *t*, hinge-tooth ($\times \frac{2}{3}$; after Chenu).

by several species in the British Eocene beds and one in the Pliocene, but only survives today in deep waters in the Atlantic Ocean and West Indies (Fig. 23), although many of the fossil species (even up to the Pliocene) must have lived in shallow waters. We take *P. fidicula* of the Inferior Oolite as an example of a fossil form.

The shell is extremely thin, so much so that when an internal mould is found this requires careful examination to determine whether any of the shell is still adherent or not. Internally nacreous, the shell is equivalve, and strongly inequilateral, the umbones being far forward and prosogyral. The valves gape at the posterior end, the long siphons not being fully retractile; consequently the interior is always filled with rock-matrix and the posterior end has a rough, broken appearance where the matrix within joined that without. The presence of the gape indicates that the animal could burrow and the protruding siphons ensured its food supply whilst in the burrow. The surface is marked with numerous oblique, slightly curved radiating costae, dying away in the postero-dorsal area. (In some species the marking is more elaborate, with fewer and coarser costae and tubercles, and differing in different areas.) The shell-markings are shown by the internal mould as well as by the outer shell. There is a short, opisthodetic external ligament. In the fossil it is difficult to establish whether teeth are present or not; but in the recent species there is only a vestigial tooth. The shell is isomyarian and sinupalliate, though the impressions are too feeble for this to be determined in many cases.

The ten genera of Lamellibranchia here described form an inadequate sample of a very extensive class, for which from time to time various methods of classification have been proposed. The classification here adopted is that of the French palaeontologist Douvillé,

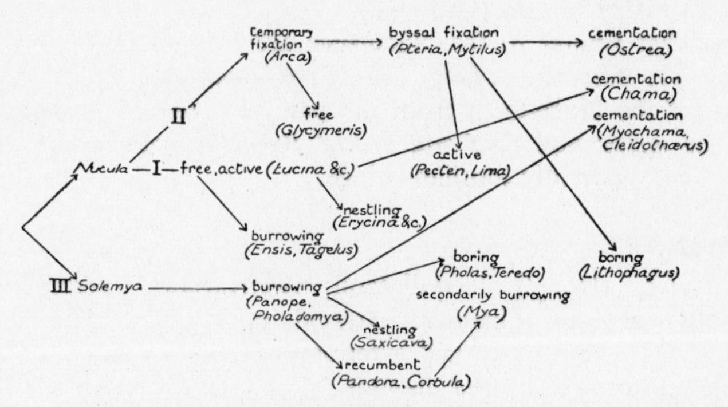

FIG. 24.—Adaptative Radiations in Lamellibranchs (from *Proc. Malac. Soc.*, xx, 1933). For "nestling" (against *Saxicava*) read "Boring."

according to whom the main lines of divergence of the lamellibranchs were determined by original differences in habits of life. There are three primary branches—active, sessile, and burrowing—so that we have here what has been termed by vertebrate palaeontologists an **adaptative radiation.** But the members of any lineage are not rigidly confined to one groove; in each lineage are cases of re-adaptation to one of the other modes of life, re-adaptations which lead sometimes to a close imitation of one type by another (**convergence**), but which can never obliterate or reverse the effects of ancestral history, which are shown by such shell-characters as those of the hinge (see Fig. 24).

A ACTIVE BRANCH

I. Nuculacea

Equivalve; isomyarian; internally nacreous; hinge-line primitively taxodont; without distinct cardinal area; surface rarely patterned otherwise than by concrescent striae. Ligament external (*Ctenodonta*) or internal (*Nucula*, *Nuculana*). Integripalliate (except *Nuculana* and *Yoldia*). Ordovician to Recent.

Nuculana [*Leda*] (Ord.–Rec., Fig. 25*a*) and *Yoldia* differ from *Nucula* in having the mantle drawn out into distinct though small siphons. This greater development of the posterior region results in a more central umbo, while there is a slight pallial sinus, and in *Nuculana* a tendency to more or less rostration (drawing out of the posterior end).

Ctenodonta may have some resemblance in form to *Glycymeris*, but is far earlier in time (Ord.–Sil.).

Somewhat similar to the Nuculacea are two Silurian–Devonian groups—(1) *Actinodonta* (Fig. 25*b*), *Lyrodesma*, etc., which show a change of the hinge-teeth from the parallel series of *Nucula* to a radiating series; and (2) *Cardiola* and *Buchiola*, with strong radial costae and apparently toothless hinge.

II. Naiadacea

Equivalve, strongly inequilateral; isomyarian; integripalliate; internally nacreous. Ligament external, opisthodetic. Hinge-teeth variable, often lamellar. Shell exterior rarely with other than concrescent striae. Freshwater habitat. [Devonian–Permian?] Trias.–Recent.

From Mesozoic times to the present day the genus *Unio* shows a hinge which is transitional between the actinodont and heterodont types: its formula may be given as 2*a*, *b*, 4*a*, *b*/1, 3*a*, *b*, 5*b*. Numerous other genera are associated with it in the family Unionidae.

Other freshwater mussels in the Newer Palaeozoic are provisionally united as a family Anthracosiidae: *Archanodon* is Devonian (Upper Old Red Sandstone), *Carbonicola* (Fig. 25*c*) and *Anthracosia* provide valuable zonal species for the Lower Westphalian Coal Measures, and *Palaeomutela* is found in the Permian of the Northern Dvina basin (North Russia) and the equivalent Lower Beaufort Series of South Africa. It is doubtful whether these Palaeozoic forms had a nacreous interior. Their range of shape is similar to that of the Unionidae. Other Coal Measures mussels may belong to the Mytilacea (see p. 65).

III. **Prae-Heterodonta**

Equivalve, inequilateral, isomyarian, integripalliate; internally nacreous or porcellanous. Hinge-teeth consisting of cardinals without anterior or posterior laterals. Ligament opisthodetic. Shell markings various. Devonian–Recent.

There are two groups—(*a*) **Trigoniacea,** with nacreous interior and dental formula 2, 4*a*/3*a*, *b* (Carb.–Recent); and (*b*) **Prae-astartacea,** with porcellanous interior and dental formula 2, 4*b*/3*a*, *b* (Dev.–Jur.).

The Trigoniacea are a small but important group. *Schizodus* (Carb.–Perm.) and *Myophoria* (Trias.) show a gradual change from smooth and prosogyral, with tooth 2 simple, to the highly-patterned and opisthogyral *Trigonia* (Fig. 15) with tooth 2 bifid. The geological and geographical distribution of this genus have already been mentioned.

The Prae-Astartacea form a less compact and less familiar group. It includes *Megalodon* (Devonian and Rhaetic) and *Pachyrisma* (Jurassic), large forms in which the hinge-region is short and high, with large, coarse teeth. Another more familiar form is *Cardinia* of the Lias (Fig. 25*d*, *d'*), resembling in shape the freshwater mussel, but marine and non-nacreous. We have already seen how in some brachiopods (*Ornithella*, see p. 8) the continuance of growth in thickness after growth in length and breadth has almost stopped leads to a change of shape in old age (**gerontic** form). A similar thickening is seen in aged individuals of *Cardinia*, but the allied genus *Hippopodium* takes on the gerontic condition while still young, and continues growing in thickness through the rest of its life, acquiring a very clumsy and heavy form, in which the earliest-formed portions of the valves lie back to back instead of closing on one another. This early assumption of senile characters is generally considered to indicate old age in the race or lineage, and in this, as in other cases, it is commonly followed by racial extinction.

IV. **Heterodonta**

Equi- or inequivalve; isomyarian; integri- or sinu-palliate; internally porcellanous. Hinge-lamellae differentiated into anterior-lateral and cardinal teeth, with frequent development of posterior-laterals behind the shortened ligament. Ligament opisthodetic, typically external, in some groups internal (resilium). Shell-markings various. Jurassic to Recent.

This very large and important order falls into two series, in each of which a number of suborders or superfamilies may be recognized.

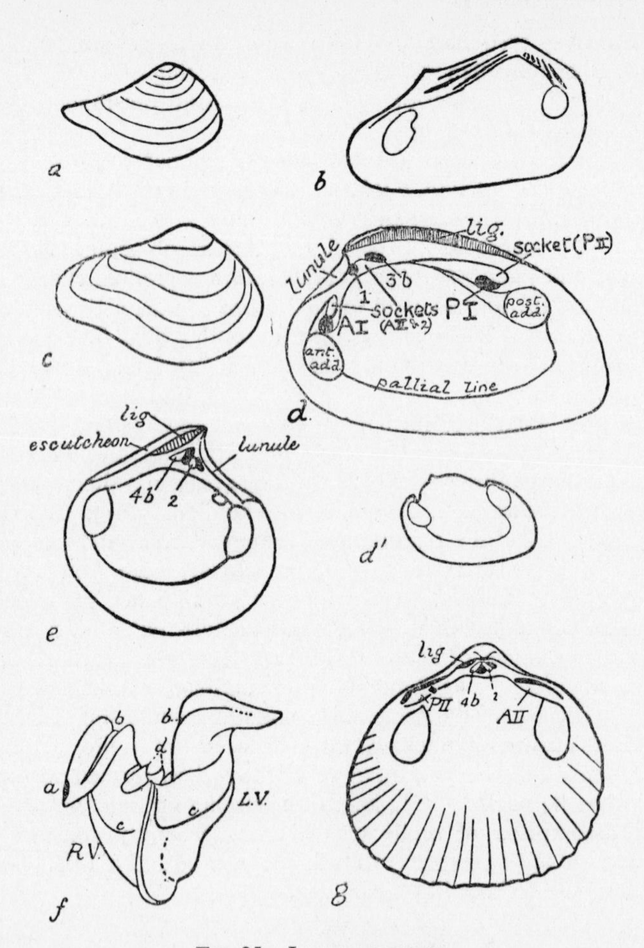

FIG. 25.—LAMELLIBRANCHS

a, *Nuculana lacryma* (J. de C. Sowerby), Inferior Oolite; right side (× 3; after Goldfuss); *b*, *Actinodonta cuneata* Phillips, Silurian. Interior of left valve; teeth white, sockets black (× $\frac{3}{4}$; after Phillips); *c*, *Carbonicola pseudorobusta* Trueman. Coal Measures, Upper Carboniferous, right side (× $\frac{1}{2}$; after Hind); *d*, *Cardinia listeri* (J. Sowerby), right valve (× 1); *d'*, *C. ovalis* (Stutchbury), internal mould (× $\frac{1}{2}$; after Quenstedt); *e*, *Astarte semisulcata* (Leach) Norwich Crag, Pleistocene; interior of left valve; teeth white, sockets black (× $\frac{5}{6}$ after S. V. Wood); *f*, *Diceras minor* Deshayes, Upper Jurassic; anterior view of internal mould (Natural size). *a*, Attachment; *b*, *c*, grooves on mould corresponding to myophoric laminae; *d*, moulds of tooth-sockets; *L.V.*, left, and *R.V.*, right valve (after S. P. Woodward); *g*, *Cardium parkinsoni* J. Sowerby, Red Crag, Pleistocene; interior of valve (× $\frac{1}{2}$; after S. V. Wood).

A. WITH TEETH OF LUCINOID TYPE: 2, 4*b*/3*a*, *b*

1. **Cardiacea.**—With radial costae, at first (*Protocardia*) in siphonal region only, later on whole surface. Umbones fairly central. Anterior and posterior laterals well developed.

The common cockle, *Cardium* (Fig. 25*g*), has a rounded outline, a central umbo, with well-defined and separated anterior-lateral, cardinal and posterior-lateral teeth; the whole surface is marked by strong radial costae. In the Mesozoic *Protocardia* radial ornament is confined to the posterior end (i.e., to the part secreted by the siphonal part of the mantle) the rest being concentrically marked (compare the case of *Trigonia*). In the Aralo–Caspian region, from Miocene to Recent, are found *Adacna* and *Limnocardium*, cockles adapted to a freshwater or brackish habitat, with long siphons and a pallial sinus.

2. **Rudistes.**—Highly aberrant fixed forms, probably derived from Cardiacea. Inequivalve. Fixed either by left valve (normal) or right (inverse). Upper Jurassic and Cretaceous.

In the Upper Jurassic of north-east France is found *Diceras* (Fig. 25*f*), having both valves twisted into a loose spiral, and fixing itself indifferently by either valve. This is the forerunner of a great series of the most remarkably modified of all lamellibranchs. In the Lower Cretaceous there are genera fixed, some by the left valve, the majority by the right: in either case the fixed valve tends to become conical or cylindrical in shape, while the free valve takes on the character of a lid or operculum. The ligament becomes deeply internal and finally disappears, the valves being no longer hinged but the free valve sliding up and down. The whole shell gradually loses all resemblance to ordinary lamellibranchs. *Requienia* still has both valves spirally coiled, but the right valve is quite flat, so that the whole shell has a strange resemblance to a gastropod with a spiral operculum. In the Upper Cretaceous the forms fixed by the right valve develop into still more extraordinary forms, mimicking the rugose corals of the Palaeozoic. Such is *Hippurites*, the right valve of which forms a cylinder a foot or more in height, of which the animal only occupies the uppermost portion, the lower part of the cavity having been cut off, as growth proceeded, by a series of calcareous partitions (like the tabulae of a coral or the septa of a cephalopod). The left valve fits on as a lid, and the hinge-teeth and adductor muscles are strangely modified. These *Hippurites* grew in reefs, like corals, and they (as well as their Lower Cretaceous predecessors) are almost restricted to tropical and sub-tropical latitudes in the sea called Tethys, the forerunner of the present Mediterranean; only a few stragglers being found in British strata.

3. **Lucinacea.**—Concentric shell-markings in some cases combined

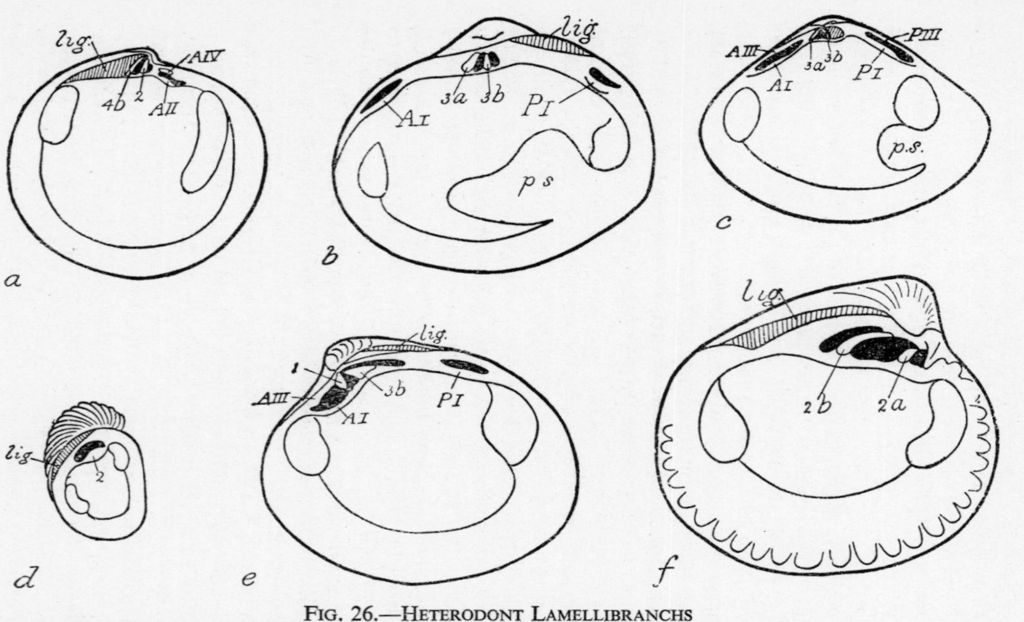

FIG. 26.—HETERODONT LAMELLIBRANCHS

In all, teeth are white, sockets black; *a*, *d*, *f*, interiors of left valves; *b*, *c*, *e*, interiors of right valves. *a*, *Lucina* (*Codakia*) *sp.*, Recent (× ½); *Codakia* differs from the typical *Lucina* in the position of the ligament, which has become internal (a resilium); *b*, *Macoma obliqua* (J. Sowerby), Pliocene (× ¾); *c*, *Mactra antiquata*, Spengler, Recent (× ½); *d*, *Chama squamosa* Solander, Eocene (× ½); *e*, *Arctica* [*Cyprina*] *islandica* Linné, Pliocene (× ½); *f*, *Cardita* [*Venericor*] *planicosta* (Lamarck), Eocene (Natural size, small specimen). *a*, *c*, original; *b*, *d*, *e*, *f*, after S. V. Wood.

with radial. Trias. to Recent. Forms resembling *Lucina* (but hinge unknown) occur in the Silurian and Devonian. The chief genera are the nearly circular *Lucina* (Fig. 26*a*), varying from lenticular to globose, with a long and narrow anterior adductor muscle scar (numerous subgenera, Trias.–Rec.), the stout *Unicardium* (Trias.–Cret.) and *Corbis* (Jur.–Rec.) in which fine radial costae are seen in the depressions between the strong concentric ridges.

4. **Chamacea.**—Fixed forms, inequivalve, closely analogous to the simpler Rudistes, but derived at a later period probably from Lucinacea.

The only genus is *Chama* (Cret.–Rec., Fig. 26*d*), which is fixed commonly by the left valve (some species by the right). The umbones are strongly prosogyral, the fixed valve larger and the free valve lid-like. The exterior has a scaly appearance.

5. **Astartacea.**—Concentric (striate or costate) ridges, never radial; valve-margins usually crenulate. Ligament external (except in *Crassatella*). Integripalliate. Owing to the strong development of 3*b*, and the occasional presence of a small 5*b*, the hinge may have a Cyrenoid aspect. Trias.–Recent.

The chief genera are the nearly circular *Astarte* (Fig. 25*e*) with numerous subgenera, the trapezoidal *Opis* (Mesozoic) and *Crassatella* (Cret.–Rec., Fig. 18).

6. **Mactracea.**—Differing from most Lucinacea in having tooth 2 bifid, in being sinu-palliate and having the ligament internal (resilium). Concentric, striate shell-markings.

Mactra (Eoc.–Rec., Fig. 26*c*) is oval in form. *Ensis* (the razor-shell, Eoc.–Rec.) has taken to burrowing, and so has acquired the elongated tubular shape common among Desmodonts (see later).

7. **Tellinacea.**—With very large pallial sinus. Shell-surface with concrescent striae.

Tellina (Paleoc.–Rec.) is a warm-water shell, rather elongate and compressed, sometimes rostrate, with lateral teeth; *Macoma* (Eoc.–Rec., Fig. 26*b*) is a cold-water form, more rounded and without lateral teeth. *Donax* (Eoc.–Rec.) is opisthogyral, and has a short posterior side (like *Nucula*).

B. WITH TEETH OF CYRENOID TYPE: 2*a*, *b*, 4*b*/1, 3*a*, *b*

1. **Cyprinacea.**—Concentric markings rarely more than striate; valve-margins smooth. Umbones anterior. *Cyprina* (Fig. 25*e*), a sub-circular shell with beaks curved forward, crowding out the anterior cardinal tooth, abundant in the colder seas (Jur.–Rec.); *Trapezium* (*Cypricardia*), a somewhat four-sided form (Jur.–Rec.); and *Isocardia* (Mio.–Rec.), with highly spiral umbones.

2. **Cyrenacea.**—Freshwater habitat. Concentric striate markings;

valve-margins smooth; integri or sinu-palliate. Umbones fairly central.

Anterior and posterior lateral teeth well developed. The chief genera are *Cyrena* (Jur.–Rec., Fig. 16) and *Corbicula* (Eoc.–Rec.).

3. **Carditacea.**—Radial costae; valve-margins crenulate; umbones anterior, leading to a lengthening of the teeth 2*b* and 3*b*. The chief genera are *Cardita* (Jur.–Rec.) and *Venericor* (Cret.–Rec., Fig. 26*f*), the latter being distinguished by the great height and thickness of the hinge-plate and large size of the teeth.

4. **Veneracea.**—Concentrically ribbed, striate or occasionally radial costae also. Valve-margins smooth or crenulate; umbones anterior. Sinu-palliate.

Many genera have been distinguished in this superfamily, and there has been much confusion in nomenclature. *Venus* has a scaly surface, crenulated valve-margins, lateral teeth obsolete, and a small pointed pallial sinus (Mio.–Rec); *Macrocallista* is smoother, with smooth valve-margins, teeth 3*a* and 1 close together and parallel, anterior-laterals well developed, large rounded pallial sinus (Paleocene–Rec.); *Paphia* [*Tapes*] differs from the latter in not having anterior-laterals, and the cardinals are radiating (Paleoc.–Rec.); *Dosinia* [*Artemis*] resembles compressed species of *Lucina* in shape, but has a Cyrenoid hinge and very pointed pallial sinus (Paleoc.–Rec.); *Cordiopsis* (Eoc.–Rec.) has already been described (Fig. 19).

B. SESSILE BRANCH

Fixation may be temporary or permanent, by a byssus or by cementation of the shell. The development of a byssus tends to the abortion of the anterior part of the body including the anterior adductor muscles, to the development of ears, and to an inequivalve condition.

I. Arcacea

Equivalve; isomyarian; internally porcellanous; hinge-line more or less taxodont; large cardinal area; ligament external and amphidetic. Radial or concentric ridges. Fixation never more than temporary.

The members of this suborder have generally been classified with the Nuculacea as Taxodonta, but the simple row of similar teeth, as seen in *Glycymeris* (Cret.–Rec., Figs. 12, 13) is found only in late genera, earlier forms having a more complex plan, which, taken with the amphidetic cardinal area, indicates a relationship to the suborder Pteriacea, from which the Arcacea diverge in (*a*) having lost the nacreous interior, and (*b*) being almost always equivalve.

One of the earliest and longest-lived genera is *Cucullaea* (Dev.–Rec.), in which the teeth are short and vertical under the centrally-placed umbo, but splay out fore and aft into horizontal lamellae, and there is a raised flange to the posterior adductor (showing in the mould as a narrow groove). In *Parallelodon* (Dev.–Tertiary) it is the anterior scar which has a flange, and the far forward position of the umbo draws out the posterior teeth into long horizontal laminae occupying three-quarters the length of the hinge. In *Grammatodon* (Carb.–Cret.), the hinge is somewhat intermediate between *Cucullaea* and *Parallelodon*. *Barbatia* (Jur.–Rec.) has a straight hinge with small teeth vertical under the umbo and gradually becoming oblique towards each end (radiating upwards); while in *Arca* (Jur.–Rec., Fig. 27*a*) they are all short and vertical.

II. Dysodonta

Usually inequivalve; hetero- or mono-myarian; commonly with anterior and posterior "ears," and a right anterior byssal notch. Teeth absent or insignificant. Ligament amphidetic, internal in some forms.

1. **Pteriacea.**—Inequivalve (except *Perna* and *Gervillia*); aniso-myarian; internally nacreous; with ears; typically with a byssal notch below the right anterior ear; hinge-line straight, without teeth or with vestiges of an actinodont type. Silurian to Recent.

A very large and varied suborder, in which there is a general tendency (a) to the inequivalve condition, the left valve being the more convex; (b) to the lengthening of the hinge-line by the formation of "ears" or "wings"; (c) to the disappearance of the anterior adductor. The internal shell layer is nacreous, the outer prismatic; cardinal area narrow and ligament amphidetic; hinge-teeth very feebly developed. *Pteria* [*Avicula*] (Devonian? to Recent) differs from *Meleagrinella* (already described) in its more oblique shape and well-marked left anterior ear (Fig. 27*c*, *c'*); *Aviculopinna* (Carb.–Perm.) and *Pinna* (Jur.–Rec.) are acutely triangular in shape; *Conocardium* (Dev.–Carb.) is equivalve, inflated, the posterior ears combining when the valves are shut to form a sort of tube (Fig. 27*b*). These genera are all costate. *Perna* (Trias.–Rec.), *Gervillia* (Trias.–Eoc.), and *Inoceramus* (Jur.–Cret.) have the ligament partly sunk into the hinge-line, and fixed into a long series of ligament-pits (Fig. 27*d*, *d'*); these genera are concrescently ribbed, as a rule: the two first are equivalve, without curved beaks, while *Inoceramus* is inequivalve, and has beaks incurved. This last attains great size and thickness in the Upper Cretaceous, just before extinction, showing senile characters similar to *Hippopodium*.

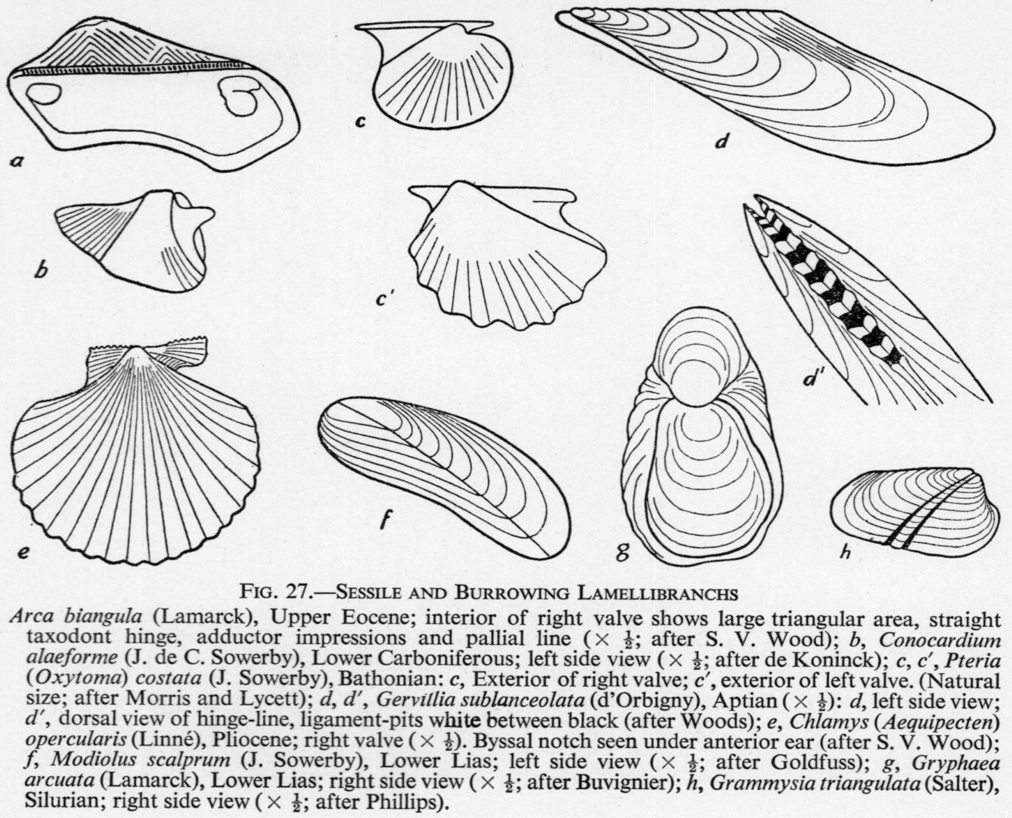

FIG. 27.—SESSILE AND BURROWING LAMELLIBRANCHS

a, *Arca biangula* (Lamarck), Upper Eocene; interior of right valve shows large triangular area, straight taxodont hinge, adductor impressions and pallial line (× ½; after S. V. Wood); *b*, *Conocardium alaeforme* (J. de C. Sowerby), Lower Carboniferous; left side view (× ½; after de Koninck); *c*, *c′*, *Pteria* (*Oxytoma*) *costata* (J. Sowerby), Bathonian: *c*, Exterior of right valve; *c′*, exterior of left valve. (Natural size; after Morris and Lycett); *d*, *d′*, *Gervillia sublanceolata* (d'Orbigny), Aptian (× ½): *d*, left side view; *d′*, dorsal view of hinge-line, ligament-pits white between black (after Woods); *e*, *Chlamys* (*Aequipecten*) *opercularis* (Linné), Pliocene; right valve (× ½). Byssal notch seen under anterior ear (after S. V. Wood); *f*, *Modiolus scalprum* (J. Sowerby), Lower Lias; left side view (× ½; after Goldfuss); *g*, *Gryphaea arcuata* (Lamarck), Lower Lias; right side view (× ½; after Buvignier); *h*, *Grammysia triangulata* (Salter), Silurian; right side view (× ½; after Phillips).

2. **Anomiacea.**—Inequivalve; monomyarian; thin-shelled with an aragonite nacreous lamellar layer; without ears; byssal notch (at least in early life) converted into a perforation (which may afterwards close up); no hinge-teeth; ligament amphidetic, internal.

Anomia (Jur.–Rec.), left valve cup-like, right valve flat with the perforation for the more or less calcified byssus persisting throughout life. *Placunopsis* (Jur.) resembles *Anomia*, but the right valve is not perforate and is fixed by cementation. *Placenta* [*Placuna*] (Mio.–Rec.), the "window-pane oyster," has both valves very flat and perforation closed.

3. **Pectinacea.**—Usually inequivalve; monomyarian; interior lamellar, sub-nacreous; with thin external ligament and thicker resilium, both amphidetic; teeth on either side of resilium, well developed or vestigial; usually with ears, sometimes with a byssal notch. Commonly with radial costae. Most genera fixed, by byssus or cementation, by the right valve, a few active swimmers. Silurian to Recent.

These shells resemble the Pteriacea in (a), being typically inequivalve, (b) commonly having the hinge lengthened by "ears," (c) having an amphidetic area and ligament, and (d) having a byssus, and consequently in many cases a notch under the right anterior ear. On the other hand, the ligament tends to sink into the position of a resilium, between the teeth; the anterior adductor is entirely lost, making the shell monomyarian; and the interior layer is sub-nacreous. The chief genera are *Chlamys* (Trias.–Rec.), both valves rather flat, with good byssal notch (Fig. 27*e*); *Pecten* (Eoc.–Rec.), with very convex right and flat left valve, and no byssal notch; *Pterinopecten* (Dev.) and *Dunbarella* (Carb.), with long straight hinge and posterior ears not sharply marked off: all these have feeble hinge-teeth, and radial costae. *Posidonia* (Sil.–Jur.) has a shorter hinge-line with practically no ears, and strong concentric ribbing. These last three genera are very close to the Pteriacea. In *Spondylus* (Jur.–Rec.) the cardinal area is large, especially on the right valve, which is attached; the two subequal hinge-teeth are large, curved (**isodont**) and smooth; the surface is radially costate and spiny. *Plicatula* (Trias.–Rec.) is flatter and without the large area; attached by umbonal region of right valve, the impression of the surface of attachment showing through on the free valve (as in the oyster); isodont teeth and sockets crenulated. *Lima* (Carb.–Rec.) is equivalve, inequilateral, toothless or nearly so; triangular ligament-pit between umbo and hinge-line; no byssal notch, but a slight gape for byssus when present. Although belonging to the "Sessile" group, *Pecten* and *Lima* are the most active of all lamellibranchs; *Pecten* is fixed in early life but later it swims by a pulsatory movement of its valves.

4. **Ostreacea.**—Inequivalve; monomyarian; interior lamellar, sometimes sub-nacreous; hinge-line short, without teeth, with thick amphidetic ligament; without ears or byssal notch; with concentric-striate ridges, or costate in some cases coarsely. Fixed by left valve (distinction from preceding Dysodonta).

These are the oysters (Fig. 21). Their very short hinge-line makes the amphidetic ligament shorter than high and usually triangular in shape. The subdivision into genera is unsatisfactory, as the characters that have been relied upon are such as recur again and again on independent lines of descent. Thus strongly costate forms have been named *Lopha*; forms with the beaks curved spirally backward, *Exogyra*; these have a flat right valve. Those with a concave right valve and large left valve with an overhanging left umbo, the shell being fixed only in early life and breaking off by its own weight later, *Gryphaea* (Fig. 27*g*). But these are probably all false or **polyphyletic** genera, consisting of a number of species belonging to different oyster-stocks which have assumed similar characters by convergence. True genera consist of species derived from the same immediate stock. Polyphyletic genera can be recognized by differences in the ontogeny of their several species. *Gryphaea* in particular is a series of senile forms, the change of growth-direction being similar to that seen in the huge forms of *Inoceramus* or *Hippopodium*.

Ostrea is abundant from the Jurassic onwards; it is marine or estuarine. *Gryphaea* is mainly Jurassic; *Exogyra*, Upper Jurassic and Cretaceous; both are marine.

5. **Mytilacea.**—Equivalve; heteromyarian; interior nacreous; umbo anterior; hinge-line short, without teeth, or with cardinal teeth 2*a*, *b*/1; ligament opisthodetic; no ears or byssal notch, but a slight byssal gape; shell elongated in an oblique direction. Concrescently or partly radially costate. Marine or fluviatile.

This suborder is typified by the common marine mussel, *Mytilus*, which leads a sedentary existence fixed by byssal threads throughout adult life; it lives in communities between the tide-marks. The shell is always elongated in an oblique direction and tends to a triangular shape, the apex being formed by the umbo, which is at or near the anterior end of the hinge-line and is never conspicuous, the base of the triangle being the posterior border. The area is generally obscure, but apparently amphidetic, but the ligament is opisthodetic. Hinge-teeth are wanting or indistinct. Early genera such as *Modiolopsis* (Ord.–Sil.) and *Myoconcha* (Carb.–Cret.) are less heteromyarian than the rest. *Modiolus* (Dev.–Rec., maximum in Jurassic, Fig. 27*f*) is more quadrilateral than *Mytilus*, the beaks not being so far forward. *Lithophaga* [*Lithodomus*] (Carb.–Rec.) is flask-shaped, and bores into corals, thick shells such as *Perna*, or rocks on the sea-bottom: moulds of its borings sometimes puzzle the fossil-collector. *Dreissena* (Mio.–Rec.) and *Congeria* (Eoc.–Plio.) are freshwater forms, having a small external ligament and the anterior adductor borne on a myophoric plate. Some of the Coal Measures freshwater shells—*Naiadites* and *Anthraconauta*—may belong to this suborder.

C. BURROWING BRANCH

Desmodonta

Usually equivalve, tending to an oblong shape; often gaping; isomyarian; internally nacreous or porcellanous. Hinge-line simple, commonly one cardinal tooth in each valve; ligament opisthodetic and, or internal. With long siphons; usually sinu-palliate. Surface markings concrescently striate, rarely radial.

A number of Palaeozoic genera belong to this branch, but, owing to imperfect preservation, are not capable of being further classified: such are *Orthonota* (Sil.), *Grammysia* (Sil.–Dev, Fig. 27*h*.), and *Edmondia* (Carb.–Permian). The Mesozoic and later forms can be more definitely grouped.

1. **Anatinacea.**—Equi- or inequivalve; hinge with one cardinal tooth in each valve, opisthodetic ligament, with or without resilium; internally nacreous (with some exceptions); sinu-palliate.

Pleuromya (Trias.–Lower Cret.) is elongate-oval, with fairly strong concentric ridges, and the right valve slightly overlaps the left along the hinge-line. *Pholadomya* (Jur.–Rec.) has been already described (Fig. 23); it is almost the only Anatinacean with radial costae, and is divided into a number of subgenera according to shape and surface patterns. *Thracia* (Trias.–Rec.) is inequivalve, the right valve being the larger; *Panope* (Cret.–Rec.) has a thick shell, gaping at both ends; *Hiatella* [*Saxicava*] (Paleoc.–Rec.) is a rock-borer, with the pallial line represented by a series of disconnected marks. The last two genera have non-nacreous shells.

2. **Myacea.**—Equi- or inequivalve; ligament internal, on a horizontal resiliophore; internally porcellanous; sinu-palliate.

Corbula (Trias.–Rec.) has already been described (Fig. 22). *Mya* (Olig.–Rec.) differs in being almost equivalve, and gaping at both ends; the anterior adductor is long and narrow, the pallial sinus deep and the resilium lies horizontally below the inner surface of the right umbo, being carried on a horizontal process of the left valve.

3. **Adesmacea.**—Borers in stone or wood, with highly-modified shell, in which the ligament becomes obsolete with the structural although not functional separation of the two valves; various acces-

sory shelly structures are added to the valves until finally the latter form an insignificant part of the skeleton.

Pholas (Jur.–Rec.) the "piddock," bores mechanically into hard and soft rocks; it has thin, brittle but hard valves with costae provided with sharp corrugations which assist in boring. The inner layer of the shell is reflected over the hinge-line and covers the umbo. The hinge-teeth are in the form of a ball in each valve, the two serving as a double ball and socket joint. The valves are widely separated in front where a sucker-like foot protrudes, motivated by muscles attached partly to a curved rod projecting inside the valves ventral to the umbo. Boring is affected by contractions of the anterior and posterior adductor muscles, the ball and socket joints permitting a see-saw-like motion of the valves. *Teredo*, the "ship-worm" (Jur.–Rec.) only bored in wood, as it now does in submarine timbering: it has two small, short valves which are of little service as a covering to the body. The two valves of the young animal each become divided into 2 or 3 regions, the anterior and middle regions develop rows of sharp costae, the hinge teeth become converted into balls and sockets as in *Pholas* but additionally a second ball-joint develops near the ventral edge of the middle region so that dorsal and ventral articulations come into play as by a rotary movement, the animal bores into the wood, the shavings of which the animal eats. The mantle deposits a calcareous lining to the tube occupied by the animal with its immensely long siphons, and there are also a pair of small calcareous plates (pellets) which close the mouth of the tube when the siphons are retracted.

Short Bibliography

MOLLUSCA GENERALLY

ADAMS, H. and A.—*Genera of Recent Mollusca*, vols. i and ii (text), vol. iii (plates) (1858). [Classification out of date, but nomenclature fairly sound; few direct references to fossils.]

BÖGGILD, O. B.—"The Shell Structure of the Mollusks," *Mém. Acad. roy. Sci. Danemark* (9), ii, 231 (1930).

CHENU, J. C.—*Manuel de Conchyliologie et de Paléontologie Conchyliologique*, 2 vols. (1859). [Figures a number of species to each genus.]

COSSMANN, M., et PEYROT, A.—"Conchologie Néogénique de l'Aquitaine," *Soc. Linnéene Bordeaux* (1909–32).

COSSMANN, M., et PISSARRO, G.—*Iconographie Complète des coquilles fossiles de l'Eocène des environs de Paris* (1904–13).

DALL, W. H.—"Contributions to the Tertiary Fauna of Florida," pts. 1–6, *Trans. Wagner Free Inst. Sci., Philadelphia* (1890–1903).

DAVIES, A. M.—*Tertiary Faunas*, vol. i (1935) London (Murby).

DESHAYES, G. P.—*Coquilles Fossiles des Environs de Paris*, 3 vols. (1824–37).

FISCHER, P.—*Manuel de Conchyliologie* (1880–7).
GRANT, U. S., and GALE, H. R.—"Catalogue of the Marine Pliocene and Pleistocene Mollusca of California," *Mem. San Diego Soc. Nat. Hist.* (1931).
NEWTON, R. B.—"British Oligocene and Eocene Mollusca," *Brit. Mus. Cat.* (1891).
WOODWARD, B. B.—*The Life of the Mollusca* (1913).
WOODWARD, S. P.—*Manual of the Mollusca* (1851–6). Still very useful, though much out of date in terminology and classification.
WRIGLEY, A.—Numerous papers in *Proc. Malac. Soc.* and *Proc. Geol. Assoc.*, on Tertiary Gastropods and Lamellibranchs (1925–46).

MONOGRAPHS OF THE PALAEONTOGRAPHICAL SOCIETY, DEALING ENTIRELY OR PARTLY WITH LAMELLIBRANCHIA

ARKELL, W. J.—"British Corallian Lamellibranchia" (1929–37).
COX, L. R. and ARKELL, W. J.—"Survey of the Mollusca of the British Great Oolite," (1948–50).
HIND, W.—(1) "*Carbonicola, Anthracomya* and *Naiadites*" (1894–96). (2) "Carboniferous Lamellibranchia," 2 vols. (1896–1905).
KING, W.—"Permian Fossils" (1850).
LYCETT, J.—"British Trigoniae" (1872–83).
MORRIS, J., and LYCETT, J.—"Great Oolite Mollusca," with Supplement (1851–63).
TRUEMAN, A. E., and WEIR, J.—"A Monograph of British Carboniferous non-marine Lamellibranchia," Part I (1946, continuing).
WHIDBORNE, G. F.—"Devonian Fauna," (1889–1907). Cephalopoda and Gastropoda in vols. i and iii, Lamellibranchia in vols, ii and iii.
WOOD, Searles V.—"Crag Mollusca," vol. ii Bivalves (1851–61). (2) "Eocene Bivalves" (1861–77).
WOODS, H.—"Cretaceous Lamellibranchia," 2 vols. (1899–1913).

LAMELLIBRANCHIA

The works listed above are all useful for Lamellibranchia. Out of the enormous mass of literature available, the following are selected as of special value:

BERNARD, F.—"Développement de la Coquille chez les Lamellibranches," *Bull. Soc. Géol. France* (3), xxiii–xxv (1895–7).
COX, L. R.—"Evolutionary History of the Rudists," *Proc. Geol. Assoc.*, xliv, 379 (1933).
DAVIES, J. H. and TRUEMAN, A. E.—"Revision of the non-marine Lammellibranchs of the Coal Measures . . . ," *Quart. Journ. Geol. Soc.*, lxxxiii, 210 (1927).
DOUVILLÉ, H.—(1) "Classification des Lamellibranches," *Bull. Soc. Géol. France*, 1913; (2) many papers on Lamellibranchs, especially the Rudistes, in *Bull. Soc. Géol. France* for many years past.
EAGAR, R. M. C.—"A study of a non-marine lamellibranch succession in the *Anthraconaia lenisulcata* zone of the Yorkshire Coal Measures." *Phil. Trans. Roy. Soc.* (B), ccxxxiii, 1–54 (1947).
HIND, Wheelton.—"The Lamellibranchia of the Silurian rocks of Girvan," *Trans. Roy. Soc. Edinb.*, xlvii (1910).
JACKSON, H. T.—"Phylogeny of Pelecypoda: Aviculidae and their allies," *Mem. Boston Soc. Nat. Hist.*, iv, no. 8 (1890).

KITCHIN, F. L.—"Jurassic Fauna of Kutch: Lamellibranchiata," *Palaeont. Indica* (9), iii, pt. 2 (1903).

LEITCH, D.—"A Statistical Investigation of the Anthracomyas of the basal *similis-pulchra* zone in Scotland," *Quart. Journ. Geol. Soc.*, xcvi, 13 (1940).

NEWELL, N. D.—"Late Palaeozoic Pelecypods: Pectinacea," *State Geol. Surv. Kansas*, x (Text and Plates), 1937–8.

STENZEL, H. B.—"Successional speciation in Paleontology: the case of the oysters of the *sellaeformis* stock." *Evolution*, iii, 34 (1949).

WEIR, J., and LEITCH, D.—"Zonal distribution of non-marine Lamellibranchs in the Coal Measures of Scotland." *Trans. Roy. Soc. Edinb.*, lviii, 697 (1936).

WRIGLEY, A.—"Observations on the Structure of Lamellibranch shells," *Proc. Malac. Soc. London*, xxvii, 7–19 (1946).

III

THE GASTROPODA

THE Gastropoda, or Snails, are much more active animals than the Lamellibranchia, most of them crawling about in search of their food, which they take into the mouth by means of the **radula**—a rasping tongue set with chitinous teeth. The head and foot are permanently outside the mantle-chamber, which is relatively small, with a small aperture and in nearly all cases lies on one side of the body. This small feature is part of a profound asymmetry which affects all parts of the body except the head and foot. In the mantle-chamber of primitive genera are a pair of gills much like those of *Nucula*; these are reduced to one as a part of the asymmetry of higher forms, and disappear altogether in the air-breathing forms, the whole mantle-chamber then serving as a lung-sac. Except in these last, a ciliary mechanism exists for the purpose of respiration, and in many it also serves to remove the excrement from the mantle-chamber, while in a few it even collects microscopic food, as in lamellibranchs, though it cannot convey it to the mouth, which lies outside the mantle-chamber. The gastropod shell like that of the lamellibranch is external but it is distinguished by not being divided into right and left valves: it is therefore described as **univalve***. The simplest form of this shell is an elliptical cone, widely open at the base; but in the great majority of cases the shell is a cone coiled in a helicoid spiral.

In many cases there is a horny or calcareous plate (**operculum**) which closes the aperture of the shell when the animal withdraws into it. This, however, is never hinged to the shell, and the idea that it represents a second valve like that of lamellibranchs has no justification whatever.

Gastropods are the only class of Mollusca that has representatives which inhabit the dry land, as well as others that live in fresh and salt waters. They are known from all the geological systems, but it was only in the Cainozoic era that they became really abundant. From Eocene times to the present day, they constitute, with lamellibranchs, the main part of the invertebrate fauna preserved as macrofossils.

* As this edition was in press, a notice appeared (*Nature*, 185, 1960, 749–51) concerning a rare gastropod with a bivalved shell living in the Inland Sea of Japan; attention was drawn to a possible relative, also smooth, in the Paris Basin Eocene.

1. **Emarginula reticulata** (Fig. 28), found fossil in the crags of East Anglia and still living in British seas, is an example of an almost symmetrical gastropod. The shell is shaped like a cone, with an elliptical base, and the apex bent over in the plane of the major axis, which thus appears to be a plane of symmetry. More careful examination with a lens shows, especially in youthful examples, that the apex is curled in a spiral to one side, so that there is no perfect plane of symmetry. The side on which the spiral lies is the right side, the direction in which the apex points being posterior (with reference to the animal's anatomy). The anterior margin has a deep, parallel-sided slit. The outer surface is patterned by alternately strong and weak radial ribs, crossed by concentric ridges (growth-lines), except the apex, which is smooth. Internally there is seen a horseshoe-shaped scar, marking the position of the muscles by which the animal pulls the shell tightly down upon the surface to which it clings by its broad, muscular foot. The opening of the horseshoe marks the position of the head.

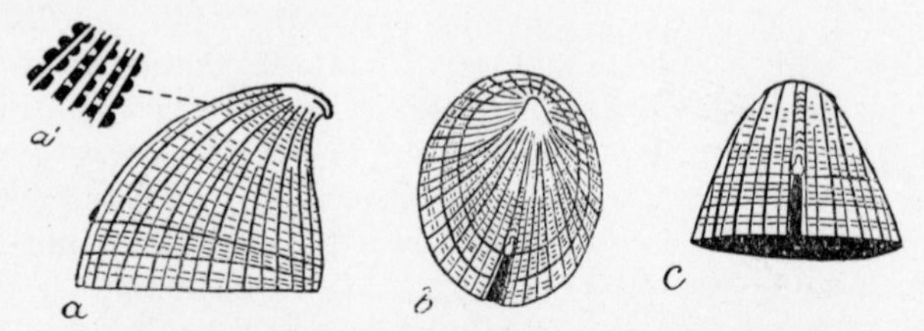

FIG. 28.—EMARGINULA RETICULATA J. Sowerby, Pliocene

a, Left side view: *a'*, small portion of surface greatly enlarged to show the surface pattern which is only diagrammatically shown in the other figures; *b*, view from above; *c*, anterior view. Slit, in *b* and *c*, black; at its upper end is seen the callus, and above this the band (× 2; original).

The slit at the anterior margin indicates the importance of ciliary mechanism in this genus, for it lodges a process of the mantle containing the exhalent aperture. Shell-growth in gastropods takes place as in lamellibranchs, at the shell-margin and by thickening of the internal layer: the apex of the shell corresponds to the umbo. The presence of a marginal notch, however, causes an interruption to the continuity of the region of growth. Hence to prevent the notch from becoming deeper and deeper as growth proceeds it is filled up at the inner end by a deposit of shell-substance like that of the inner layer (**callus**): in time this forms a band from the apex to the slit that in some species of *Emarginula* is very prominent.

This marginal slit, in one form or another, is characteristic of an important series of gastropods, whose generally primitive character is confirmed by the fact that they formed a much larger proportion of the Palaeozoic than of the modern gastropod fauna.

2. **Turritella imbricataria** (Fig. 29) is a very common fossil in the Barton Beds of the Hampshire coast, and the Bracklesham Beds of the Selsey peninsula. Its shape approaches that of a very acute cone, but it is formed of a tube coiled in a spiral, and increasing steadily in size from the apex of the cone downwards, which is the direction in which growth takes place. Each turn of the spiral is called a **whorl,** and all the whorls except the last collectively form the **spire.** Lines drawn at a tangent to opposite sides of the spire will meet beyond the apex at an angle of 11°, which is the **spiral angle** of this particular species. The

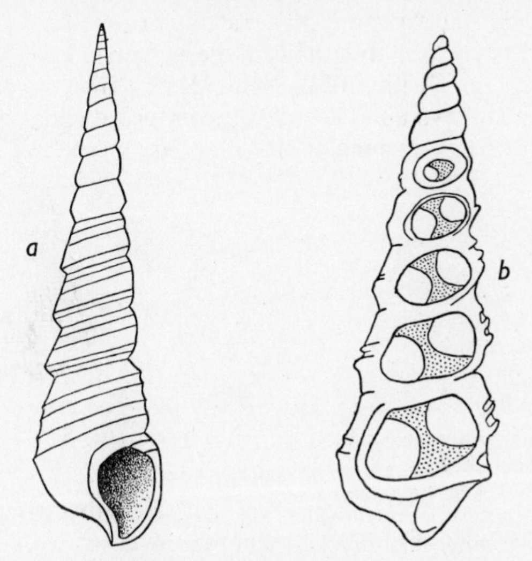

FIG. 29.—TURRITELLA IMBRICATARIA Lamarck, Eocene
a, Surface details omitted from upper half of spire; *b*, vertical section (not quite median) of an allied Recent species. Columella stippled (Natural size; original).

spiral line which marks the visible junction of one whorl with the next is called the **suture.** The side of each whorl forms an apparently convex curve, but this is seen under a lens to be made up of eight concave curves, the ends of which correspond to seven more or less conspicuous ridges which run spirally round the shell from apex to base, bearing on them a vast number of fine and close-set tubercles. Between and parallel to these ridges are much finer lines, from four to twelve in each interval. These are all crossed by very fine growth-lines, parallel to the margin of the aperture.

The **aperture** of the shell is oval. When the shell is placed in the usual position chosen for illustrating gastropods—apex upwards*—the aperture is seen on the right-hand side of the base, this being a right-handed spiral or **dextral** shell, as is the case with the great majority of gastropods. The margin of the aperture is called the **peristome**; it is divided into **inner lip** (near the middle line) and **outer lip.** Living species of *Turritella* have a horny operculum, which fits into the aperture when the animal is completely withdrawn into the shell; but this, being horny, is not found fossil.

If a vertical section is cut through the axis of the spiral, it is seen that the inner faces of the whorls are united into a solid pillar extending from base to apex—the **columella.**

The outline of the whorls and detail of the surface pattern are

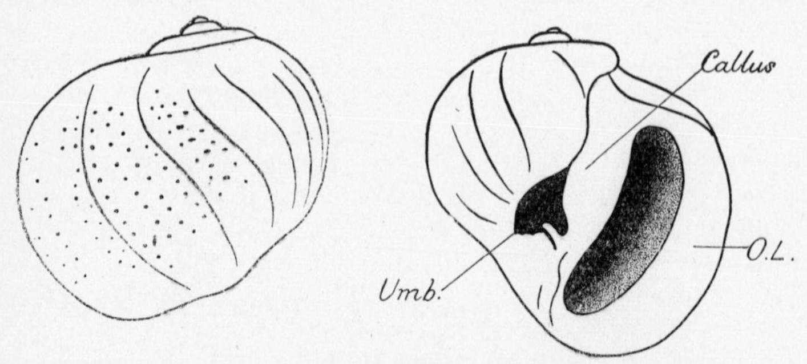

FIG. 30.—NATICA MULTIPUNCTATA S. V. Wood, Pliocene
O.L., Outer lip; *Umb.*, umbilicus (Natural size; after S. V. Wood).

specific characters of *T. imbricataria*: all the other characters described (except where common to all or many gastropods) are generic characters of *Turritella.* A shell of this shape, with a long spire, and a last whorl not conspicuously larger than the last but one, is described as **turreted.** The possession of a solid columella is denoted by the term **imperforate**; the simple outline of the peristome is indicated by the term **holostomatous.** These terms will be more clearly understood when the contrary terms are illustrated.

3. **Natica multipunctata** (Fig. 30) is a common fossil in the masses of accumulated shells found in East Anglia and known as Crag, and the species still lives at the present day, though only in warmer seas, as the Mediterranean. The shape is unlike that of *Turritella*, the last whorl forming the greater part of the external surface, the spire being very low, and having a spiral angle of 120°. The whorls are highly

* This is the accepted position in English works, but in French works it is usual to place the apex downwards.

convex, nearly semicircular in outline, but flattened or even slightly concave near the suture. The large size of the last whorl is not due to an abrupt increase of sectional area, but to the mode of coiling, the greater part of each whorl being concealed by the next: this can be realized either by following in imagination the extension of the last whorl which would result from further growth, or by making a vertical section through the axis of coiling. Such a section would show that instead of being united centrally into a columella the whorls are coiled around a central cavity, the **umbilicus,** the presence of which makes this a **perforate** shell.

The aperture is nearly semicircular, the inner lip being nearly straight. In the adult animal, a fold of the mantle extends over the inner lip, and secretes an extension of the inner layer of the shell, which more or less completely closes over the umbilicus. This deposit is called **callus.**

The surface of *N. multipunctata* is nearly smooth, being only marked by fine lines of growth and by the punctations to which it owes its trivial name. In the Crag specimens the lines of growth are commonly irregular, some—often several near together—being more prominent than the rest. As in lamellibranchs and brachiopods, such irregularities denote interruptions to the steadiness of growth, possibly periods of starvation or of a greatly varying supply of calcareous matter.

Species of the genus *Natica* are world wide, frequenting sandy bottoms in shallow water. In time they range from at least the Cretaceous period, with allied genera extending back to the Triassic. They are carnivorous, using their radula to bore circular holes in lamellibranch shells, as well as those of other gastropods, through which they feed on the animal within. Shells thus bored are common among the fossils of the Crags and of the Miocene of Touraine. *Natica*, however, is only one of the genera which bores in this way, and it is itself sometimes a victim to this mode of attack.

4. **Cerithium serratum** (Fig. 31) of the famous "Calcaire Grossier," the Paris building-stone, of Eocene age, is a turreted shell, differing from *Turritella*, not only in its more elaborate surface pattern but in the form of its aperture, the anterior end of which is produced into a well-marked channel, the **anterior canal.** The shell attains a length of 75 mm. (3 in.). The spiral angle is about 20°. The whorls are flat-sided; in the first 10 mm. the surface markings are weak and then they suddenly become strong. These consist in each whorl of an upper row of prominent compressed tubercles, 12 or 13 to a whorl, just below the upper suture; a row of very small tubercles, about 30 to a whorl, three-quarters down the side of the whorl; and a row of rather

larger tubercles, 25 to a whorl, just above the lower suture. These rows are crossed by fine lines of growth, which on the last whorl are seen to have the form of a reversed **S,** though on the whorls of the spire the lower limb of the **S** is concealed. These growth-lines are parallel to the margin of the outer lip. The aperture is pear-shaped; its long axis is at 27° to the axis of the spire; the anterior canal is nearly straight, but with a slight twist which tends to bring it more in line with the spire-axis, about two-fifths the length of the aperture proper, and nearly half its width. The shell is imperforate. Having an anterior canal it is **siphonostome,** *Natica* and *Turritella* being **holostome.**

FIG 31.—CERITHIUM SERRATUM Lamarck, Eocene
A.C., Anterior canal; *I.L.*, inner lip; *O.L.*, outer lip; *S*, suture. Details omitted near apex (× ½; after Deshayes).

Many species differing in details of shell pattern from *C. serratum* will equally well illustrate the genus in the broad sense. The original *Cerithium* has now been split up into numerous smaller genera. Recent species are all marine, and though the group is world wide, the most typical species are tropical. The genus is known fossil from the Cretaceous onwards, and had a wide geographical distribution until Eocene times, after which it declined in abundance.

5. **Rimella rimosa** (Fig. 32) is found in the Barton Beds of the Hampshire coast, and belongs to a family which is nowadays almost confined to tropical seas. In shape it differs from turreted shells in that the anterior part of the last whorl (and of other whorls, though

this is not seen externally) is drawn out to such an extent that the last whorl almost repeats the shape of the spire in inverted position: such a shell is spindle-shaped or **fusiform.** The aperture and anterior canal take part in this drawing-out. The outer lip is thickened, and the posterior end of the aperture is continued into a long narrow channel (**posterior canal**) which runs along the spire in close contact with it. This canal is analogous to the notch of *Emarginula* and its allies, but does not indicate any near affinity with them.

The surface pattern consists of numerous slightly-curved vertical ridges (corresponding to growth-lines), the intervals between which

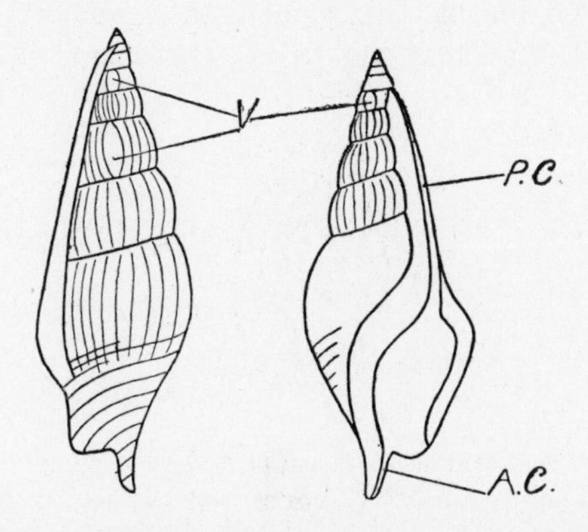

FIG. 32.—RIMELLA FISSURELLA (Lamarck), Eocene
This species is very close to *R. rimosa* (Solander). *A.C.*, Anterior canal; *P.C.*, posterior canal; *V*, varices (× $\frac{3}{2}$; after Deshayes).

are marked by finer spiral lines, which become broader and flatter as they cross the vertical ridges.

Careful examination of a well-preserved specimen shows two things: first, that the actual apex is obtuse, and the first four whorls are quite smooth; second, that at intervals the place of about three of the vertical ridges is taken by a much broader and more conspicuous elevation. The elevations, called **varices,** agree in character with the thickened outer lip of the aperture; they are in fact a series of outer lips successively abandoned as the shell grew larger. It is evident that a shell with an elaborate aperture like that of *Rimella* cannot possibly maintain the form of its aperture by steady growth as *Natica* or *Turritella* can do with its simple aperture. It must either form its aperture once for all and abandon further growth, or it must alternate between periods of rapid shell-growth without the specialized aperture, and periods in which no growth takes place and the elaborate aperture is in full use. The latter is the case with *Rimella*: in place of the occasional irregularities of the growth-lines of *Natica* there are periodical growth-periods. What each of these represents in actual time cannot be stated with certainty, but it is probably less than a year.

It should be noticed that none of the varices of *Rimella* show any sign of an old posterior canal. This does not mean that posterior canals were not formed until the adult stage, since immature specimens are found having a posterior canal: it therefore means that each posterior canal is actually absorbed during the growth-period that follows its abandonment. This suggests the possibility that in some genera all the special structures at the aperture might be resorbed

FIG. 33.—TRIVIA AVELLANA (J. Sowerby), Pliocene
A.C., Anterior canal; *P.C.*, posterior canal ($\times \frac{3}{2}$; original).

when growth restarts. These considerations have a bearing upon some features found in Cephalopoda (pp. 109, 112).

6. **Trivia avellana** (Fig. 33), found living in British seas and fossil in the Crag, is a small hemi-ellipsoidal shell in which no spiral coiling is recognizable externally, because the last whorl completely envelops and hides the spire. Such a shell is said to be **convolute.** The aperture is very narrow, nearly parallel sided and slightly curved, with slight anterior and posterior canals; the outer lip is thickened. The whole surface is traversed by ridges disposed more or less at right angles to the aperture but without the steady parallelism of the spiral ridges of a turreted shell: these ridges extend over the edges of both lips, which are therefore said to be toothed.

Trivia is a genus of the family Cypraeidae, the Cowries. The typical *Cypraea* is a tropical genus, which includes the money-cowry and many large and richly coloured species. It differs from *Trivia* in having the outer surface of the shell smooth and polished by lustrous callus (termed **enamel**) deposited by lobes of the mantle which extend out and lap over it. In the youthful shell the spire is visible though very short, and only as the adult stage is reached does the shell become convolute.

7. **Fusinus porrectus** (Fig. 34), of the Barton Clay, is a many-whorled, fusiform shell. The aperture is pear-shaped, and is drawn out into a narrow, straight anterior canal, nearly three times the length of the aperture itself. The whorls are very convex in outline, marked by nine spiral costae of unequal strength, crossed by a large number of vertical ribs (12 to a whorl except in some of the earliest whorls); where they cross, the spiral costae are particularly prominent. In the part of the last whorl corresponding to the anterior

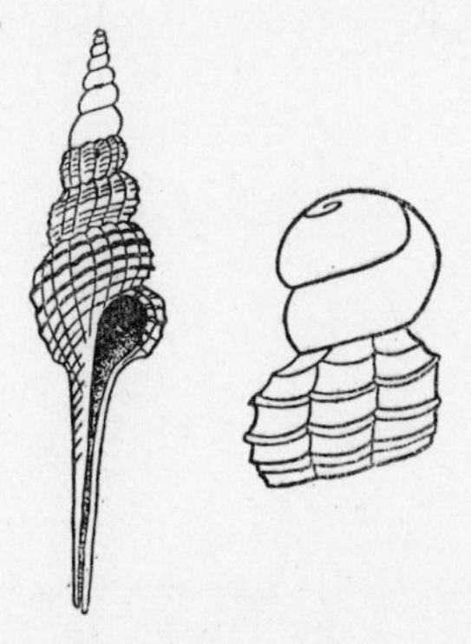

FIG. 34.—FUSINUS PORRECTUS (Solander), Eocene
Left-hand figure, natural size (surface pattern omitted from earlier whorls); right-hand figure, protoconch and first post-embryonic whorl ($\times$ 18; original).

canal (a part concealed in the other whorls) the vertical ribs die away and the spiral ridges are numerous and more uniform.

If the shell is perfect at the apex, the first two whorls are seen to be quite different from the rest—smooth and globose, coiled on an axis inclined at an angle to that of the rest of the shell. This early stage, so markedly different from the adult stage, is termed the **nucleus** or **protoconch.** Some writers have inferred that in certain brachiopods the earliest-formed shell suggests by its shape the ancestral brachiopod, and in the process of growth, changes gradually into the adult form. Attempts have been made to apply the principle of palingenesis (recapitulation of ancestral evolution) to gastropods. These have not been successful, because closely allied forms may have very different protoconchs. The change from larval shell to adult is abrupt, and corresponds to a change in the mode of life.

Fusinus is a marine genus, now confined to sub-tropical seas, but, as in so many other cases, it is found in Britain in the Eocene. Its earliest record appears to be Cretaceous.

8. **Planorbis (Planorbina) euomphalus** (Fig. 35) is a common fossil in the Headon Beds (Lower Oligocene) of the Isle of Wight, a freshwater formation. It is a **discoidal** shell, i.e., one surface is almost perfectly flat, except for a central depression: the other surface is convex, with a much wider central hollow. When this hollow is cleared of rock-matrix, the protoconch can be seen in the centre: this surface is therefore the upper surface, and consequently the shell is **sinistral** since the aperture fronts one on the left-hand side. This sinistral character is confirmed by the anatomy of living species of *Planorbis*.

The number of whorls is small. The shell is thin and fragile, and

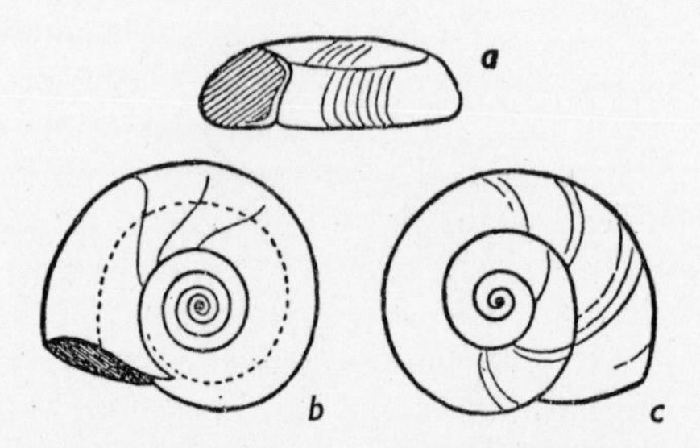

FIG. 35.—PLANORBIS EUOMPHALUS (J. Sowerby) (Headon Beds, Lower Oligocene, Isle of Wight)
a, apertural view; *b*, basal view (flat); *c*, apical view (hollowed). (Very slightly enlarged: a few of the growth-lines indicated.)

smooth except for growth-line striations. The aperture for two-thirds of its extent has an elliptical outline, but in contact with the next inner whorl this changes to a concave curve, meeting the ellipse nearly at a right angle.

Planorbis is first known in the freshwater Purbeck Beds at the top of the Jurassic system. Species of this genus are still abundant in streams and lakes of the Northern Hemisphere.

The classification of the Gastropoda is an unsatisfactory matter for the palaeontologist. The anatomical features which zoologists have found to be the best basis for a natural classification leave no mark upon the shell, and there are not, as there are in lamellibranchs, important shell-characters that can be made a basis for broad divisions. If gastropods were entirely or mainly extinct, we should be driven to make a shell-classification, but it would be obviously artificial. As it is, there are but few extinct genera which are not obviously related to recent forms, so the framework of the zoological classification is accepted and extinct genera are placed with their nearest allies.

The fundamental division is based upon a difference in the nervous system: in one main subclass, the Streptoneura, the twisting of the body during the growth of the individual which brings the anus to a position near the head, has twisted the nerve-loop which supplies the gills into a figure-of-8; in the other, the Euthyneura, the loop remains straight. The latter group falls naturally into two divisions, the Opisthobranchia, marine forms with gills, and the Pulmonata, terrestrial and freshwater forms in which the mantle chamber is converted into a lung-sac. These two latter groups are so distinct that many systematists raise them to the rank of subclasses, and for palaeontological purposes we will do so here. No shell character can be pointed to as distinctive of either of these three subclasses; nevertheless, anyone who is familiar with gastropod shells generally would feel little hesitation in placing a genus he had never seen before in its proper subclass. He would not find it easy to state his reasons, as they would consist in combinations of characters, positive and negative, which could only be stated with many qualifications. A remarkable discovery (1957) in deep waters off the Pacific coast of Mexico of a small oval gastropod *Neopilina* belonging to a primitive group thought to have become extinct in Devonian times makes it desirable to consider a fourth subclass, the Monoplacophora.

SUBCLASS: **MONOPLACOPHORA**

This group includes the Tryblidioidea which have a cone-shaped shell with forwardly directed apex and a horseshoe-shaped pattern of paired muscle-scars on the shell interior. The anus is located at the rear of the shell and there is no trace of torsion. Cam.–Recent. *Scenella* (Cam.), *Tryblidium* (Ord.–Sil.).

SUBCLASS: **STREPTONEURA**

ORDER: **Aspidobranchia**

This order is divided into suborders on radula structures but palaeontologically they may be divided into series, according to the (a) presence or (b) absence of the lateral slit (described in *Emarginula*).

(*a*) In the Fissurellidae (Carb.–Rec.) *Fissurella* resembles *Emarginula* in shape, but the slit becomes closed in at an early age by the union of the margin and forms a "key-hole" in front of the apex; as the shell grows this perforation becomes larger, by resorption of the shell, until it occupies the whorl apex and is far from the margin. In the superfamily **Pleurotomariacea** (late Cam.–Rec.) the important genus *Pleurotomaria* (Trias.–Cret., Fig. 38*a*) is spirally coiled and has a slit, the filling in of which gives rise to a spiral band, towards which the lines of growth are indented in a V-like manner. Some species are trochiform, others turbinate (see below for these terms). The geological history of the Pleurotomariacea is similar to that of *Pholadomya* or *Trigonia* among lamellibranchs: world-wide in its distribution until the Cretaceous period, it then became restricted and now survives only in the seas of Japan and the East and West Indies.

The **Bellerophontacea** (Cam.–Trias.) typified by *Bellerophon* (Ord.–Trias., Fig. 38*b*), contains the few gastropod genera in which the

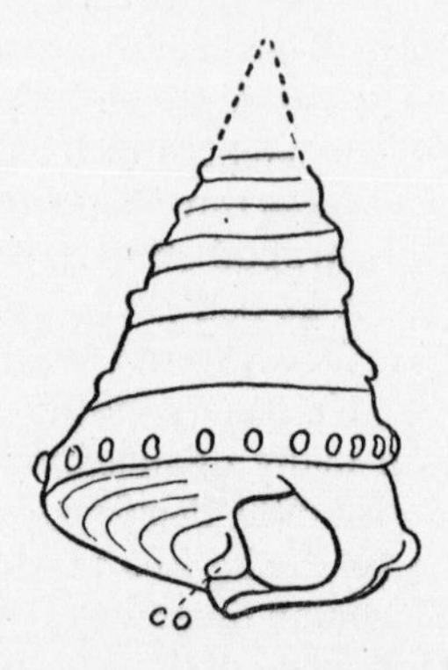

FIG. 36.—TROCHUS (TECTUS) CRENULARIS Lamarck, Middle Eocene, Paris Basin *co*, columella (Original, slightly reduced).

shell is coiled symmetrically, so that it resembles externally a cephalopod shell (but has no internal septa): the last whorl more or less completely envelops the others, leaving either a very narrow umbilicus on each side, or none. This symmetrically coiled shell is taken as evidence of a free-swimming life. The superfamily **Euomphalacea** (Cam.–Cret.) with *Euomphalus* (?Sil.–Perm., Fig. 38*c*) have a discoidal shell with a low flat spire. In shells of these two superfamilies, the slit is reduced to a slight notch, but there is a well-marked band.

(*b*) Among genera without a lateral slit, the most important are members of the **Trochonematacea** (Cam.–Cret.) and the **Trochacea** (Trias.–Rec.): *Trochus* (Mio.–Rec.) a conical shell with a fairly flat base and an aperture wider than high, this shape being described as **trochiform** (Fig. 36); *Turbo* (U. Cret.–Rec., with allies from M.–Jur., Fig. 38*d*) differing from *Trochus* in having a convex base and a nearly circular aperture, so that if placed apex downwards it resembles a spinning-top, hence described as **turbinate** (Fig. 38*d*). Among **Neritacea** (Dev.–Rec.) is *Nerita* (Jur.–Rec.), globose, imperforate with semicircular aperture and thick, often grooved, lips; *Theodoxus* [*Neritina*] (Jur.–Rec.), similar but with thin shell and lips, a fluviatile snail (Fig. 38*e*). Two important groups of shells with a high narrow spire are the **Loxonematacea** (Ord.–Jur.) and the **Subulitacea** (Ord.–Cret.). In the Ordovician is a short-lived low spired group the **Macluritacea.** *Maclurites* has a dextral shell with a depressed spiral and flattened base. The **Patellacea** (Sil.–Rec.) include *Patella*, the limpet.

ORDER: **Ctenobranchia**

Divided into a number of suborders on a basis of radula-structure, convenient palaeontological division is into holostome and siphonostome forms, though this is not necessarily a natural grouping for the siphonal notching of the shell may be acquired in different lineages.

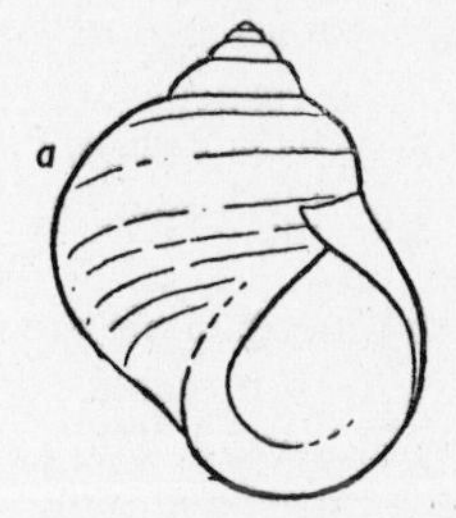

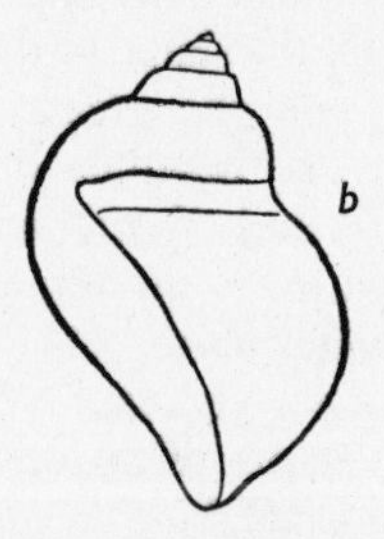

FIG. 37.—LITTORINA LITTOREA (Linné), Recent, Britain
a, apertural view; *b*, side view, showing labral profile (Original, slightly enlarged).

(*a*) **Holostome.**—*Epitonium* [*Scalaria*] (Trias.–Rec.) is turreted, with circular aperture, strong, sharp vertical ridges and finer spiral lines. *Architectonica* [*Solarium*] (Cret.–Rec.) is conical, with obtuse spiral angle, umbilicus very wide and deep, and four-sided aperture. The protoconch is a left-handed (sinistral) spiral, projecting downwards into the umbilical cavity. A shell in which there is this difference between larval and adult portions is said to be **heterostrophic.**

Littorina (Eoc.–Rec., Fig. 37), the periwinkle, somewhat resembles *Turbo* in form, but is not nacreous. *Natica* and *Turritella* have already been described. *Viviparus* [*Paludina*] (Carb.–Rec.), is turbinate with rounded apex, rounded whorls and a subcircular aperture: it is a freshwater form (Fig. 38*g*). A single record from the Lower Carboniferous of Yorkshire makes it one of the earliest known freshwater gastropods, others being members of allied families—*Pila* [*Ampullaria*] from Nevada, and *Bernicia* from Northumberland—all Lower Carboniferous.

The family Capulidae is an interesting case of degeneration: the members have taken to microscopic food and a fixed habit; in consequence the shell reverts to a more or less limpet-like form. There is therefore some uncertainty as to whether certain fossils belong to this family or to the Patellidae. *Capulus* (Fig. 38*f*) shows a trace of spiral curvature at the apex. *Calyptraea* (Cret.–Rec.), the bonnet-limpet, or Chinaman's hat, is depressed-conical, with a spiral shelf in the interior; *Platyceras* (Cam.–Trias.), limpet-like, with spirally-rolled apex, has been found adhering to the ventral surface of crinoids, on which it was in some degree parasitic. The Capulidae attained their acme in the Silurian and Lower Devonian, particularly of Bohemia, where many strange forms occur, including some which are high-conical and almost cylindrical.

Thiara [*Melania*] (Cret.–Rec.) and *Melanopsis* (Cret.–Rec.) are freshwater shells. *Thiara* (Fig. 38*h*) is turreted but distinguished from *Turritella* by the shape of the whorls, which gave a step-like outline to the shell, and the greater strength of the vertical costae, which develop strong tubercles or spines. *Melanopsis* is **subulate** (fusiform but broad and short) and has an anterior canal, thus being transitional to the siphonostome group.

(*b*) **Siphonostome.**—*Cerithium* has been described already. Closely allied to it is *Potamides*, which at present lives in mangrove swamps and the estuaries of tropical and sub-tropical rivers. This genus may be distinguished from *Cerithium*, if the aperture be uninjured, by the different appearance of the anterior canal, which is abruptly truncated. Recent specimens are further distinguished by their opercula, and by the presence of a dark-coloured **periostracum**: this is an external, non-calcareous layer of the shell (rarely preserved in fossils), usually developed in freshwater shells as a protection against the more highly solvent power of fresh water. *Potamides* had a much wider geographical distribution in Eocene and Oligocene times than now, being common in the strata of the Isle of Wight. These two genera are members of the family Cerithiidae, mainly Cainozoic, but first known in the late Cretaceous. It is preceded in time by similar forms with a feebler anterior canal—family Procerithiidae, mainly Jurassic but extending to Upper Cretaceous, where its last members overlap the earliest Cerithiidae.

In the Mesozoic rocks there is found another family, the Nerineidae, some members of which have great superficial resemblance to *Cerithium* in form and shell sculpture, so that they can only be distin-

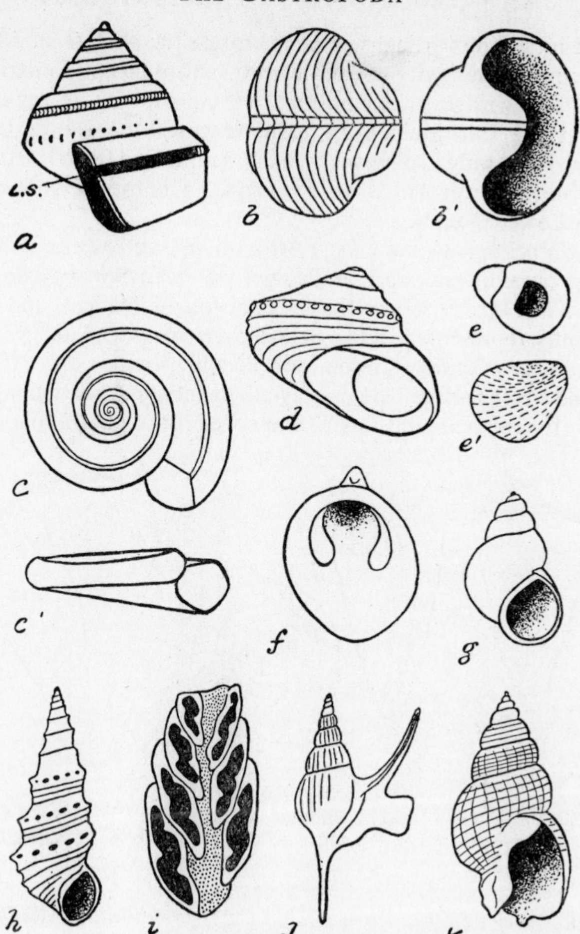

FIG. 38.—GASTROPODS

a, *Pleurotomaria mosensis* Buvignier, Upper Jurassic (× ½; after Buvignier); *L.S.*, Lateral slit; *b*, *b′*, *Bellerophon hiulcus* (J. de C. Sowerby), Lower Carboniferous (× ½; after de Koninck); *c*, *c′*, *Euomphalus pentangulatus* (J. Sowerby), Lower Carboniferous (× ½; after Goldfuss); *d*, *Turbo* (*Bolma*) *baccatus* Defrance, Miocene, Touraine (× 2; original); *e*, *e′*, *Theodoxus apertus* (J. de C. Sowerby), Oligocene (× ⅘; after Edwards); *f*, *Capulus ungaricus* (Linné), Pliocene (× ½; after S. V. Wood); *g*, *Viviparus lentus* (Solander), Oligocene (× ¾; after Deshayes); *h*, *Thiara* (*Brotia*) *inquinata* (Defrance), Eocene (× ¾; after Deshayes); *i*, *Nerinea moreana* (d'Orbigny), Upper Jurassic; vertical section, apex broken off, umbilical cavity dotted (× ½; after Buvignier); *j*, *Dicroloma parkinsoni* (Mantell), Gault (× ½; after Starkie Gardner); *k*, *Hinia* [*Nassa*] *reticosa* (J. Sowerby), Pliocene (× ¾; after S. V. Wood). In several cases the shell sculpture is omitted from the earlier whorls for the sake of clearness of outline.

guished by taking vertical sections through the shell (Fig. 38*i*). It is then seen that the cavity of each whorl is more or less constricted by internal spiral ridges (not confined to the columella as any are that occur in the Cerithiidae). In extreme cases these ridges are so developed that only a narrow space of complicated form is left for the snail's body; for instance, in *Ptygmatis*, a common genus in the English Lower Oolites.

Rimella belongs to the family Strombidae, other genera of which have the outer lip expanded and sometimes branching in a finger-like manner. The family ranges from Cretaceous to Recent, and though abundantly represented in the British area up to and during Eocene times, it is now confined to the Indo-Pacific province, the Mediterranean and West Indies. The closely-allied family Aporrhaidae, however, with similar digitations to the outer lip (e.g., *Dicroloma*, Fig.

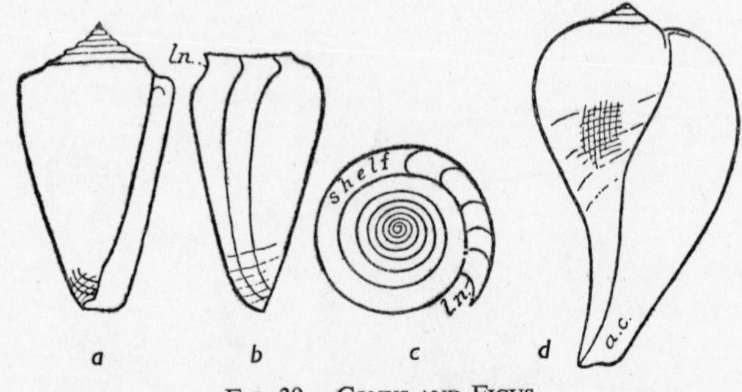

FIG. 39.—CONUS AND FICUS
a, b, c, Conus ponderosus Brocchi, Miocene, Vienna Basin; *l.n.*, lateral notch; × 1; *d, Ficus ficus* (Linné), Recent; *a.c.*, anterior canal; × ¾ (Original).

38*j*), and most abundant in the Jurassic, is essentially Atlantic in its modern distribution, and the species *Aporrhais pespelicani* is common in the British seas.

Trivia and the Cypraeidae have already been mentioned. Allied to them is *Ficus* [*Pyrula*] (Cret.–Rec., Fig. 39*d*), another of the forms, now exclusively tropical, found in the British Eocene: a pear-shaped form with very short, obtuse spire, very large aperture with thin outer lip, and with a surface marked by vertical and spiral lines.

Another group of siphonostomes is represented by the whelks, *Buccinum* and *Nassa* (Upper Tertiary and Recent), the former of boreal, the latter of general distribution. These are elongate-oval imperforate shells with spires of moderate length, wide apertures and a short interior canal; the surface is marked by coarser vertical and finer spiral lines. In *Buccinum* the canal is in line with the long axis of the aperture; in *Nassa* (Fig. 38*k*) it makes a considerable angle with it. The latter genus is also distinguished by crenulations on the interior of the outer lip. Another allied genus is *Neptunea* [*Chrysodomus*], a boreal form, nearly smooth, notable for its left-handed species which abound in the Red Crag (Fig. 40*a*).

Murex (Mio.–Rec.) differs from the whelks, firstly in having a very rounded aperture and a very long and narrow canal, the edges of which bend over until they all but meet, forming a "split-tube"; secondly, in the thickening of the mouth-border, from which and along the anterior canal there often arise long spines which repeat the split-tube structure. These mouth-borders persist as varices, commonly of great regularity in arrangement, there being three to a whorl, so that those of successive whorls continue one another in a vertical (or very slightly oblique) line up the spire; it has been stated that in these forms one turn of the spire represents a year's growth. In other cases the varices are more numerous. In *Typhis* (Eoc.–Rec.) the anterior canal and spines become perfectly tubular. *Murex* is now widely distributed except in the cold seas, *Typhis* is more closely confined to warm waters, but is one of many that give the warm-water stamp to the faunas of the British Eocene.

Fusinus has been described; allied to it is *Clavilithes* (Fig. 40*b*), an abundant Eocene form, now represented by a single Polynesian species. It differs from *Fusinus* in being smooth and in the shape of the whorls, which have rather flat sides terminated above by a horizontal shelf; *Sycostoma* (Eocene), is another smooth-shelled form, differing in its much shorter spire and less clearly defined canal.

The Volutidae (Cret.–Rec.) consist of somewhat fusiform shells, though without the sharp demarcation of the anterior canal seen in *Fusinus*, the aperture being long and somewhat parallel-sided. Their special feature is the development of spiral plaits or ridges on the columella. The commonest fossil genus is *Athleta* (Cret.–Rec., Fig. 40*c*), a richly sculptured form, with several subgenera, in the British Eocene, of which *Volutospina* is the commonest. *Scaphella* (Fig. 40*e*), a smooth form with very rounded apex, is found in the British Crag deposits, and now lives in warmer seas.

Another family of fusiform shells with aperture shaped as in the Volutidae is the Turridae [Pleurotomidae] (Cret.–Rec.), with the posterior canal taking the form of a notch at a little distance from the suture (Fig. 40*d*), recalling the lateral slit of *Pleurotomaria*, and like that causing an inflection of the growth-lines, but not closed by a special secreted band. Numerous genera and species are found in the British and other Tertiaries, and in the warm seas today, one of

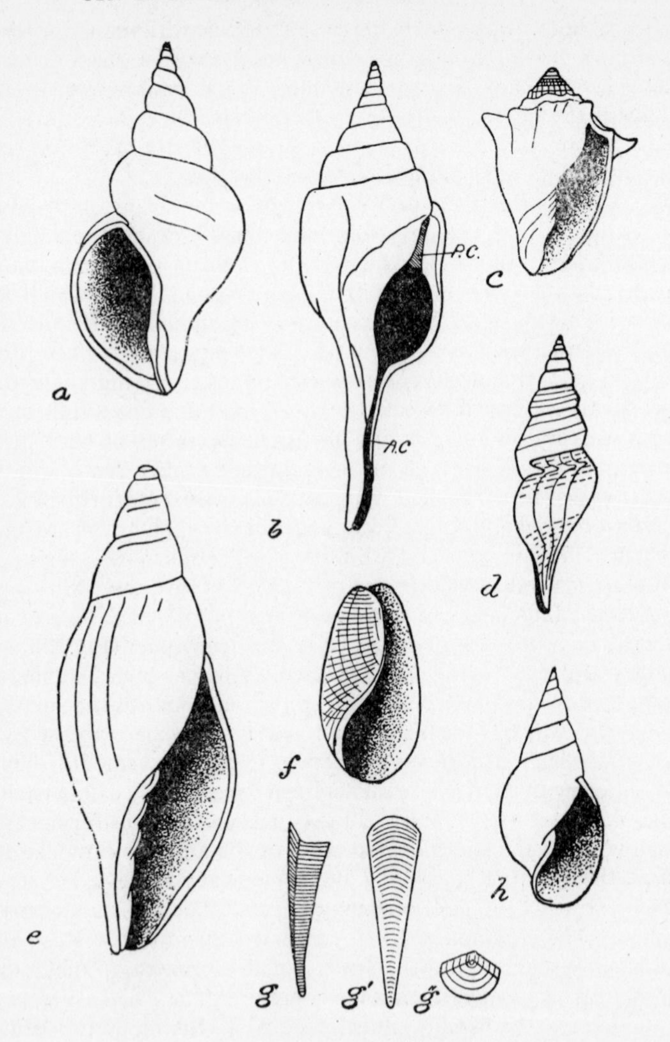

FIG. 40.—GASTROPODS

a, *Neptunea contraria* (Linné), Red Crag, Pleistocene (× ½; after Harmer); *b*, *Clavilithes longaevus* (Lamarck), Eocene (× ½; after Deshayes); *A.C.*, Anterior canal; *P.C.*, posterior canal. *c*, *Athleta* (*Volutospina*) *athleta* (J. Sowerby), Eocene (× ½; after Deshayes); *d*, *Turricula transversaria* (Lamarck), Eocene (× ½; after Deshayes); *e*, *Scaphella lamberti* (J. Sowerby), Pliocene (× ½; after S. V. Wood); *f*, *Scaphander edwardsi* (J. de C. Sowerby), Eocene (× ½; after Dixon); *g*, *Hyolithes princeps* Billings, Lower Cambrian: *g*, Side view; *g′* back view; *g″* operculum (× ⅝; after Walcott); *h*, *Lymnaea fusiformis* (J. Sowerby), Oligocene (× ⅝; after Edwards).

which, *Turricula* [*Surcula*] is shown in Fig. 40*d*. *Conus* (Ecc.-Rec., Fig. 39*a*, *b*, *c*), with very short, depressed spire, narrow, parallel-sided aperture, and posterior canal close to the suture, has a similar distributional history, but is more closely restricted to tropical seas today.

Brief mention must be made here of a specialized group of Ctenobranchs, the Heteropoda, whose shells are very thin and transparent and symmetrical in their coiling. Both these changes from typical gastropod characters indicate adaptation to a pelagic life—that is, one of swimming in the open sea. Transparency is the most effective form of camouflage for a pelagic animal which has nowhere to take cover, and the lop-sided shell, which does very well for a bottom-crawler, would be a serious impediment to active swimming. Heteropoda are very abundant in the Mediterranean and other warm seas; but fossil examples are rare. The symmetrical character of the shell of *Bellerophon* led to its being at one time placed among the heteropods, but its shell is much thicker, and the presence of a slit indicates its Aspidobranch affinities.

SUBCLASS: **OPISTHOBRANCHIA**

The **Opisthobranchia** are exclusively marine; they are divided into two suborders, Tectibranchia and Nudibranchia. The latter are shell-less and unknown fossil. The former are divided into a number of families, but on a basis of shell-form they may be roughly grouped into (1) more or less turbinate forms, such as *Actaeon*, distinguished from Streptoneura of similar shape by the strongly-marked folds on the columella, and (2) forms such as *Bulla* and *Scaphander* (Fig. 40*f*), with a more or less completely convolute shell, distinguished from *Cypraea*, etc., by the shape of the aperture (narrow posteriorly, widened anteriorly).

There is also a pelagic group known as the Pteropoda, occupying a similar position among opisthobranchs to that of the Heteropoda among Streptoneura, their shells having become superficially symmetrical, and very thin and translucent, by adaptation to a swimming life in the open sea. Pteropods abound in all the modern oceans to such an extent that where there is moderately shallow water far from land their shells form an important proportion of the foraminiferal ooze on the sea-floor—so-called pteropod ooze (at greater depths their thin shells dissolve before reaching the bottom). In the Cainozoic rocks of the Mediterranean region, similar pteropod-bearing deposits are known, and some pteropod shells have been described from the Cretaceous; but the entire absence of pteropods from the pelagic Chalk of Europe, where they might particularly be expected, seems to show that this specialized group was then at the very beginning of its existence; and the Jurassic and Triassic rocks contain no certain remains of pteropods. Yet in the Devonian rocks there are shells indistinguishable from those of the modern pteropod *Styliola*; and from thence downwards to the Cambrian are found a number of genera which, though less closely similar to modern forms, are yet more like pteropods than anything else (*Hyolithes*, Fig. 40*g*). In several Cambrian *Hyolithes* there are indications of protruding curved tapering skeletal structures which have been compared with the skeletally unsupported swimming organs of modern pteropods. It is very unlikely that any group should have abounded in the Palaeozoic era, have lain hidden through the Mesozoic, and have again become abundant in the Cainozoic. Probably we have here a case in which adaptation to like conditions produces similarity of form: such cases are described as **convergences.** It has been suggested that the Hyolithids belong to an extinct group of the Mollusca. Other Palaeozoic supposed "Pteropods" may belong to quite other groups: *Tentaculites* (Ord.–Dev.), a ringed and tapering tube, may be an annelid; and *Conularia* (Cam.–Carb.), a 4-sided elongated pyramid, may be related to the 4-rayed medusae coelenterates.

SUBCLASS: **PULMONATA**

These are terrestrial or freshwater forms, with shells of very variable shape, but usually thin and unornamented (except by colour, which is sometimes preserved in fossils). *Planorbis* has been described; other freshwater forms are *Lymnaea*, with long pointed few-whorled spire and oval aperture (Fig. 40*h*), and *Physa*, of somewhat similar shape, but sinistral. Among land-snails are *Helix*, the common snail, *Zonites*, the cellar-snail, *Pupilla*, which has a cylindrical form, the maximum diameter being attained very early, and *Poiretia*, resembling *Lymnaea* in form.

All these genera are common in the numerous freshwater beds of the Tertiary. Previous to that freshwater deposits are scanty. The oldest, in the Devonian, have yielded no freshwater gastropods, though they contain freshwater lamellibranchs, and from the Upper Devonian of Canada some land-Pulmonates have been reported. In the Coal Measures there are a few terrestrial forms, such as *Dendropupa* (without teeth on aperture) and *Anthracopupa* (with teeth on aperture), genera almost identical in shell-characters with the *Pupilla* of today. The oldest known freshwater Pulmonates appear to be *Physa* and *Planorbis* of the Purbeck Beds, except for some very doubtful Palaeozoic records.

SOME OTHER NON-GASTROPODAN MOLLUSCA

Two small classes may conveniently be mentioned here. The **Scaphopoda** form a small class of Mollusca, agreeing with the Gastropoda in having unchambered univalve shells. They show no spiral coiling, however, and the tubular shells are open at both ends. The shape of the chief genus, *Dentalium*, is that of an elephant's tusk, i.e., a curved and tapering cylinder; it may be smooth or longitudinally ribbed. The class ranges from Palaeozoic to Recent.

The **Polyplacophora** (chitons) are another small class of the Mollusca, in which there are eight plates in a longitudinal row covering the dorsal surface and articulated with one another: they thus have some superficial resemblance to an Arthropod (Chap. V) and can roll themselves up after the fashion of some Trilobites and woodlice. They are rare fossils, ranging from Ordovician to Recent.

Short Bibliography

(See under "Mollusca Generally," p. 66)

COSSMANN, M.—*Essais de Paléoconchologie Comparée*, vols. i–xiii (1895–1925). A very detailed study of all fossil Opisthobranchs and Prosobranchs except Docoglossa, Pleurotomariidae, Fissurellidae, Capulidae, Calyptraeidae, and a few minor families, the work being incomplete through the author's death.

COX, L. R.—(1) "*Anthracopupa britannica*, sp. nov.," *Quart. Journ. Geol. Soc.*, lxxxii, 401 (1926). (2) "*Bernicia praecursor*, gen. et sp. nov., a Lower Carboniferous freshwater gastropod," *Geol. Mag.*, lxiv, 326 (1927).

DONALD, Jane [MRS. LONGSTAFF].—Many papers on Palaeozoic Gastropods, *Quart. Journ. Geol. Soc*, (1887–1933).

WEIR, J. W.—"The British and Belgian Carboniferous Bellerophontidae," *Trans. Roy. Soc. Edinb.*, lvi, 767 (1931).

WENZ, W.—*Handbuch der Paläozoologie, Bd. VI: Gastropoda, Teil* 1: *Allgemeiner Teil und Prosobranchia*, Berlin (Borntraeger), 1938. [Deals with a large part of the Palaeozoic Gastropods.]

MONOGRAPHS OF THE PALAEONTOGRAPHICAL SOCIETY

BLAKE, J. F.—"Fauna of the Cornbrash" (1905–7).

EDWARDS, F. E., continued by WOOD, S. V.—"Eocene Mollusca," pts. 2–4 (Gastropods) (1852–77).

HARMER, F. W.—"Pliocene Mollusca" (1914–24).

HUDLESTON, W. H.—"Gastropoda of the Inferior Oolite" (1887–96).

MORRIS, J., and LYCETT, J.—"Great Oolite Mollusca," with Supplement (1851–63).

REED, F. R. C.—"Ordovician and Silurian Bellerophontacea" (1920–1).

SLATER, Ida L.—"British Conulariae" (1907).

WHIDBORNE, G. F.—"Devonian Fauna" (1889–1907). Gastropoda in vols. i and iii.

WOOD, S. V.—Crag Mollusca pt. 1 (1848).

IV

THE CEPHALOPODA

THE present-day Cephalopoda (cuttle-fish, squid, octopus and nautilus) are the most highly organized class of the phylum Mollusca. They conform with gastropods in having a rasping tongue, but the crawling foot is replaced by a series of powerful arms around the mouth, and the mantle and body-wall become muscular swimming organs. While the crawling life of gastropods has made them asymmetrical, and only a few have acquired a superficial symmetry through taking to a swimming life, nearly all cephalopods retain perfect bilateral symmetry, while as active voracious animals their sense-organs have become highly developed. Only a few extinct forms are asymmetric, and these may well have been crawlers exclusively, while the rest of those known as fossils may, like the modern members of the class, have been at least capable of swimming and sometimes swimmers exclusively. The habitat of cephalopods is more restricted than that of gastropods, being exclusively marine.

Most of the modern cephalopods, such as the cuttle-fish and squid, show no sign of a shell externally, though it is to be found in a modified condition buried in the interior of the body; some, as the octopus, have none at all. Only three living genera have an external shell: that of *Argonauta*, the "paper nautilus," is of a special type peculiar to it, and need not be considered here; that of *Spirula* is almost embedded in the mantle, but is of the same general type as the completely external shell of *Nautilus* (the "pearly nautilus"). This last is a spirally-coiled univalve shell, divided internally into chambers, and the same general type is found in an enormous number of extinct Cephalopoda. As the spiral coiling introduces a complication into the structure, it will be convenient to describe first the similar but straight shell of the extinct genus *Orthoceras*, filling up the gaps in our knowledge by reference to the living nautilus. There are a great many species of *Orthoceras*, ranging in age from uppermost Cambrian (Tremadoc) to Triassic, but as none of these is so common as to be obtainable with certainty for examination and it may often be necessary to take several species to demonstrate the full structure, the description here given is generic instead of specific. It should be stated, however, that the generic name *Orthoceras* has lately acquired a restricted usage and many other generic names have been introduced for various of the genera of orthocones (see p. 105). For reasons of simplicity the term *Orthoceras* is retained here and used in a wide sense.

1. **Orthoceras.**—The shell (Fig. 41) is a cone approaching to a cylinder in some cases circular, in others elliptical in plan, and ending in a more or less blunt rounded apex. Rarely there has been found, at the apex, a more or less globular structure—the **protoconch,** or initial shell, secreted in early life. The shell (or **conch**) is

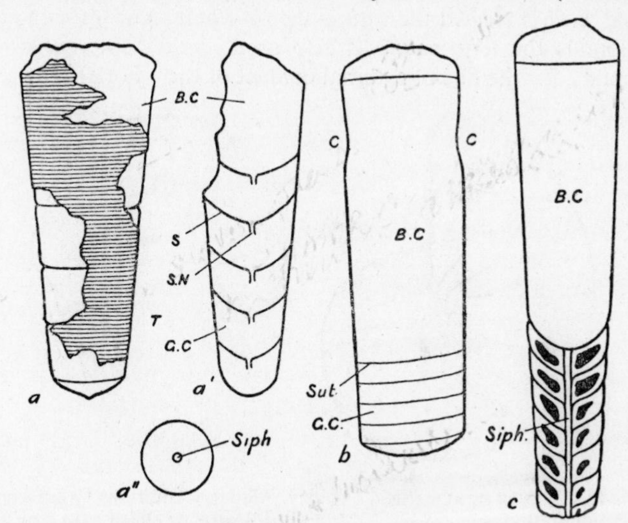

FIG. 41.—ORTHOCERAS

a, *a'*, *a''*, *O. capax* Barrande, Silurian: *a*, mould of body-chamber and three gas-chambers, part of test remaining (shown by fine striations); *a'*, median section of a similar broken specimen; *a''*, view of end broken along a septum; *b*, *O. decipiens* Barrande, Silurian, mould showing perfect body-chamber with constriction, septa more closely set than in *O. capax*; *c*, *O. capillosum* Barrande, Silurian, median section showing perfect body-chamber without constriction, gas-chambers partly filled, and continuous siphuncle (All after Barrande); *B.C.*, Body-chamber; *C*, constriction; *G.C.*, gas-chamber; *S.*, septum; *Siph.*, siphuncle; *S.N.*, septal neck; *Sut.*, septal suture; *T*, test.

divided internally into chambers by partitions (**septa**), which are convex towards the protoconch and are perforated generally in the centre by an opening around which the septum extends like a bottle-neck (Fig. 41, *S.N.*). At the opposite end from the protoconch a considerable portion of the cavity of the shell is undivided, forming the **body-chamber,** lodging the animal's body. The rest are **gas-chambers,** containing (in *Nautilus*) a gas secreted from the blood, resembling atmospheric air but with more nitrogen. Through the central perforations of the septa ran a living cord, the **siphuncle,** through all the gas-chambers back to the protoconch. While the animal was growing, new shell was continually being added to the aperture or "mouth" of the body-chamber; while at intervals the animal moved forwards in its body-chamber, and secreted behind it a new septum. Thus any given septum was formed at a rather later time than the part of the external shell with which it is in contact—a fact that must never be forgotten. The line of junction of the edge of a septum with the external shell is termed the **suture-line** or **septal suture** (for in coiled cephalopods the term **suture** is also used in the same sense as in gastropods, for the line of external contact of the whorls). The septal

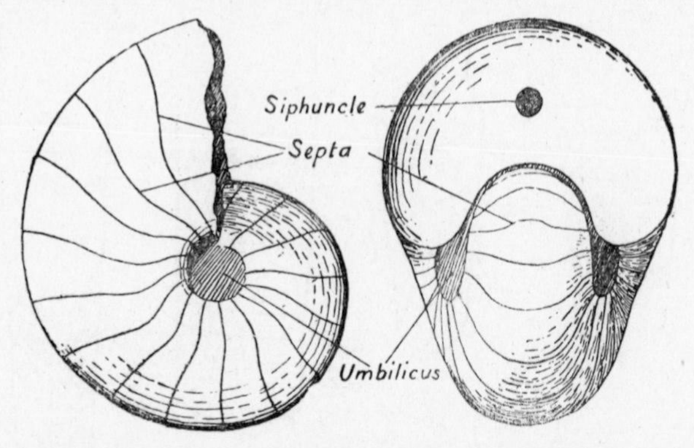

FIG. 42.—NAUTILUS ALBENSIS d'Orbigny, Albian, Cambridge Greensand Internal mould. Body-chamber at least missing. Umbilicus filled with phosphatic matrix. Side view and front view (Natural size; original).

sutures are invisible externally as in all cephalopods, but when the shell is removed the internal mould shows them.

The outer shell surface of *Orthoceras* is usually marked simply by growth-lines parallel to the margin of the aperture—usually only fine striae, at other times with stronger rings at intervals. The actual aperture of the body-chamber is rarely preserved (the thin shell, unsupported by septa, being easily crushed); but in a few perfect specimens the aperture is seen to be a simple circle, which may or may not have just behind it a circular constriction (Fig. 41*b*).

Some orthocone shells attain great length; a damaged specimen of *Rayonnoceras pergiganteum* J. S. Turner, itself 5 feet (1·53 metres) long is estimated to have been 6 feet 3 in. (1·90 metres); it was found in a Lower Namurian limestone on Holy Island, Northumberland.

2. **Nautilus albensis** is a fairly common species in the Gault, and the

following description applies to it, but, with slight alterations, would serve for many other species. It is a shell symmetrically coiled in a plane-spiral, so tightly that each whorl nearly conceals the whorl within, leaving only a very deep and narrow **umbilicus** on each side. The shape of the whorl-section thus comes to be that shown in Fig. 42 (right-hand figure). The outer surface or **periphery** is known from the anatomy of living species to be ventral, so that the inner surface must be dorsal. The outer surface shows a series of slight corrugations which are essentially growth-lines and correspond to the aperture-margin, but owing to the spiral coiling there has to be more rapid growth on the ventral side, so that the growth-lines radiate out instead of being parallel as in gastropods: in cephalopods they are

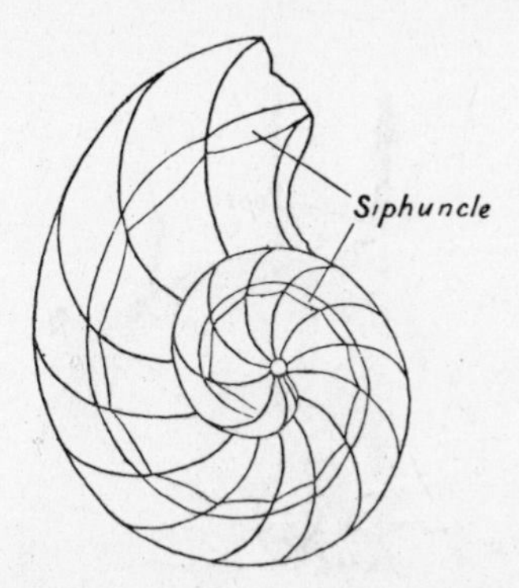

FIG. 43.—NAUTILUS EXCAVATUS J. de C. Sowerby, Inferior Oolite
Approximately median section (not accurately median at the centre) of a fragmentary specimen (Natural size; original).

therefore usually termed the **radial** lines. Those lines are very faint near the umbilicus, but become more distinct in the outer part of the lateral areas, where they swing backwards and cross the peripheral area in a curve which is concave forwards (towards the aperture). This is called the **hyponomic sinus** (cf. Fig. 50*a'*), and corresponds to an embayment of the aperture occupied by the funnel or hyponome, an organ by which the *Nautilus* ejects water from the mantle-chamber and so projects itself through the sea. This hyponomic sinus is characteristic of most Nautiloidea, though wanting in many of the straight shells. If the fossil is broken, the septa are seen to be concave forwards as in *Orthoceras*, with the siphuncle placed rather nearer the dorsal side (Fig. 42, right-hand figure, and Fig. 43). On the mould the suture-lines are seen to run radially outwards from the umbilicus for half the height of the lateral area, then to swing forwards, and to cross the peripheral area with a very slight backward curve: they thus contrast sharply with the radial lines, of which they cross four or more in their course.

3. **Asteroceras obtusum** (Fig. 44) is the index-fossil of one of the zones of the Lower Lias. Specimens of this or closely allied species are familiar to visitors to Lyme Regis. The shell is of a light yellow-brown colour, but is usually only preserved in fragments, the fossils being essentially internal moulds. The shell is spirally coiled in a plane- or bilaterally-symmetrical spiral. Each complete turn of the spiral is termed a **whorl.** The number of whorls cannot usually be counted, the

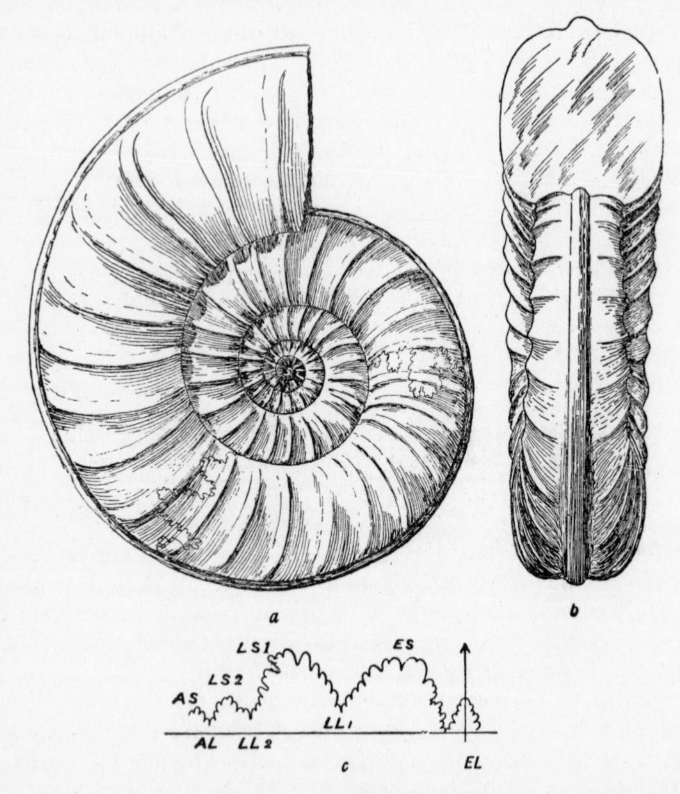

FIG. 44.—ASTEROCERAS OBTUSUM (J. Sowerby), Sinemurian (Lower Lias), Lyme Regis

a, Side view; *b*, apertural view (× ½; original); *c*, suture-line of an allied species (after d'Orbigny). The arrow denotes the middle line of the periphery and points towards the body-chamber; the horizontal line is the guide line; *EL*, External lobe; *ES*, external saddle; *LL*1, lateral lobe; *LS*1, lateral saddle; *LL*2, *LS*2, second lateral lobe and saddle; *AL*, *AS*, auxiliary lobe and saddle.

central being hidden by rock-matrix difficult to remove, but usually four or more can be counted: they become larger rather rapidly from the centre outwards. Internal moulds show that the body-chamber occupies rather less than half the last whorl: it is filled with a grey matrix of argillaceous limestone, while the gas-chambers are filled with brown crystalline calcite. The lateral region seen within the encircling last whorl is called the **umbilicus.** Unlike that of gastropods this forms a concave area on both sides equally, since the shell is a plane-, not a helicoid, spiral. The spiral line of contact between each whorl and the next is the **umbilical suture** (corresponding to the suture of a gastropod, but present on both sides of the ammonite).

A plane-spiral shell is measured as follows: the **diameter** is measured from the outer edge of the aperture (or if the shell is imperfect, of the latest part preserved) straight across the centre to the opposite edge. The measurement of a single whorl in the same direction gives its **height**; the maximum measurement at right-angles to this is its **thickness.** In an ammonite with strong ribs, such as *Asteroceras*, this thickness will vary very much according as it is measured at or between the ribs. The **width** of the umbilicus is measured between the two limbs of the umbilical suture along the same line as the diameter. These are the most important dimensions, and it is their relative proportions, rather than their absolute amounts, that are of value for comparison of one species with another. As, however, these proportions may alter in the process of growth, the actual size of an ammonite measured are, then—(1) diameter; (2) height of last whorl; (3) thickness of last whorl; (4) width of umbilicus. The first is generally stated in millimetres, the others as percentages of the first. Thus for a particular specimen of *Asteroceras obtusum*, we express these dimensions thus:

145: 36, 27–31, 41.

The 27–31 refers to the difference of thickness between and at the ribs.

For comparison we give measurements of two allied species, *Asteroceras marstonense:* 75: 39, 32–36, 33, and *Asteroceras margarita:* 120: 34, 26–29, 27.

A glance at the three formulae indicates that *A. marstonense* is thicker, *A. margarita* thinner than *A. obtusum*, while both are less widely umbilicate. Nevertheless, all three have a wide umbilicus in comparison with that of *Nautilus*: they may therefore be termed **latumbilicate,** while *Nautilus* is **angustumbilicate.**

If the specimen is placed so that the aperture faces us (apertural view, Fig. 44*b*) the shape of the whorl in cross-section is seen (see also Fig. 57*e*). It is somewhat elliptical, and the outline may be divided into six parts, corresponding to six regions on the surface of the whorl*: (1) the somewhat flattened **peripheral** or **external** area or **venter,** which bears a broad rounded **keel** in its middle line; (2) and (3) the lateral areas, one on each side, slightly convex but with a general direction approximately parallel to the plane of symmetry; (4) and (5) the paired **inner areas** or **umbilical slopes,** one on each side, which slope inwards to the sutures; and (6) the concave **dorsal** or **impressed** area, lying between the right and left umbilical sutures and in close contact with the next inner whorl. The ratio between the height of this overlapped part of an inner whorl and the height of the outer whorl which overlaps it is the **amount of indentation** of the outer whorl; while the ratio between it and the total height of the whorl of which it forms part is the **amount of inclusion** of the inner whorl. In *A. obtusum* the former is about $\frac{1}{7}$, the latter about $\frac{2}{7}$ (the height of the outer whorl being about double that of the next inner). These ratios are low when the umbilicus is wide (as in *Asteroceras*), but they increase as it becomes narrower.

The spiral line along which the peripheral area meets the lateral area is called the **peripheral shoulder**; that on which the lateral area meets the umbilical slope, is the **umbilical edge** (for descriptive purposes, as distinct from measurement, the umbilicus is taken to include the umbilical slope of the outer whorl). In *Asteroceras* these margins are not as sharply defined as in some ammonites.

The shell sculpture of *Asteroceras* consists of a series of strong ribs, which start on the umbilical slope and run almost straight outwards until near the peripheral shoulder, where they bend decidedly forwards and die away. Between their ends and the median keel, the periphery forms a shallow groove on either side. The forward bend of the ribs shows that the aperture instead of having a hyponomic sinus, as in *Nautilus*, has a median projection or **rostrum** (cf. Figs. 52*a*, *b*).

The ribs are well spaced out, their distance apart being proportionate to the height of the whorl; thus in the last whorl of the specimen figured it gradually increases from 8 mm. to 16 mm. (as measured in the middle of the lateral area). The ribs are not thickenings of the shell, but corrugations, and therefore show as well on the mould as when the shell is present.

The septal sutures (Fig. 44*c*) are very different from those of *Orthoceras* or *Nautilus*: they are thrown into a series of forward and backward curves, each of which is further puckered up or frilled. The forward curves (convex towards the body-chamber) are called **saddles,**

* In *Nautilus* (and ammonites with similar whorl-section) the first five of these areas pass indefinitely into one another.

the backward curves are **lobes.** The visible suture-line consists of (1) a median **external** or **ventral lobe,** *EL*, divided into two by a small pointed saddle, (2) a pair of **external saddles,** *ES*, on the peripheral margins, followed on each side by (3) the **lateral lobe,** *LL*1, (4) the **lateral saddle,** *LS*1, (5) the elements resulting from subdivision of the primitive umbilical lobe, *LL*2, *LS*2, *AL*, *AS*, and others hidden on the impressed area. Finally, quite invisible where covered by an outer whorl, is (6) the **internal** or **dorsal lobe**: this, like (1) is median, all the others being paired.

If a straight line is drawn from the centre of the ammonite to the point of the external lobe (**normal line** or **guide-line**), it will be noticed that all the other lobes fall short of this line, the first lateral lobe most of all.

4. **Promicroceras planicosta** is another ammonite from the zone of *A. obtusum*. It is common at Lyme Regis, but far more beautiful specimens have been found in big nodules of limestone at Marston Magna near Yeovil—the "Ammonite Marble" of Marston Marble, at one time much used for ornamental purposes, and still generally obtainable in small fragments from mineral dealers. In both localities the usual specimens are small (diameter about 2–3 cm.), the dimensions of a small one being 20: 33, 32–36, 45.

Whorl-section reniform (kidney-shaped) in inner whorls, becoming higher and less broad later. Amount of indentation one-tenth, of inclusion one-sixth (at 20 mm. diameter). When the umbilicus is free of matrix there can be seen in the centre a small **protoconch,** with six whorls around it (up to 20 mm. diameter). The first two whorls or so are quite smooth; then fine striae appear; after another whorl and a half, obscure ribs are seen, and about the end of the fourth whorl well-defined ribs appear. At the 20 mm. diameter the ribs are about 2 mm. apart, and run as follows: starting in the steep inner area they first slope backwards, then curve round to a truly radial direction on the rounded umbilical edge, from which they curve very gently to a slight backward slope on the outer half of the lateral area; at the peripheral shoulder they swing forward and cross the periphery without interruption in a forwardly-convex curve, but at the same time they become much broader and lower (as though modelled in a plastic substance and then pressed down). The fine striae on and between the ribs follow the same direction, which shows that here as in *Asteroceras* there was no hyponomic sinus to the aperture but a rostrum, though not a very prominent one. Larger specimens of allied species (genus *Xipheroceras*) show that soon after attaining 20 mm. diameter the ribs begin to swell up into tubercles at the peripheral shoulder, and in some the flattened peripheral rib tends to divide into a bundle of three ribs. In another species (*X. ziphus*) the tubercles are developed into long spines.

The body-chamber of the small specimens of *Promicroceras* occupies just over half a whorl, and the last two or three septa are usually crowded together—a feature common in the last septa of all ammonites. This suggests that the small size of these ammonites was not due to premature death, but that they were a small species.

The suture-line of *Promicroceras* is (at 20 mm. or less) like a much-simplified *Asteroceras* suture, except that there are no auxiliary lobes and saddles, and the lateral lobes are nearly as deep as the external lobe.

These four examples of cephalopods have been chosen as easily

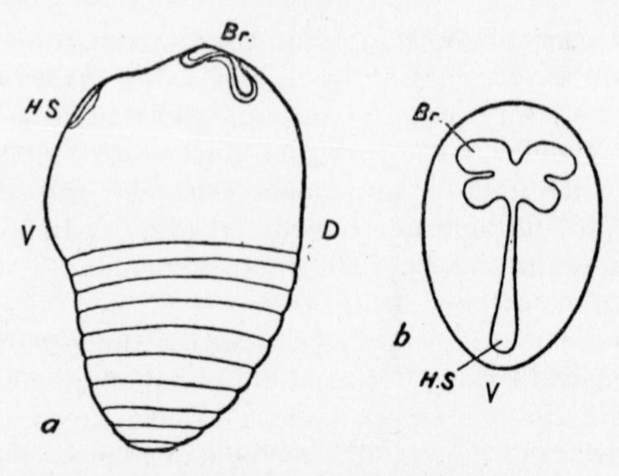

FIG. 45.—TETRAMEROCERAS OBOVATUM (Blake), Silurian (Lower Ludlow Beds) *a*, Side view; *b*, apertural view; *Br.*, Brachial portion of aperture; *H.S.*, hyponomic sinus; *D*, dorsal side; *V*, ventral side ($\times \frac{3}{8}$; after Blake).

obtainable for description, so that readers can verify the descriptions on actual specimens. The next few examples are less common, and will be described more briefly.

5. **Tetrameroceras obovatum** (Fig. 45) is one of several more or less ellipsoidal species found in the Lower Ludlow shales of Herefordshire and Shropshire. It is about 110 mm. long, 58 mm. wide, and 75 mm. high. Preserved as a mould, it is seen that about one-half of the shell is body-chamber, the rest consisting of about nine gas-chambers. The whole shows a very slight curvature, the side which we shall presently recognize as ventral being slightly concave: such a curvature is the opposite from that of *Nautilus*, where the ventral side is convex. This is expressed by saying that *Tetrameroceras* is **endogastric,** *Nautilus* **exogastric.** A more striking difference from both *Orthoceras* and *Nautilus* is the form of the aperture, which is greatly narrowed in the

centre, forming a slit only 3 mm. wide. This widens at the ventral end into an oval, at the dorsal end into a four-lobed expansion. Evidently the animal was permanently shut in its body-chamber, and could only protrude two pairs of arms for obtaining food by the larger four-lobed aperture, and the swimming funnel by the other. The latter therefore represents the hyponomic sinus and identifies the side towards which it lies as ventral. In other contemporary species there may be openings for only one pair of arms (genus *Gomphoceras*) or for three or four pairs.

6. **Meloceras** [**Cyrtoceras**] **elongatum** (Fig. 46*a*) is found in the Upper Silurian limestone of Bohemia. It is elliptical in section and curved in form, the curvature decreasing from youth to age (i.e., from apex to body-chamber), and the plane of curvature being that of the major-axis of the ellipse (compressed-elliptical). It attains a length of 80 to 90 mm. of which nearly a third is body-chamber. The septa have the simple form seen in *Orthoceras*, but are more closely set than in most species of that genus, their distance apart being only about one-sixth the major axis of the cross-section. The siphuncle is close to the convex margin, and forms a continuous calcareous tube, constricted as it passes through the septa. The shell is only patterned by fine striae, the course of which shows a very slight hyponomic sinus on the convex margin, which is therefore ventral. Thus this species is exogastric, and the siphuncle is ventral (its usual position when not central).

7. **Kophinoceras** [**Gyroceras**] **ornatum** is found in the Middle Devonian limestone of the Eifel. Compared with the last it has a stronger curvature, so that it forms a complete spiral of at least a whorl and a half, but there is no contact between the whorls. It is depressed-elliptical in section (i.e., the long axis of the ellipse is at right angles to the plane of coiling). The only evidence of shell-markings seen on the mould is a paired series of rather coarse and ill-defined tubercles along the peripheral margins, but on the shell itself the periphery is marked by longitudinal ridges and transverse striae. The septa are farther apart than in *Meloceras*; the siphuncle is in the same position.

8. **Ophidioceras simplex** (Fig. 50*b*), from the Silurian of Bohemia, is a flat-sided spiral shell, 35 mm. in diameter, of several whorls which are just in contact, except that a very small portion of the last whorl becomes straight and so separates from the next inner whorl. The aperture is constricted and **Y**-shaped, the hyponomic sinus (the stem of the **Y**) being towards the periphery, so that the shell is exogastric (as are most, but not all, of the spiral cephalopods). The sides are marked by simple ribs, interrupted by a keel on the periphery.

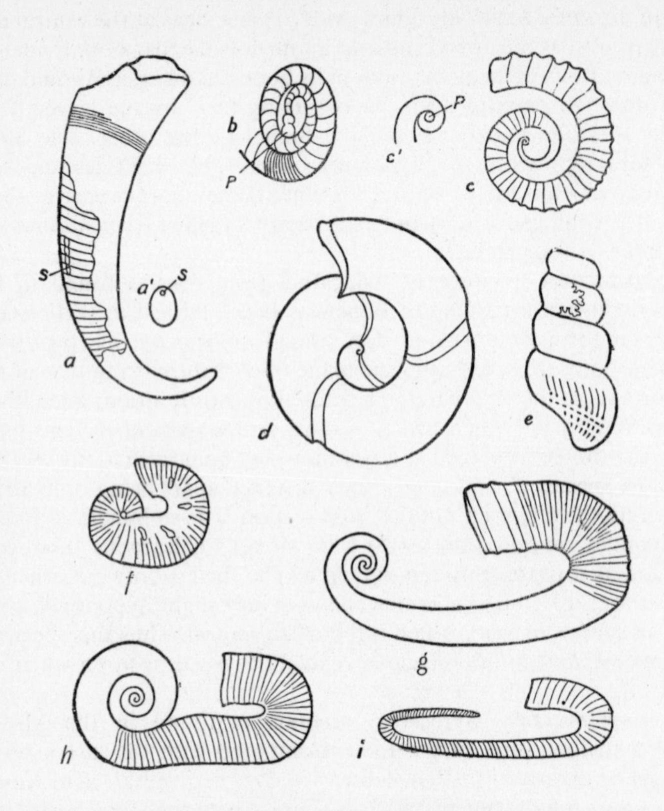

FIG. 46.—CEPHALOPOD COILING TYPES

a, a', *Meloceras* [*Cyrtoceras*] *elongatum* (Barrande), Silurian (× ½), cyrtocone: *a*, side view, striate growth-lines indicated at one place, part of test removed, showing suture-lines and part of siphuncle (*s*); *a'*, cross-section; *b*, Inner whorls of *Agoniatites fecundus* (Barrande), Middle Devonian (× 3), showing unstable early stage; *P*, protoconch; *c*, *Parapatoceras calloviense* (Morris), Callovian (Kellaways Rock) (× 2); *c'*, initial part (× 10); *P*, protoconch. This shows uncoiling from serpenticone to criocone. *d*, *Morphoceras dimorphum* (d'Orbigny), Bajocian; shell sculpture omitted; shows uncoiling, sphaerocene to serpenticone; also five constrictions; *e*, *Turrilites bechei* (Sharpe), Cenomanian; shell sculpture omitted except on last whorl, and one suture-line indicated; *f*, *Scaphites aequalis* J. Sowerby, Cenomanian; scaphiticone, beginning as sphaerocone and ending in a hook. *g*, *Lytocrioceras pulcherrimum* (d'Orbigny), Neocomian; criocone opening out, but with hook-line end; sculpture omitted except near aperture. *h*, *Macroscaphites ivani* (Puzos), Neocomian (× ¼); serpenticone uncoiling, with hook-like end; sculpture omitted except on body-chamber; *i*, *Hamites* (Albian); a generalized figure, earliest part missing. *a*, *b*, *c*, after Barrande; *c*, after Sharpe; the rest after d'Orbigny. All × ½, unless otherwise stated.

These three examples illustrate intermediate stages between a straight shell like *Orthoceras* and a tightly-coiled one like *Nautilus*. At one time (a century or more ago) the different degrees of curvature were regarded as differences of generic value, and names were given on that basis (*Orthoceras*, *Cyrtoceras*, *Gyroceras*, etc.), but it was gradually realized that genera so defined included species of widely different stocks. So a series of morphological terms has been introduced instead—**orthocone** (straight), **cyrtocone** (curved), **gyrocone** (open spiral), **ophiocone** (spiral with whorls in contact) and **nautilicone** (outer whorls completely hiding inner whorls). Among Ammonites it has been supposed that this represented a necessary evolutionary sequence, and that genera in these stages followed one another in time during the Palaeozoic era. There are also some cases where the earliest whorls are less curved than the later, and on the principle of palingenesis this would be a repetition of the history of the stock. But more exact knowledge does not confirm these older generalizations. For instance, the earliest known nautiloid genera are cyrtocones, not orthocones.

There are other cases, from the Ordovician onwards, but more frequently in the Mesozoic, and especially in the Cretaceous, which have the early whorls more curved or coiled than the later, which may be even straight. Some examples of these will now be given.

Among Ammonoids there is greater variety of **occlusal** forms (inner whorls completely hidden) than in Nautiloids, so a distinction is drawn between **sphaerocones** (globular, Fig. 50*c*; 52*k*), **cadicones** (barrel-shaped, Fig. 57*i*) and **oxycones** (with sharpened periphery, Figs. 57*c*, *d*).

9. **Parapatoceras calloviense** (Figs. 46*c*, *c*′) is a common fossil in the Kellaways Rock of Wiltshire. At first sight it appears to be a gyrocone, but close examination shows that it begins with a globular protoconch (unlike the typical ammonite protoconch which is barrel-shaped), around which the tube-like shell winds closely for more than one whorl, after which its curvature diminishes and it continues as an open (evolute) spiral. As this is a stage in an **uncoiling** process it must not be called a gyrocone: the term applied is **criocone.**

10. **Baculites** (Figs. 52*l*, *l*′) is a long, stright, compressed form, found in the uppermost Cretaceous strata. Its initial stages are usually lost, but in specimens from South Dakota they have been found to be substantially similar to those of *Parapatoceras*, the closely-coiled stage lasting rather longer. It is thus the final stage of uncoiling (**baculicone**).

Parapatoceras and *Baculites* thus afford examples of palingenesis, and by their early coiled stages are distinguishable as derivatives of normal ammonites. They are also distinguished from similarly-coiled primitive forms by their complex ammonitic suture-lines. In both these respects they illustrate the principle of **irreversibility** in evolution, according to which an organism which reverts to the condition of one of its remoter ancestors never does so entirely: it retains some features possessed by its intermediate ancestors. It should thus always be possible to distinguish degenerate from primitive forms.

Uncoiling forms, including a baculicone, occur as early as the Norian (Upper Triassic), where they are contemporaneous with the last orthocones; but it is not until the Cretaceous that they become abundant.

Just as early ophiocones commonly show a tendency to revert to the straight form in the adult, so the uncoiling forms show a tendency to coil again at the end, producing hook-like body-chambers. *Lytocrioceras* ends thus after a criocone stage, *Macroscaphites* after a serpenticone stage, while *Scaphites* passes direct from sphaerocone to hook. *Hamites* (initial stages an unstable spiral) makes two hook-like bends. All these are Cretaceous (Figs. 46 *f–i*), but in the Jurassic period *Oecoptychius* had the form of *Scaphites* (but different ornament), and there are a number of stocks which showed a tendency to uncoil in the adult (Scaphitoids).

All these forms so far described are bilaterally symmetrical, but there are a few asymmetrically-coiled forms (termed **turricones** in Ammonoids, **trochocones** in Nautiloids), resembling sinistral turreted gastropods, mainly Cretaceous (e.g., *Turrilites*, Fig. 46*e*) but with one example in the Norian; and one extraordinary form (*Nipponites*) which can best be described as a three-dimensional tangle. By analogy with the gastropods, which are mostly asymmetric and crawlers, but in which the few swimming or floating forms become symmetric; we may suppose these few asymmetric cephalopods to have been crawlers (**benthic**), while the rest were, at the least, not confined to the bottom, even if they were not habitually swimmers (**nektic**) or floaters (**planktic**).

The thin shell with its large volume of gas-chambers must have been a very light object in the animal's lifetime, enabling the animal to rise or sink in the sea with ease, as does the modern *Nautilus*. After death it might float and be carried by currents to a considerable distance. Thus the three surviving species of *Nautilus* live in the Polynesian seas between Sumatra and Fiji, but their empty shells are found on the coast of Japan and elsewhere; the three species of *Spirula* live in moderately deep sub-tropical waters, and have rarely been found alive, but the empty shells are thrown up in great numbers on the coasts of New Zealand, and are also known in the Banda Sea, the Canary Islands, West Indies, etc., and have been found on the Cornish coast. Thus it may be the case with fossil cephalopods that the geographical distribution of the shells is far wider than that of the living animals. In travelling long distances, however, such delicate shells could not escape injury; so that when we find (as we frequently do) fossil cephalopods with the most delicate structures preserved uninjured, it is reasonable to suppose that they lived where they are now found; but when shells are found showing signs of travel—broken body-chamber, abraded surface, or general damage—they may fairly be regarded as having floated from a distance.

In the process of burial on the sea-floor, the sediment would fill the body-chamber, and penetrate through the septal neck into the first gas-chamber, which it could only fill up to the level of the entry. A little might penetrate into the second chamber, but beyond that it would have great difficulty in making its way unless the shell was broken in places (though fine calcareous mud is sometimes found in the air-chambers of apparently unbroken shells). The empty chambers would rarely remain empty, however: water would permeate them and deposit materials from solution. Thus in calcareous rocks the chambers are usually filled or partly filled with crystalline calcite; in clays, with pyrite or marcasite. Where very little sediment was being deposited, they were commonly filled with calcium phosphate. As a consequence, fossil cephalopods, especially those of large size, are usually heavy objects, and it needs an effort of imagination to realize them as swimming or floating. Further, the thin shell itself may have been removed, in one or two ways: by solution in the rock, or by flaking away during extraction leaving an internal mould. When the outer shell surface is corrugated and not thickened the ridges show as well on the mould as on the shell, but the moulds are at once distinguished by their showing the septal sutures, which of course are on the inner, not the outer, surface of the shell. Sometimes when no suture-lines are visible on the surface, the shell itself has been replaced by silica or pyrites.

Among existing Cephalopoda two orders may be recognized, distinguished by the number of gills and various other features. The order Tetrabranchiata (four-gilled) includes the single genus *Nautilus*; all other living forms belong to the Dibranchiata (two-gilled). It has been usual to extend this classification to extinct forms, but while many of these are so evidently allied to the recent forms that they can safely be assigned to one or other of these orders, this is not the case with the ammonites. It is therefore advisable to recognize three orders which may be briefly characterized thus.

1. The **Nautiloidea** are the ancestral and most conservative order.

Commencing as cyrtocones towards the end of the Cambrian period, they evolve all types of shell in the Palaeozoic era, but subsequently continue without further marked change to the present time. Only a very few forms, and those quite early, show any tendency to uncoil.

Generally distinctive features are (1) the protoconch is of varied shape (globular, hemispherical, conical); (2) the septal sutures appear on the outer surface of the internal mould either as simple circles or slightly wavy or even angulated, but never frilled; (3) the siphuncle is usually near the centre of each septum, very rarely is it in contact with either margin; (4) the septal necks point away from the body-chamber; (5) there is usually a hyponomic sinus.

2. The **Ammonoidea** contrast with the steady-going Nautiloidea: they had a short life and a merry one. Possibly diverging from coiled Nautiloidea at the end of the Silurian period, they are at first not strikingly different. Some of them uncoil completely as early as the late Triassic period; others at various stages in the Jurassic; others after some uncoiling start coiling-in again; but in the Cretaceous period uncoiling becomes very general and is followed by the apparently abrupt extinction of the order (though coiled forms also persist to the last).

General distinctive features are more difficult to state than in the case of the Nautiloidea, because of the great changes through which the order passes, but the following may be given:

(1) The protoconch is usually barrel-shaped; (2) the septa are like those of nautiloids in some primitive families, but soon become folded and eventually frilled at the edges, so that the suture-lines attain a very high degree of complexity; (3) the siphuncle migrates to one of the margins, in the great majority to the periphery; (4) the backwardly directed septal necks (**retrosiphonate**) soon become replaced by forwardly projecting **septal collars** (**prosiphonate**); (5) the hyponomic sinus is present in most Palaeozoic genera, but afterwards disappears, and may be replaced by a rostrum.

3. The **Dibranchiata** vary so much in shell-structure that it is difficult to make any statement applicable to the whole order. The main feature is the general subordination of the chambered shell (**phragmocone**) to other skeletal structures. The septa are simple, the siphuncle more or less marginal and retrosiphonate.

Among the large number of Palaeozoic nautiloids we choose a few for mention.

(1) Genera retaining the straight form of *Orthoceras*, but distinguished from it by peculiar features of the siphuncle. In *Endoceras*

(Ord.) the siphuncle is nearly half the width of the shell and the septal necks are funnel-like and fit into one another, forming a continuous tube. In *Actinoceras* the siphuncle is central, and swells out in each gas-chamber, but is constricted as it passes through the septa.

(2) Orthocones or cyrtocones which acquire peculiar features when fully grown. *Tetrameroceras* (p. 98) is an instance; *Gomphoceras* (Ord.–Sil.) is a straight form with T-shaped aperture; *Phragmoceras* (Sil.) an endogastric gyrocone.

(3) Ophiocones (Figs. 50*a*, *b*) are found from the Cambrian up to the Triassic. Sphaerocones begin with *Solenocheilus* (Carb.), and *Nautilus* (Trias.–Rec.) continues this form. *Hercoglossa* (U. Jur.–Eocene) and *Aturia* (Cainozoic) are remarkable as mimicking the Devonian Clymeniina (see below), in the folding of their septa and, in the case of *Aturia*, also in the dorsal position of the siphuncle.

(4) Uncoiling forms occur among nautiloids: *Lituites* is Ordovician, *Ophidioceras* (Fig. 50*b*) is Silurian: both are ophiocones (sometimes beginning as cyrtocones) which become straightened out, the former at an early period of its life, the latter only at the end. Otherwise the Nautiloidea seem to have found stability and permanence in the sphaerocone stage.

The Ammonoidea are more interesting to the geologist than are the Nautiloidea, because of their great value as zone-fossils. This arises from the wonderful vitality of the stock, which during six geological periods threw off swarm after swarm, each evolving with rapidity and then either dying out or giving rise to a new stock repeating a similar evolution, until at last exhaustion is reached in the Cretaceous period and the whole order dies out.

Examples of some of the most highly organized ammonoids (*Asteroceras*, *Promicroceras*), as well as of uncoiling forms (*Parapatoceras*, *Baculites*), have already been described. Examples of the earlier less specialized forms will now be given.

In the Upper Devonian shales of Büdesheim in the Eifel district of Western Germany, there are found great numbers of internal moulds of small tightly-coiled cephalopod shells, in pyrites oxidized on the surface to a rich brown colour. They rarely exceed 15 mm. in diameter, but they are practically complete moulds, since they show the body-chamber as well as the septate portion. They vary very much in relative thickness and degree of involution, so that a number of different species are distinguishable, but if the suture-lines are examined they can at once be sorted into two series. An example of each will now be described.

11. **Manticoceras** [**Gephuroceras**] **retrorsum** has (in the specimen

shown in Fig. 47) a diameter of 15 mm. and its proportions are 15: 53, 57, 20. The body-chamber, so far as preserved, is not quite half the last whorl in length. The umbilical slope is very steep, almost perpendicular near the aperture, but less steep in the earlier-formed part of the last whorl. The lateral and peripheral areas seem at first sight to pass insensibly into one another, but very careful examination reveals a very faint longitudinal depression about 2½ mm. on either side of the middle line: this is more pronounced in the earlier formed part of the last whorl.

The shell, as seen in the mould, has two quite independent series of lines—(1) a series of very faint ribs, most clearly seen at the peripheral shoulder, where they are less than 1 mm. apart. They are curved with a forward concavity on the lateral area, form a lappet at the peripheral

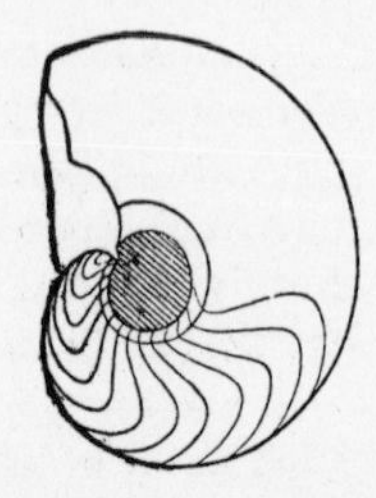

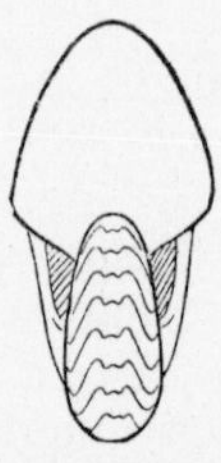

FIG. 47.—MANTICOCERAS aff. RETRORSUM (von Buch), Upper Devonian; Büdesheim, Eifel

Internal mould of an immature example; the faint surface markings omitted; umbilicus filled with matrix; numerous suture-lines behind the body-chamber (× 2; original).

margin, and then they bend backwards again and cross the periphery in a deep curve like the hyponomic sinus of Nautiloidea; these are growth-lines. (2) The other pattern, that of the innermost of the three shell layers, is a series of much finer and closer transverse, anastomosing striae, the so-called **wrinkle-layer.**

The suture-line (septal suture) is of a kind intermediate between those of a *Nautilus* and an ammonite: lobes and saddles are well defined, but entirely free from frilling (**goniatitic type**). There is a broad external lobe, divided into two by a small median saddle (itself divided by a broad shallow lobe); a wide external saddle on each side, and a rounded lateral lobe. It is the special feature of the family Gephuroceratidae that the external lobe and saddle are so wide that the latter comes to lie in the middle of the lateral area; hence many authors call it the "lateral saddle." In ontogeny the little median saddle appears late, so that there can be no doubt that the divided lobe of which it forms part is the external lobe and not a pair of lobes (Fig. 55*b*).

Although the Eifel specimens are never much larger than the specimen described, in beds of the same age in New York State there have been found specimens of a closely-allied species which grew to a fairly large size: in these the hyponomic sinus disappeared in the adult and was replaced by a forward projection or rostrum (Fig. 48).

FIG. 48.—MANTICOCERAS RHYNCHOSTOMA (J. M. Clarke), Upper Devonian; Big Sister Creek, Erie County, New York State, U.S.A.
Adult, with rostrum and lappets. Last suture-like only shown ($\times \frac{1}{6}$; after Clarke).

12. **Tornoceras simplex** is said to attain a diameter of 75 mm., but the Büdesheim specimens are usually small, and the size and proportions of such a one, stated in the manner already explained, are 14·5: 55, 47·5,?. The umbilicus is either entirely closed, or too small to be measured accurately.

The body-chamber extends to nearly a whole whorl in complete

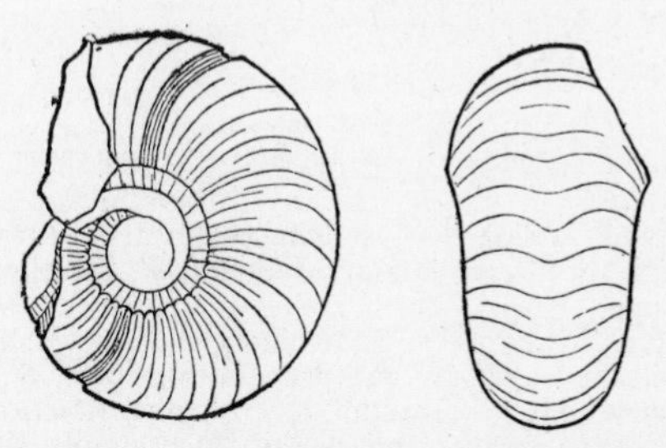

FIG. 49.—GASTRIOCERAS CARBONARIUM (von Buch), Lower Coal Measures, Lancashire
Immature; two constrictions shown; the growth striae indicate a hyponomic sinus (right-hand figure) (Natural size; original).

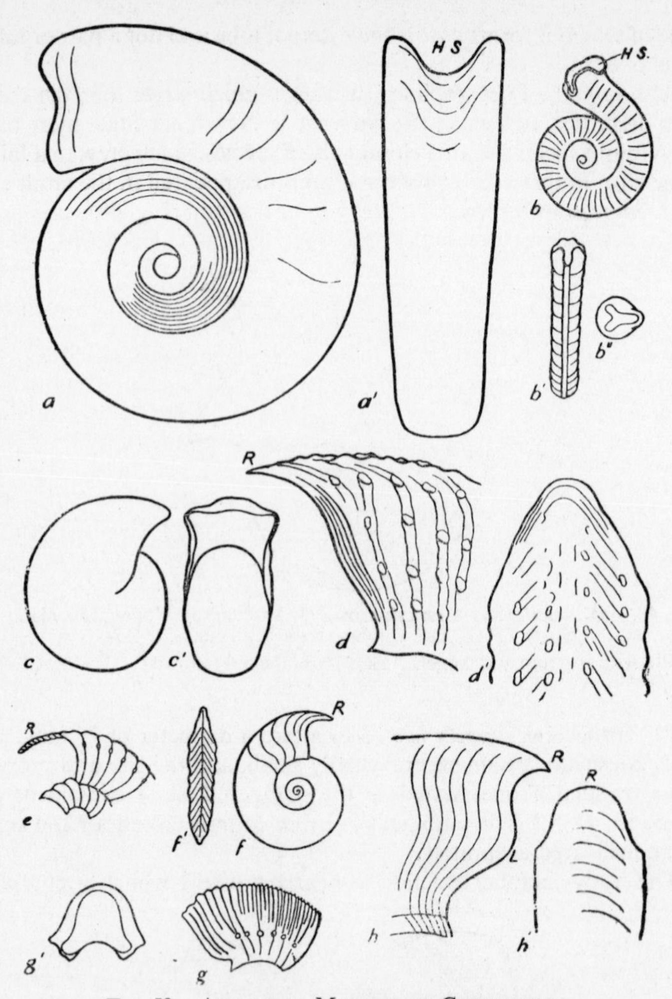

FIG. 50.—APERTURAL MARGINS OF CEPHALOPODA

a, *Discitoceras leveillianum* (de Koninck), Carboniferous Limestone, Clane, Ireland, ophiocone; *a*, Side view, showing perforated centre, there are indications of the radial and longitudinal ornament; *a′*, Peripheral view, showing hyponomic sinus (After Foord and Crick). *b*, *Ophidioceras simplex* (Barrande), Silurian, Lochkow, Bohemia; *b*, Side view, showing ophiocone finally uncoiling, and aperture with hyponomic sinus; *b′*, Peripheral view; *b″*, apertural view (After Barrande). *c*, *Arcestes distinctus* (Giebel), Norian (Alpine Trias); sphaerocone; no distinct hyponomic sinus, rostrum, or lappets (After Quenstedt). *d*, *d′*, *Protrachyceras archelaus* (Laube), Ladinian (Alpine Trias); apertural region, showing broad rostrum (After Mojsisovics); *e*, *Pleuroceras pseudocostatum* (Hyatt), Upper Pliensbachian (Middle Lias); long crenulate rostrum (After Quenstedt). *f*, *f′*, *Pleydellia subcompta*

examples. The mould is nearly smooth, but shows growth-lines (radial lines) whose course is similar to that of *Manticoceras*. The suture-line, however, is quite different: the external lobe is narrow and undivided; on each side of it is a rounded external saddle, a rounded lateral lobe and a rounded lateral saddle (Fig. 55*c*). It is the lobe that comes in the middle of the lateral area, so that the contrast with *Manticoceras* is striking.

13. **Gastrioceras carbonarium** (Fig. 49) and an allied species *G. listeri* occur in some abundance in the Bullion Mine marine band which forms the "roof" of a coal-seam in the Lower Coal Measures of Lancashire and Yorkshire. Specimens of the former have been found as much as 160 mm. in diameter, but a more usual size is about 30 mm. and one of these small specimens gives the following proportions (the query to the last item denotes difficulty of measurement owing to the umbilicus being filled with tough carbonate matrix):

30: 55·5, 60, 40?

The inner area falls very steeply from the umbilical edge. The periphery is arched and cannot be marked off from the lateral area. The greater part of the surface is marked by very fine striae, which on the lateral area are straight and radial but cross the periphery in a forwardly-concave curve, indicating a much shallower hyponomic sinus than in the Devonian forms already described. The striae are not uniform, but are arranged in bundles, which in passing from the periphery towards the umbilicus take on the character of ribs and at the umbilical margin project slightly as blunt tubercles.

At intervals of about half a whorl there occur grooves or **constrictions,** 1 mm. or more in width, following exactly the course of the growth-lines. The constrictions are analogous to the varices of gastropod shells: each one was originally formed just behind the apertural margin (as in *Orthoceras decipiens*, though in that species there is only one), and it may be presumed that its formation was followed by a period of cessation of growth, after which renewed growth went on rapidly for half a whorl, when another constriction was formed.

Most specimens preserve the shell, but even where it is chipped

---

FIG. 50.—APERTURAL MARGINS OF CEPHALOPODA (*continued*).

(Branco), Upper Toarcian; showing relation of radial line to apertural margin (After Buckman). *g*, *g′*, *Cadomites exstinctus* (Quenstedt), Upper Bajocian; showing tendency to constriction of aperture, absence of distinct rostrum and lappets, and relation of radial line to apertural margin (After Quenstedt). *h*, *h′*, *Brasilia effricata* (S. Buckman), Lower Bajocian, Bradford Abbas, Dorset; showing very broad rostrum, no definite lappets, and relation of radial line to apertural margin (After Buckman). (All × ½) *H.S.*, Hyponomic sinus; *L*, lappet; *R.*, rostrum.

away no suture-lines are visible. This is because the body-chamber was over a whorl in length: only on the inner whorls of broken specimens can the suture-line be seen (Fig. 55*d*). It resembles that of *Tornoceras* in having a lateral lobe and saddle, but the lobe is pointed instead of rounded; while the external lobe is divided by a median saddle which is much more prominent than that of *Manticoceras retrorsum*.

These three Palaeozoic species are conveniently described as "goniatites," being characterized by suture-lines with simple lobes and saddles.

GENERAL CHARACTERS OF AMMONOIDS

The body-chamber of the typical ammonoids varies from less than half a whorl to over two whorls in length. The aperture of nearly all

FIG. 51.—LUNULOCERAS BRIGHTI (Pratt), Callovian (Lower Oxford Clay), Christian Malford (Wilts)
Rostrum and lappets (Natural size; original).

Devonian forms has a hyponomic sinus on the periphery, but some of the goniatites (*Manticoceras*, Fig. 48) lose this, and begin to develop a slight forward projection (**rostrum**) in its place. In the Jurassic and Cretaceous periods this rostrum may become very prominent, until in some case it projects as a narrow rod with a length greater than the height of the whorl (Figs. 50*c–f*). Sometimes the rostrum may project outwards as a horn, or even be curled backwards (Fig. 53). Genera with a prominent rostrum are found onwards to the Cretaceous period, but a pair of lateral protuberances (**lappets**) (already seen in *Manticoceras*, Fig. 48) become much commoner in the Jurassic (Fig. 51), and as these increase the rostrum diminishes until it may disappear altogether (Figs. 52*a–h*). Genera with large lappets are

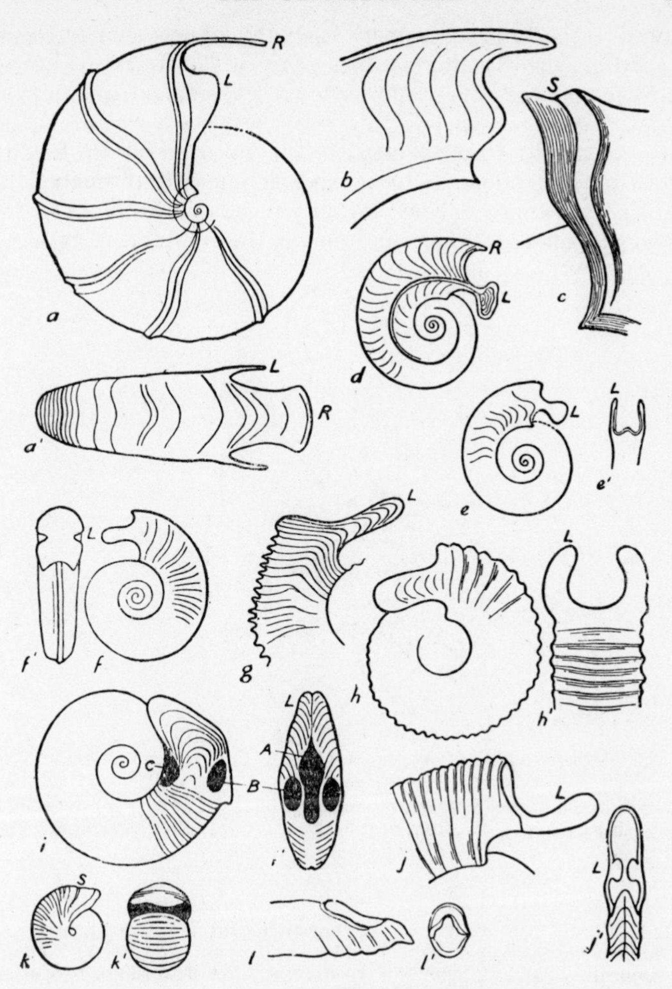

FIG. 52.—APERTURAL MARGINS OF CEPHALOPODA

a, a′, *Phylloceras mediterraneum* (Neumayr), Upper Bajocian, near Digne (Basses Alpes); showing rostrum and lappets (unusual in Phylloceratidae) and a number of ribs which are parallel to apertural margin if rostrum and lappets are excluded ($\times \frac{1}{3}$; after Haug); *b*, *Hildoceras sp.*, Lower Toarcian (Upper Lias) rostrum, lappet, and radial line. (After Wright.) *c*, *Lucya cavata* (S. Buckman), Lower Bajocian (*concava* zone), Bradford Abbas (Dorset); constriction behind aperture. (After Buckman.) *d*, *Oppelia fusca* (Quenstedt), Upper Bajocian, Oeschingen (Württemberg); lappet constricted at base; relation of radial line to apertural margin well shown. (After Quenstedt.) *e, e′*, *Fontannesia curvata* (S. Buckman), Middle Bajocian, Bradford Abbas (Dorset); very feeble rostrum constricted (spoon-like) lappets. (After Buckman.) *f, f′*, *Witchellia deltafalcata* (Quenstedt), Bajocian; rostrum confined to the Jurassic. In many cases the lappets tend to close in the aperture, the extreme case being that of *Ebrayiceras* (Fig. 52*i*), where the aperture is even more completely subdivided than in the Silurian *Tetrameroceras*.

The constrictions already described in *Gastrioceras* are found in various other ammonoids, sometimes at intervals throughout life (Fig. 46*d*), sometimes only in very early stages of growth. As we have seen in gastropods, all specializations of the apertural margin must

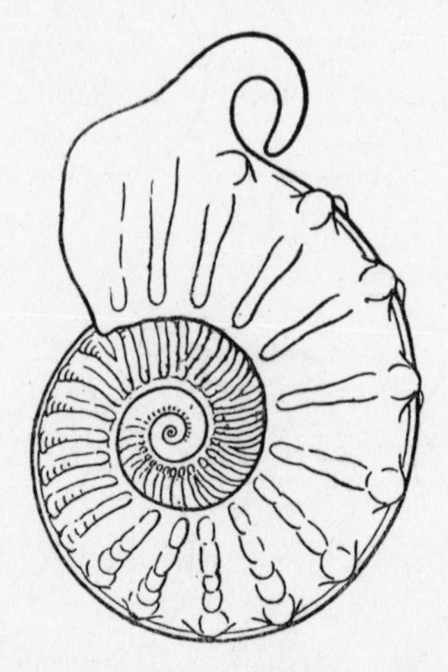

FIG. 53.—MORTONICERAS cf. ROSTRATUM (J. de C. Sowerby), Albian (Upper Gault)

Shell markings omitted in centre; bifurcating ribs replaced later by single multi-tuberculate ribs. Rostrum bent backwards (× ⅜; after Buvignier).

FIG. 52.—APERTURAL MARGINS OF CEPHALOPODA (*continued*).

almost lost, lappets bent inwards to constrict aperture. (After Quenstedt.) *g*, *Kosmoceras elizabethae* (Pratt), Callovian (Lower Oxford Clay), Christian Malford (Wilts); long lappets, no rostrum; relation of radial line to apertural margin. (After Quenstedt.) *h*, *h′*, *Otoites sp*., Bajocian, Lauffen (Württemberg); incurved lappets, no rostrum; no evidence of lappets in radial line. (After Quenstedt.) *i*, *i′*, *Ebrayiceras pseudoanceps* (Ebray), Upper Bajocian, St. Honoré-les-bains (Nièvre, France); lappets and bifurcated rostrum uniting to divide aperture into five orifices—*A*, median hyponomic orifice; *B*, paired opening, probably for eyes; *C*, paired opening for arms. (After H. Douvillé.) *j*, *j′*, *Binatisphinctes comptoni* (Pratt), Callovian; long incurved lappets, no rostrum, a constriction shown some distance behind aperture; radial line parallel to constriction, not to apertural margin. (After Quenstedt.) *k*, *k′*, *Sphaeroceras brongniarti* (J. Sowerby), Bajocian; sphaerocome, aperture depressed without rostrum or lappets. (After Quenstedt.) *l*, *l′*, *Sciponoceras* [*Baculites*] *baculoides* (Mantell), Cenomanian, Chardstock (Somerset). (After Crick.) All × ½, except where otherwise stated. *R*, Rostrum; *L*, lappet; *S*, constriction.

have interfered with the simple growth of the shell, and we must suppose, either that they were not formed until the shell had attained its full size, or that, if formed, they had to be resorbed before growth could continue. Either of these suppositions may be true in particular cases. Where constrictions occur in the adult the apertural margin is parallel to the constriction, not to the growth-line as usual (Fig. 54).

In the body-chamber of some ammonites there may be found a pair of symmetrical, subtriangular calcareous plates, which together form a heart-shaped body (**aptychus**). Similar aptychi occur apart from ammonites, and there are even certain Jurassic shales in the Alps which are full of aptychi, but no ammonites are found: as the aptychi are made of calcite, while the shells are composed of the less stable

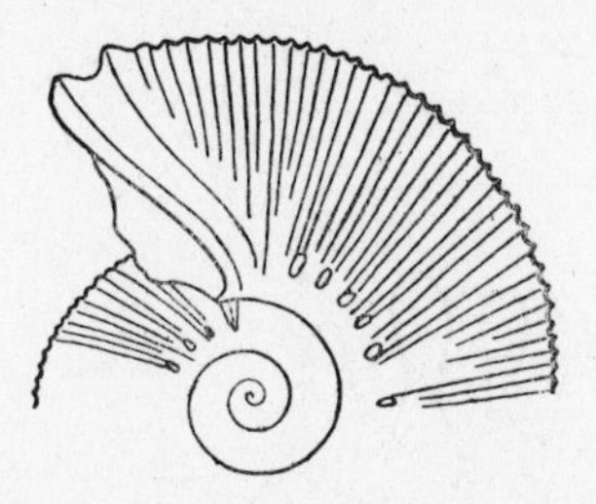

FIG. 54.—OLCOSTEPHANUS ASTIERANUS (d'Orbigny), Neocomian
Plain apertural margin, oblique to radial line ribs ($\times \frac{1}{2}$; after d'Orbigny).

aragonite, this is not surprising. In one unique example, from the Inferior Oolite of Dundry Hill, Somerset, the aptychus was found in its natural position, closing in the aperture of the shell, so that it evidently acted as an operculum. The two halves of the aptychus were permanently united in *Scaphites* (**synaptychus**); in the Liassic families Asteroceratidae and Amaltheidae and some Palaeozoic genera, a single horny plate (**anaptychus**) served as an operculum but this plate was not in all cases capable of complete closure of the aperture; many ammonoids and the Clymeniina (a group of late Devonian ammonoids in which the siphuncle is dorsal) possibly had no operculum of any kind.

The septa of ammonoids always show a tendency to folding at the margin, giving rise to more or less complexity in the suture-line. Whatever may be the meaning of this folding, puckering, and frilling, the various resulting patterns of suture-line are of the greatest value in tracing affinities and classifying ammonoids, so that careful attention must be paid to its details. The number of lobes and saddles is extremely small at first (Fig. 55), but tends to increase, some families having a very large number already in the Devonian. In the typical families of the later Mesozoic the number tends to keep within certain limits, and to bear some relation to the degree of involution of the shell. The following description applies especially to those genera in which this relation holds, and needs some modification to apply to the forms with either very few or very many lobes and saddles.

Beginning at the periphery, there is a median **external lobe** (also

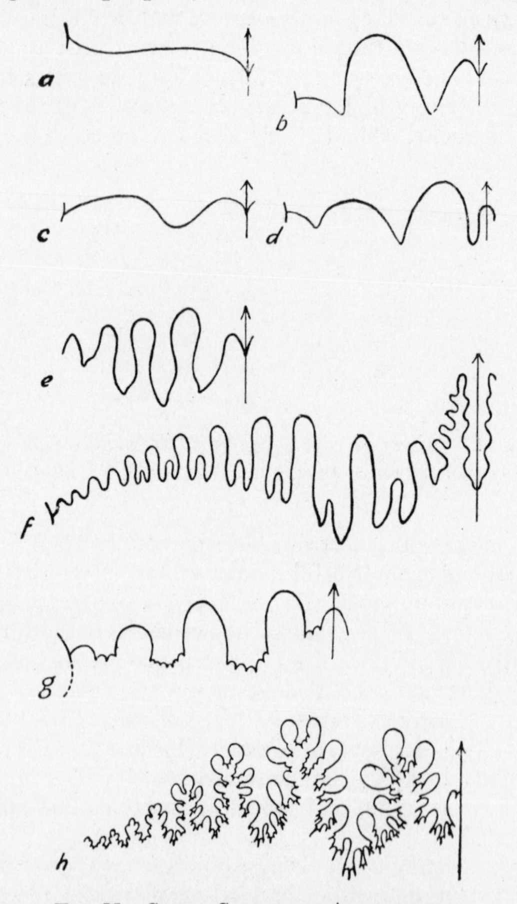

FIG. 55.—SEPTAL SUTURES OF AMMONOIDEA

Each figure has the umbilical margin on the left and the middle line of the periphery on the right. The arrow points towards the body-chamber. *a*, *Anarcestes*, Devonian; *b*, *Gephuroceras*, Devonian; *c*, *Tornoceras*, Devonian; *d*, *Gastrioceras*, Carboniferous; *e*, *Prolecanites*, Carboniferous; *f*, *Medlicottia*, Permian; *g*, *Paraceratites*, Triassic; *h*, *Phylloceras*, Jurassic. *a–f*, after Crick: *g*, after Perrin Smith; *h*, after d'Orbigny.

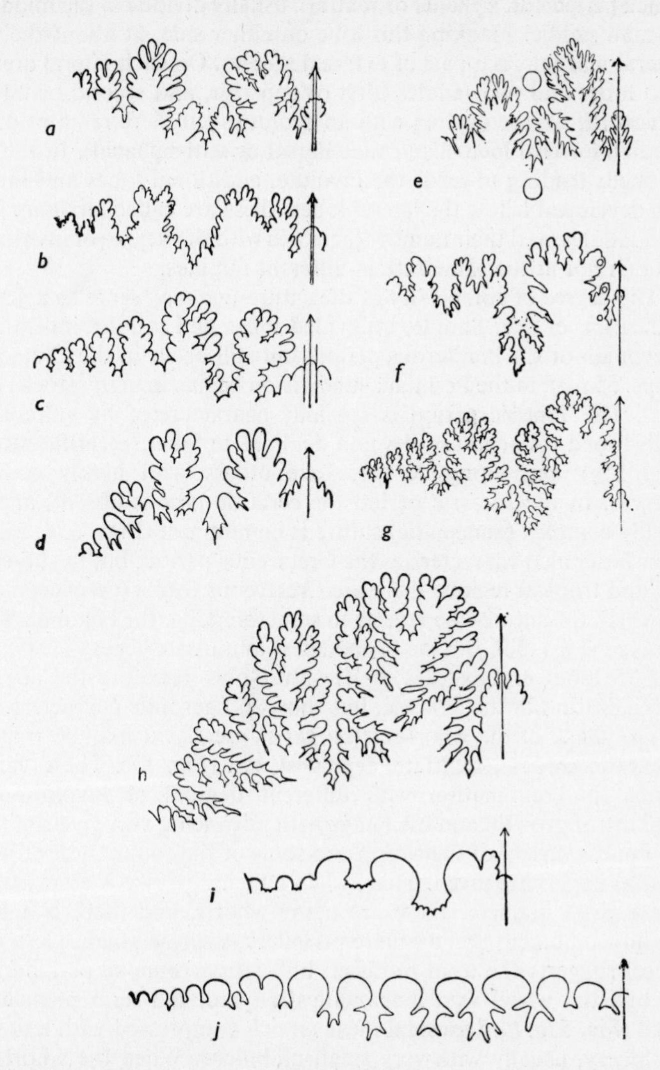

FIG. 56.—SEPTAL SUTURES OF JURASSIC AND CRETACEOUS AMMONIDEA
Arrangement as in Fig. 55. Circles in *e* denote the position of tubercles. *a*, *Pleuroceras*, Upper Pliensbachian; *b*, *Grammoceras toarcense*, Upper Toarcian; *c*, *Oppelia*, Upper Jurassic; *d*, *Caloceras*, Hettangian; *e*, *Eoderoceras*, Lower Pliensbachian; *f*, *Kosmoceras*, Callovian; *g*, *Hoplites*, Lower Cretaceous; *h*, *Perisphinctes*, Upper Jurassic; *i*, *Tissotia*, Upper Cretaceous; *j*, *Metengonoceras*, Upper Cretaceous. *a–h*, after d'Orbigny; *i*, after H. Douvillé; *j*, after Hyatt.

called **peripheral, siphonal** or **ventral**), usually divided in the middle by a small saddle. Flanking this lobe on either side, at about the peripheral margins, is a pair of **external saddles.** On each lateral area are two lobes and two saddles (first or superior, and second or inferior **lateral**). If we are dealing with an evolute shell there remains only a median **internal lobe** (also called **dorsal** or **anti-siphonal**). In the case of shells tending towards the involute, additional lobes and saddles are developed below the lateral lobes: these are called **auxiliary** lobes and saddles, and their number increases with the degree of involution, though not at the same rate in different families.

The degree of complexity of the suture-line may serve as a general indication of age. Simple, undivided lobes and saddles indicate the Devonian or Carboniferous period, but still occur in the Cretaceous (Figs. 55*a–e*); in the Permian, subdivision begins in many stocks (Fig. 55*f*). The Triassic period is specially characterized by ammonoids with broad rounded saddles and denticulate lobes (**ceratitic** sutures, Fig. 55*g*), but alongside these are others with highly complex sutures. In the Jurassic period the ceratitic type is absent, and the highly complex (**ammonitic**) suture is commonest (Figs. 55*h*, 56*a–h*). This latter also characterizes the Cretaceous period, but in sub-tropical and tropical regions there are Cretaceous forms (**pseudoceratites**) in which the suture-line reverts to something like the common Triassic type (Figs. 56*i, j*). One of the latest ammonite genera (*Indoceras*) has 75 lobes and saddles in its suture-line, recalling the acme of specialization in the Triassic, but shows no gerontic characters.

The shape of the whorl-section varies greatly: it may be **rounded, quadrate, cordate, sagittate, depressed,** etc. (Fig. 57). These various forms, in combination with different degrees of involution, of rapidity of growth, and of change with advancing age, give rise to an enormous variety of shell-forms, to some of the commonest of which special names are given. Thus, when increase of size is slow, so that for a given diameter there are many whorls, and there is a wide, shallow umbilicus and a square periphery, the resemblance to a cart-wheel suggests the term **rotiform**; but with a rounded periphery, as though the wheel were pneumatically-tyred, the term **planulate** is used (Fig. 57*a, b*). **Discoidal** means much compressed with a narrow periphery, usually with very small umbilicus. When the whorls are depressed, with broad periphery and rather deep umbilicus, the ammonite if placed on its side suggests the shape of a crown, especially if (as in common) it has a row of tubercles or spines along the lateral margin: such an one is **coronate** (Fig. 57*f*). The term **oxycone** is applied to discoidal forms and those approaching them in which the periphery is acute or oxynote (Fig. 57*c*, *d*); and **cadicone**

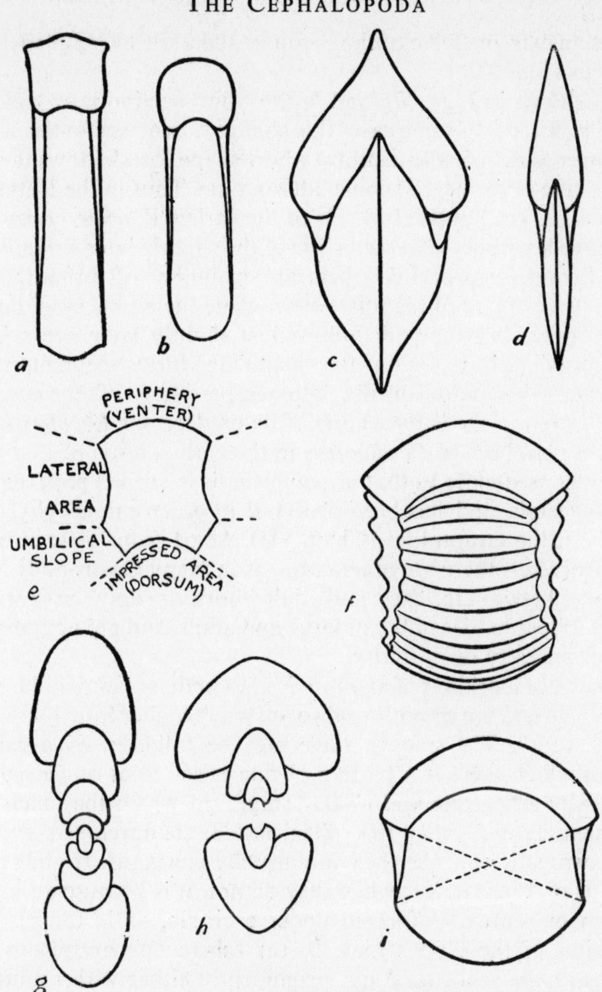

FIG. 57.—WHORL-SHAPE IN AMMONOIDEA

All except *e*, *g*, and *h* are apertural views. *a*, *Microderoceras*, latumbilicate, rotiform, whorl-section quadrate; *b*, *Dactylioceras*, latumbilicate, planulate, whorl-section rounded; *c*, *Chamoussetia*, angustumbilicate, lenticular oxycone, whorl-section cordate; *d*, *Amaltheus*, angustumbilicate, discoidal oxycone, whorl-section sagittate; *e*, Diagram based on *Aspidoceras* to explain areas of whorl-surface; *f*, *Erymnoceras*, coronate, whorl-section greatly depressed (periphery and umbilical area meet, excluding any lateral area); *g*, *Perisphinctes*, section showing early whorls of depressed crescentic shape, changing later to compressed and planulate; *h*, *Quenstedtoceras*, section showing early compressed whorls, changing later to broad and somewhat depressed; *i*, *Cadoceras*, a cadicone, greatly depressed whorls, deep conical umbilicus (indicated by dotted lines). *a*, *b*, *c*, *d*, *f*, *i*, after d'Orbigny; *g*, after Dumortier and Fontannes; *h*, after R. Douvillé.

denotes an extreme form of the coronate type, with very deep conical umbilicus (Fig. 57*i*).

As is shown in Figs. 57*g* and *h*, the whorl-section may change as growth proceeds. In some cases this change is contrary to the concept of palingenesis, since the youthful whorl-shape foreshadows the adult whorl-shape of forms that come later in time. Thus in the Hettangian (Lower Lias) the youthful whorls of the earlier *Waehneroceras* show the grooved periphery characteristic of the slightly later *Schlotheimia*. In the Lower Pliensbachian there are a number of forms the early whorls of which are of planulate type, while the adults swell out into sphaerocones; but they are followed in slightly later strata by the "capricorns" which retain the planulate form throughout life. Similarly in the Callovian, the flattened periphery of *Kosmoceras* is foreshadowed in the early whorls of its predecessor *Kepplerites*, and the acute periphery of *Cardioceras* in the early whorls of *Cadoceras*. This reverse sequence to that of palingenesis is termed **proterogenesis** or **caenogenesis,** and has been observed in other animal phyla, particularly in the Graptolites (Chap. VII). Apart from these two antithetic processes, there are other cases, as in many Gastropods, where differences between the early and adult whorls are only the expression of the different modes of life of larva and adult, and are no indication either of ancestry or posterity.

The surface markings of ammonoids fall under a few definite heads. Primitive genera are smooth, and so are many higher forms, especially in early youth. The smooth stage may be followed by a **capillate** stage, in which there is a pattern of fine raised lines on the surface; and this by **subcostate** and **costate** stages, in which the much more prominent features called **ribs (pilae, costae)** are developed: these are really corrugations of the shell, and usually appear as strongly on the mould as on the exterior; when they do not, it is because an internal partition or septum is formed under each rib, as in the planulate ammonites of the Upper Lias. In the **tuberculate** or **spinous** stage portions of the shell stand out prominently either with rounded or pointed ends. A shell which is tuberculate in youth may become costate or even smooth on the body-chamber.

The ribs run in a straight or curved direction from the umbilicus to the periphery; their course is fairly constant except towards the aperture, and is a good guide to the affinity between different species. This course is termed the **radial line** of the species, and it bears a close relation to the form of the apertural border. Thus, if there is a rostrum and lappets without any constriction behind, the radial line will be more or less sickle-shaped or **falciform** (Figs. 50*f*, *h*; 52*d*), but if there is a constriction it will be straight (Fig. 52*j*). A rostrum without lappets gives rise to a radial line like an inverted **L**, or Greek **Γ** (**gammi-radiate,** Figs. 44*a*; 50*d*, *e*).

The ribs sometimes cross the periphery without any change of direction (Figs. 52*h*, *k*); in some shells there is more or less of a forward bend, owing to the general presence of a rostrum (Fig. 50*f*). The absence of a forward bend in the ribs does not, however, prove the absence of a rostrum (Fig. 52*a*′); it may mean that rostra were produced at intervals and resorbed. In many forms the ribs are interrupted on the middle of the periphery, either by an elevated **keel** (**carinate** periphery, Fig. 44*b*) or by a median groove (**sulcate** periphery, Figs. 50*d*′; 52*i*) or by a keel flanked by grooves (**carinati-sulcate**).

The radial shell markings are in a few families crossed by longitudinal (spiral) lines.

DIBRANCHIATA

The most familiar fossils of this order are those known as *Belemnites*, the most typical of which are found in the Jurassic and Lower Cretaceous systems, though forms but little different also occur in the Upper Triassic and Upper Cretaceous. A typical belemnite shell consists of (1) a **phragmocone** (chambered shell) which is an orthocone or cyrtocone with a calcareous globular protoconch, a siphuncle along the ventral margin and a forward prolongation of the dorsal region, the **pro-ostracum** (Figs. 58*a*, *b*, *b*′) and (2) a solid **guard** or **rostrum,** more or less cigar-shaped, with a conical hollow (**alveolus**) at the blunt end for the reception of the apex of the phragmocone. The guard is composed of small prisms of calcite arranged radially at right angles to the surface and with concentric growth rings; it is more commonly found preserved than the phragmocone, and is so resistant to wear and tear that belemnite guards may be found as derived fossils in gravels. In a few cases, sufficient traces of the body of the belemnite have been found to show that the skeleton was entirely internal, the animal growing out of all proportion to the growth of the "shell," so that an ever smaller portion of the body remained within the body-chamber, and the mantle gradually enveloped the whole shell. Like its relative the modern cuttle-fish, the animal possessed an ink-sac and a number of arms around the mouth, though these were furnished with horny hooks not known in the cuttle-fish; there were eight arms with probably two longer than the rest and without hooks (Fig. 58*a*).

In some of the earliest (Upper Triassic and Lower Jurassic) forms the phragmocone extends nearly to the end of the guard (Fig. 58*c*), and in these cases it is probable that the former was still an external shell in early life and only later became enveloped in the mantle, which then began to secrete a guard around it. But in most later forms the alveolus only extends a short distance into the guard, which probably began to be formed as early as the phragmocone.

The phragmocone when found alone is apt to be mistaken for an Orthocone nautiloid with a marginal siphuncle, but it can be distinguished, (1) by the three layers in the shell wall, (2) there is no body

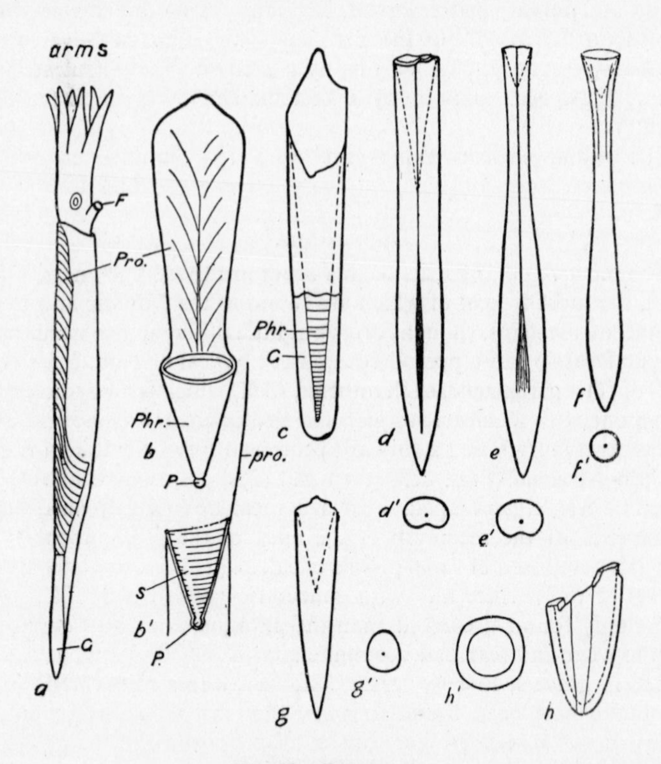

FIG. 58.—BELEMNITES

a, Restoration of the living Belemnite (modified from d'Orbigny); *F*, Funnel; *G*, guard; *Pro*., pro-ostracum; *b*, *b'*, Restoration of phragmocone, in front view and section (after Crick); *P*, Protoconch; *Phr*., phragmocone; *S*, siphuncle; *c*, *Atractites philippii* Hyatt and Smith, Carnian, California; guard (*G*) partly broken open, showing phragmocone (*Phr*.). (After Hyatt and Smith). *d*. *Belemnopsis bessinus* (d'Orbigny), Bathonian; *e*, *Hibolites hastatus* (Blainville), Callovian (Oxford Clay); *f*, *Hastites clavatus* (Blainville), Lower Pliensbachian; *g*, *Prototeuthis acutus* (Miller), Sinemurian; *h*, *Coeloteuthis excavatus* (Phillips), Sinemurian. *d–h*, Ventral views: dotted lines indicate the alveolus, unbroken lines the grooves on the surface; *d'–h'*, Cross-sections, the dot marking the axis; *d*, after d'Orbigny; *e–h*, after Phillips. (All × ½, except *a*.)

chamber, and (3) by the fact that the growth-lines (**conothecal striae**), which form its only surface markings, curve forwards on the dorsal region in correlation with the presence of a pro-ostracum (Fig. 58*a*).

The chief features serving to distinguish species among belemnite-guards are (1) the general shape, (2) the disposition of grooves on the surface, (3) the depth and apical angle of the alveolus. The two main shapes are the **lanceolate** and the **hastate**: in the former the diameter is constant for the greater part of the length, and the posterior end is conical; in the latter the diameter increases from the front end for some distance backwards. In cross-section either form may be circular, compressed (laterally) or depressed (dorso-ventrally): to distinguish between the latter in the absence of the phragmocone, it is sufficient to determine the plane of symmetry of the alveolus which is the median plane. The alveolar angle varies between 12° and 32°.

A few belemnite-guards are quite free from grooves; others have a number of short apical grooves; but the majority have either a median ventral groove or a pair of dorso-lateral grooves.

The classification of belemnites is in a confused and unsatisfactory, condition. Only morphological classifications have been proposed, except for limited groups of species, and little notice has been taken of ontogeny, although the mode of growth of the guard (by accretion) makes changes in shape and grooving recognizable.

The following is an imperfect statement of the classification most generally accepted. An asterisk denotes genera always accepted as distinct, the remainder being usually lumped in the comprehensive genus *Belemnites*.

A. GUARD NOT GROOVED

Atractites,* Triassic, Scythian to Rhaetic (guard nearly cylindrical, alveolus nearly as long as guard, Fig. 58*c*).

Prototeuthis, Sinemurian–L. Pliensbachian (guard acutely conical, alveolus half length of guard), e.g., *Bel. acutus* (Fig. 58*g*).

B. GUARD WITH VENTRAL GROOVE

Coeloteuthis, Sinemurian (short, cylindro-conical, alveolus nearly as long as guard, groove very wide), e.g., *Bel. excavatus* (Fig. 58*h*).

Belemnopsis, Bajocian–Neocomian (long, cylindrical, with acute apex, groove long, deep and narrow, alveolus one-third length of guard), e.g., *Bel. bessinus* (Fig. 58*d*).

Pachyteuthis, Oxfordian–Neocomian (stout, cylindrical, with conical apex, groove long and shallow, alveolus from one-third to two-thirds length of guard), e.g., *Bel. excentricus* (Fig. 59*a*).

Cylindroteuthis, Callovian–Neocomian (long, cylindrical, groove deepest near apex, becoming indistinct forwards, alveolus about one-fourth length of guard), e.g., *Bel. oweni*, *B. magnificus* (Fig. 59*b*).

Hibolites, Lower Bajocian–Cenomanian (hastate), e.g., *Bel. hastatus* (Fig. 58*e*).

Belemnitella,* Senonian–Maestrichtian (cylindrical, with rounded and acuminate apex, alveolus one-third to one-half length of guard, ventral slit in guard along more than half the length of alveolus), e.g., *Bel. mucronata* (Fig. 59*c*).

Actinocamax,* Cenomanian–Senonian (very like the last, but the part of the guard around the alveolus is imperfectly calcified, and is always more or less destroyed, e.g., *A. plenus*, *A. verus* (Fig. 59*d*).

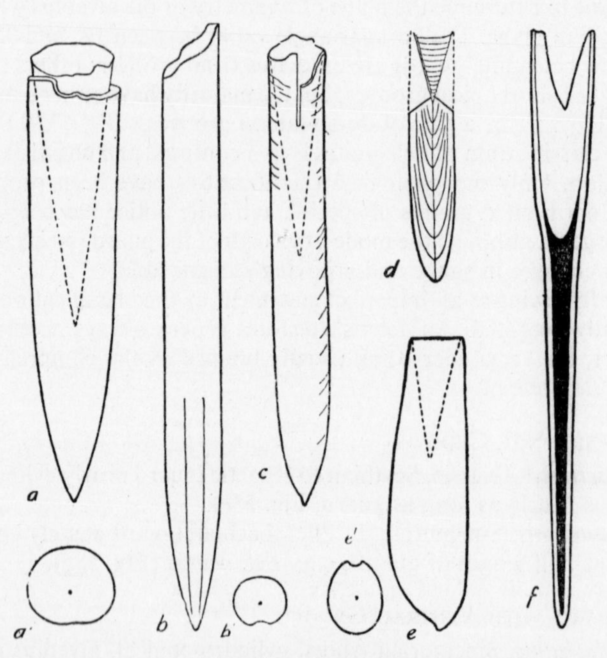

FIG. 59.—BELEMNITES

a, *Pachyteuthis explanatoides* (Pavlow), Lower Cretaceous. *b*, *Cylindroteuthis magnificus* (d'Orbigny), Upper Jurassic. *c*, *Belemnitella mucronata* (d'Orbigny), Senonian (Upper Chalk), showing alveolar slit and groovings of surface. *d*, *Actinocamax verus* (Miller), Senonian (Upper Chalk); dotted lines indicate the destroyed part of the guard. *e*, *Duvalia lata* (Blainville), Neocomian; some species of *Duvalia* are more compressed than this (*e′*). *f*, *Dactyloteuthis acuarius* (Schlotheim), Toarcian; longitudinal section, showing sudden change of shape and cavity (black). *a–c*, Ventral views; *d, f*, sections; *e*, side view; *a′–e′*, Cross-sections. All × ½ (*a*, *b*, after Pavlow; *c*, *e*, *f*, after d'Orbigny; *d*. after Crick).

C. WITH DORSO-LATERAL GROOVES

*Aulacoceras,** Triassic, Ladinian to Norian (phragmocone much longer than the small hastate guard).

Belemnites, s.s. [*Passaloteuthis*] Lower Pliensbachian (also slight ventral furrow).

Hastites, Lower Pliensbachian–Neocomian (hastate), e.g., *Bel. clavatus* (Fig. 58*f*).

Pseudobelus, Lower Bajocian–Cenomanian (lanceolate, with very deep lateral grooves), e.g., *Bel. bipartitus.*

D. WITH SEVERAL APICAL GROOVES

Dactyloteuthis, Upper Pliensbachian–Upper Toarcian (cylindro-

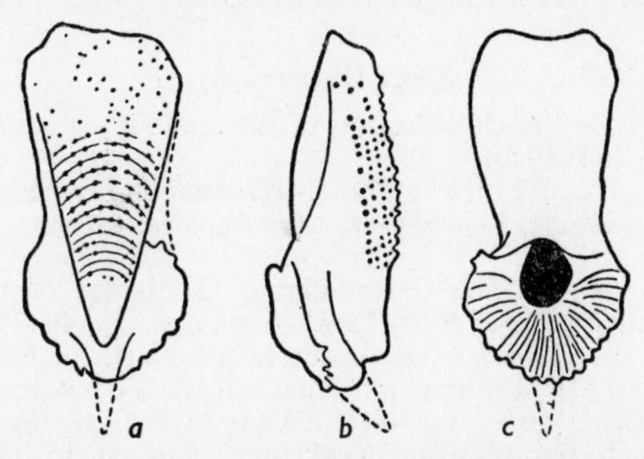

FIG. 60.—BELOSEPIA SEPIOIDEA Blainville, Middle Eocene, Bracklesham *a,* dorsal view; *b,* side view; *c,* ventral view (After R. B. Newton).

conical up to a certain age, then suddenly elongating into a long tubular apex), e.g., *Bel. acuarius* (Fig. 59*f*).

Megateuthis, Bajocian (with similar change of shape), e.g., *Bel. giganteus.*

E. WITH MEDIAN DORSAL GROOVE

Duvalia, chiefly Neocomian of Mediterranean province (much compressed), e.g., *Bel. dilatatus, B. latus* (Fig. 59*e*).

Belemnoids are known from the Permian of Greenland and lately they have been described from the Mississippian (Viséan to Namurian) of Arkansas and Utah. The typical Belemnoids die out at the end of Mesozoic time, except in Australia, where they are recorded from Eocene strata. In Europe, there are in the Eocene several forms, mostly very rare as fossils, which may be their descendants strangely modified before final extinction. Greater interest attaches to two collateral stocks which led to living forms. One of these is the Eocene

Belosepia (Fig. 60), which forms a link between the belemnites and the recent cuttle-fish *Sepia*. The siphuncle is greatly widened, the phragmocone is transitional from its typical form to the "cuttle-bone," and the guard is in process of reduction to the little spike of a modern *Sepia*. The other transitional form is the Miocene *Spirulirostra*, in which the phragmocone is a more decided cyrtocone than in any belemnite: this leads to the recent *Spirula*, a gyrocone in which all trace of guard has disappeared.

Alongside the typical Belemnoids of the Mesozoic runs an allied stock (*Phragmotheuthis*, Trias.; *Belemnoteuthis*, Upper Jurassic) in which the guard is a very thin investment of the phragmocone. By reduction of the phragmocone and guard with retention of pro-ostracum, this may have led to the modern squid, *Loligo*.

Short Bibliography

ARKELL, W. J.—"A classification of the Jurassic Ammonites," *Journ. Paleont.*, xxiv, 354 (1950).

ARKELL, W. J. and 7 other authors.—*Treatise on Invertebrate Paleontology. Part L. Mollusca, Cephalopoda. Ammonoidea.* 490 pp. (1957) Kansas Univ. Press.

BISAT, W. S.—(1) "The Carboniferous Goniatites of the North of England and their zones," *Proc. Yorks. Geol. Soc.*, xx, 40 (1924). (2) "The Goniatites of the *Beyrichocheras* Zone in the North of England," *ibid.*, xxii, 280 (1934). (3) "The junction faunas of the Viséan and Namurian," *Trans. Leeds Geol. Assoc.*, vi, 10 (1950). (4) "The Goniatite succession at Cowdale Clough, Barnoldswick, Yorkshire," *ibid.*, vi, 155 (1952).

BISAT, W. S. and HUDSON, R. G. S.—The Lower *Reticuloceras* (R1) Goniatite succession in the Namurian of the North of England," *Proc. Yorks. Geol. Soc.*, xxiv, 383 (1943).

BUCKMAN, S. S.—*Yorkshire Type Ammonites*, 2 vols. (1909–19), continued as *Type Ammonites* (1919–30).

D'ORBIGNY, A. A.—*Paléontologie Française. Terrains Jurassiques, vol. i, Céphalopodes* (1842–51); *Terrains Cretacés, vol. i, Céph.* (1840–2).

GROSSOUVRE, A. DE.—"Les Ammonites de la Craie Supérieure," *Mém. Carte géol. France* (1894).

HYATT, A. (1) Genesis of Arietidae, *Smithsonian Contrib.* No. 673 (1889); (2) Pseudoceratites of the Cretaceous, *Monograph U.S. Geol. Surv.*, xliv (1903); (3) various papers in *Proc. Boston. Soc. Nat. Hist.* (from 1875).

MILLER, A. K.—"The last surge of the Nautiloid Cephalopods," *Evolution*, iii, 231 (1949).

MILLER, A. K. and FURNISH, W. M.—"The Classification of the Paleozoic Ammonoids," *Journ. Paleont.*, xxviii, 685 (1954).

MOORE, E. W. J.—"The Carboniferous Goniatite genera *Girtyoceras* and *Eumorphoceras*," *Proc. Yorks. Geol. Soc.*, xxv, 388 (1946).

MOJSISOVICS, E.—A series of monographs on Triassic Cephalopoda published by the various learned societies of Vienna. For the chief see *Geol. Mag.* (1908), p. 189.

NAEF, A.—*Die Fossilen Tintenfische* (1922) Jena.

QUENSTEDT, F. A.—*Ammoniten des Schwäbischen Jura* (1885–8) Stuttgart.

REESIDE, J. B., Jr.—"The Scaphites, an Upper Cretaceous Ammonite group," *Prof. Pap. U.S. Geol. Surv.*, clB, 21 (1928).

ROMAN, F.—*Les Ammonites Jurassiques et Crétacées: Essai de genera* (1938) Paris.

SMITH, J. Perrin.—(1) "Carboniferous Ammonoids of America," *Monograph U.S. Geol. Surv.*, xlii (1903); (2) "Middle Triassic Marine Invertebrate Faunas of North America," *Prof. Pap. U.S. Geol. Surv.*, lxxxiii (1914); (3) "Upper Triassic Marine Invertebrate Faunas of North America," *Prof. Pap. U.S. Geol. Surv.*, cxli (1927); (4) "Lower Triassic Ammonoids of North America," *Prof. Pap. U.S. Geol. Surv.*, clxvii (1932).

SPATH, L. F.—(1) "Evolution of the Cephalopoda," *Biol. Reviews*, viii (1933); (2) "Phylogeny of the Cephalopoda," *Pal. Zeitschr.*, xviii (1936); (3) "Revision of the Jurassic Cephalopod Fauna of Kachh (Cutch)," *Palaeont. Indica* (n.s.), ix, pts. 1–6 (1927–33).

TRUEMAN, A. E.—"The Ammonite Body-chamber . . . ," *Quart. Journ. Geol. Soc.*, xcvi, 339 (1940).

WRIGHT, C. W.—"A Classification of the Cretaceous Ammonites," *Journ. Paleont.*, xxvi, 213 (1952).

PALAEONTOGRAPHICAL SOCIETY'S MONOGRAPHS

ARKELL, W. J.—(1) "Corallian Ammonites" (1935–48). (2) "Bathonian Ammonites" (1951–8).

BUCKMANN, S. S.—"Inferior Oolite Ammonites," Part I (1887–1907).

CASEY, R.—"The Ammonoidea of the Lower Greensand" (1960–).

DONOVAN, D .T.—"Synoptic Supplement to Wright's 'Lias Ammonites' " (1954).

EDWARDS, F. E.—"Eocene Mollusca: I, Cephalopoda" (1849).

FOORD, A. H.—"Carboniferous Cephalopoda of Ireland" (1897–1903).

PHILLIPS, J.—"British Belemnites" (1865–1909). Incomplete.

SHARPE, D.—"Fossil Mollusca of Chalk of England. Part I, Cephalopoda" (1853–7).

SPATH, L. F.—"Ammonoidea of the Gault" (1923–43).

SWINNERTON, H. H.—"Cretaceous Belemnites" (1936–55).

WRIGHT, C. W. and WRIGHT, E. V.—"Survey of fossil Cephalopoda of the Chalk" (1951).

WRIGHT, T.—"Lias Ammonites" (1878–86).

BRITISH MUSEUM (NAT. HIST.) CATALOGUES

FOORD, A. H.—Fossil Cephalopoda, vols. i and ii (Nautiloidea) (1888–91).

FOORD, A. H., and CRICK, G. C.—Fossil Cephalopoda, vol. iii (Goniatites) (1897).

SPATH, L. F.—Fossil Cephalopoda, vol. iv (Triassic Ammonoids) (1934).

SPATH, L. F.—"Catalogue of the Liassic Ammonite Family Liparoceratidae" (1938).

SPATH, L. F.—Fossil Cephalopoda, vol. v (Triassic Ammonoids (II) (1951).

V

THE TRILOBITA AND OTHER ARTHROPODA

1. **Calymene blumenbachii** (Fig. 61), the "Dudley Locust," is a familiar fossil, beautifully preserved in the Silurian limestone of Dudley in the Midlands Black Country, which was formerly exploited as a flux for the iron-furnaces, but that is no longer economic. This animal was of very different construction from those hitherto considered. The main part of its body is seen to be composed of a row of very short, broad divisions (**thoracic segments**), thirteen in number, all alike except for a slight diminution in size backwards. In front is a semicircular head-shield (**cephalon**), which shows suggestions of segments like those behind; and at the hind end is a tail-piece (**pygidium**) showing very well-marked segmentation (division into somites), but the segments are dorsally fused together. From end to end run two nearly parallel grooves, which divide the whole body into a median strongly-arched portion (**axis** or mesotergum) and lateral flatter portions (**pleural regions**). The existence of these three portions of the body suggested the name **Trilobita** for the group of fossils to which *Calymene* belongs.

The trilobites are quite extinct and unknown other than from the Palaeozoic rocks. There are, however, abundant living forms with so similar a general structure of skeleton that they can safely be referred to the same phylum. This is the great phylum Arthropoda, which includes the Insects, the Myriapods (centipedes and millipedes), the Chelicerata (scorpions and spiders), the Crustacea (lobsters, crabs, and woodlice), and some minor and unfamiliar classes. In all these, the body is metamerically segmented (divided into a longitudinal series of more or less similar **somites**), and encased in an exoskeleton of chitin which may be hardened by a deposit of calcium carbonate and phosphate; a number of the somites at the anterior end are inferred to have been more or less completely united into a head; and on the ventral surface is a paired series of jointed limbs (usually one pair to each somite) which fulfil the functions of locomotion (walking or swimming), the seizing of food, and (in aquatic forms) respiration.

Shell-growth in the Arthropoda is totally unlike that of Brachiopoda and Mollusca. The exoskeleton is not secreted by a mantle but by the whole surface of a complex jointed body. It is a continuous cuticle, thickened and hardened to form a number of rigid pieces (**sclerites**) but remaining thin between, where flexibility is necessary. Marginal growth is therefore impossible. The growing animal at intervals bursts and casts off its whole exoskeleton, and then secretes a new and larger one: this is the process of **ecdysis** or **moulting.**

The limbs or appendages of Arthropods are usually of the greatest importance for classification. Unfortunately the limbs of trilobites are very rarely found. The ventral exoskeleton of aquatic Arthropods has always much more of the thin cuticle and smaller sclerites than the dorsal, but in trilobites it seems to have been nearly all thin cuticle, which easily decayed and allowed the limbs, themselves thin, to be lost. The part adjoining the margins of cephalon, thoracic segments and pygiduim, however, is as thick and hard as the upper surface with which it is continuous: this is called the **doublure** and is of a width which differs in various genera.

Calymene blumenbachii is sometimes found lying flat, sometimes rolled-up in the way in which the modern woodlouse rolls itself up when alarmed—that is to say, with the ventral sides of cephalon and pygidium in close contact (compare Fig. 67*e*). Comparison of specimens in the two conditions shows that neither cephalon nor pygidium is flexible but that the somites of the intermediate region (commonly called the **thorax**) are movable on one another by pivots at the sides, their dorsal portions sliding partly over one another when the animal stretched itself out, and becoming fully exposed when it rolled up, while the converse is the case with the ventral region.

In an ordinary-sized specimen 50 mm. long, the cephalon measures about 16 mm. in length, the thorax 27 mm., and the pygidium 7 mm. Measurements recorded below refer to such an individual, but members of the species are known with nearly double these dimensions.

Each thoracic or free segment consists of a central arched portion (**axis**) and a pair of lateral **pleurae.** They have their outer ends reflexed as doublure. The first thoracic segment articulates under the hind end of the cephalon and its axis is about 11 mm. wide, that of the last about 8 mm.: each axial "ring" consists of a prominent arch, measuring 2 mm. from front to rear, and an anterior sunken area, of which about 1·5 mm. is visible in an enrolled specimen, but which almost disappears in a straight one. The pleurae vary in width from about 12 mm. in the first segment to 10 mm. in the last; about 4 mm. from the axis where the segment takes a slight backward bend (**fulchrum**) each bears a prominent little forward projection or **fulcral**

process, fitting into a socket in the segment in front, and serving as a pivot. In enrolled specimens the part of the pleura external to this fulchrum is largely concealed beneath the segment in front. A well-marked transverse groove divides each pleura into a smaller anterior and larger posterior portion.

The cephalon shows the same trilobed character as the thorax: the axis or central part is called the **glabella,** the lateral, the **cheeks** or **genae.** Evidences of segmentation are shown chiefly by the glabella; there is, however, one well-marked cross-furrow (neck or **occipital furrow**) near the posterior margin, marking off both in the axis and

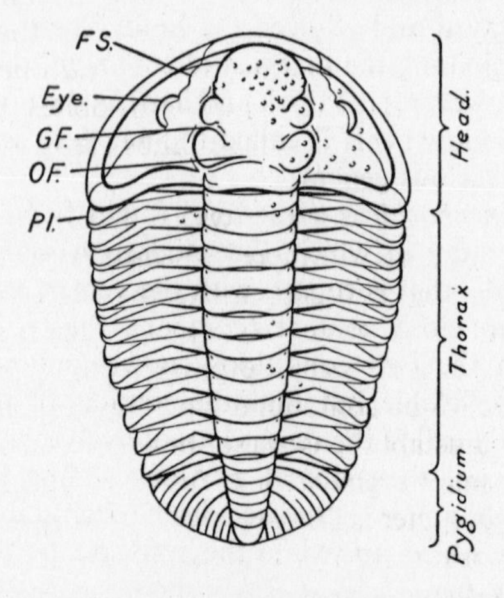

FIG. 61.—CALYMENE BLUMENBACHII, Brongniart, Wenlock Limestone
FS., Facial suture; *GF.*, glabellar furrow; OF., occipital furrow; *Pl*, pleura (Natural size; original).
For "Head" read "Cephalon."

genae a posterior portion which closely resembles a thoracic segment, but is immovably fixed to the rest of the cephalon. The other signs of segmentation are three pairs of grooves (**glabellar furrows**), more or less transversely placed, but not crossing the centre of the glabella, which is finely tuberculate. The most anterior of these is very slight, the second longer or deeper, the third still longer and deeper. The two latter bend obliquely backward, so as mark off rounded lobes joined to the rest of the glabella by a constricted base. These three furrows and the occipital furrow have been taken to indicate that the cephalon contains five united somites.

With the disappearance of flexibility in the cephalon, why have traces of the grooves between the somites been retained? The reason can be understood by examining the methods of muscular attachment in Arthropoda. Where powerful muscular action is necessary the inner face of the exoskeleton does not provide a sufficient area of attachment, and to increase that area a portion of the cuticle becomes, as it were, pushed into the interior of the body. Such an inward process is called an **apodeme,** and in the thorax of trilobites apodemes are formed in pairs near the antero-lateral margin of each ring. The lateral furrows of the glabella occur in corresponding positions and evidently mark the apodemes of the cephalon. Some trilobites have glabellar furrows much more like those of the thoracic axis than in the case of *Calymene*; in others they diverge still more from such a condition. Since apodemes serve principally for the attachment of the limb-muscles, this may indicate that in *Calymene* the cephalic limbs are more specialized than in some trilobites, but less so than others. Of the nature of the cephalic limbs (jaws, etc.) of trilobites we know very little.

The glabella of *Calymene* narrows gradually forwards. It ends bluntly, and in front of its end there is a raised border, which margins the whole cephalon, but is slightly arched up in front of the glabella.

The parts of the genae within this cephalic border are only slightly convex. Each carries a crescentic eye, opposite the second glabellar lobe. The eyes are of the compound type, found only in arthropods: instead of one adjustable lens there are many, of fixed curvature, each focusing a small segment of the field of vision on a retinula. The compound character is not well seen in *Calymene*, as the surface of the eye is smooth.

On each cheek there is seen a fine dividing suture (**facial suture**), which bisects the **genal angle** (outer end of posterior margin of cephalon), runs first obliquely forward, and then transversely, until it reaches the posterior margin of the eye; here it curves round the inner margin of the eye, and continues straight forward to the front margin of the cephalon, where it continues over the edge of the doublure. On the ventral surface a transverse **rostral** suture unites the right and left facial sutures; two short lateral connective sutures with the rostral suture bound a median ventral portion of the doublure, called the **rostrum** or **rostral plate.**

The most probable explanation of the facial suture is that it is a line of easy separation to facilitate the moulting (ecdysis) of the cephalon, and especially of the eye, the lenses of which are formed from the cuticle and like the rest of the exoskeleton must be moulted. The portion of each cheek lying external to the suture is called the **free cheek** or **librigena,** the remainder is the **fixed cheek** or **fixigena,** and the glabella and the occipital ring together with its attached fixed cheeks constitute the **cranidium.** Loose cranidia and free cheeks are often found, probably detached in the process of moulting. In all trilobites which possess both eyes and facial suture, the eye lies on the free cheek in contact with the suture, and the adjacent part of the fixed cheek has a protuberance fitting into the outline of the eye, called the **palpebral lobe.**

If the suture cuts the lateral margin in front of the genal angle, it is described as **proparian**; if shifted so as to cut the posterior margin, it would be **opisthoparian.** That of *Calymene* is on the verge between the two.

Separated from the reflexed hind margin of the rostral plate by the **hypostomal suture** there is a somewhat oblong plate with indented posterior margin. This is called the **hypostoma,** and formed a ventral front lip to the mouth (Fig. 65). The hypostome itself has a doublure on all but its front edge.

The pygidium of *Calymene* is a nearly semicircular plate the free posterior margin having much less curvature than the anterior margin which articulates under the thirteenth thoracic segment. This pygidium also has a reflexed doublure. The axis tapers backwards to a rounded termination, a little short of the actual end of the pygidium: its anterior two-thirds are divided into six well-marked "rings," the remainder is unsegmented. There are six pairs of dorsally fused pleurae, which from front to rear show an increasing backward curvature, until the last pair run backwards almost parallel to one another. The margin is evenly curved. The pygidial pleurae are covered with little tubercles.

Very little is known of the limbs of *Calymene*: traces of them have been found in sections of enrolled specimens.

2. **Dalmanites vulgaris** (formerly called *Phacops caudatus*) is another famous Dudley trilobite (Fig. 62). The most important differences between it and *Calymene* are these: The marginal border of the cephalon is drawn out into an obtuse point in front and into a pair of long **genal spines** behind, reaching beyond the middle of the thorax. The glabella is much wider in front, instead of narrower; its lateral furrows are nearly straight and do not mark off rounded lobes as in *Calymene*, but converge towards the middle line, leaving a much narrower smooth central area. The eyes are raised well above the surface of the cheeks and they are very large, usually well preserved, and their compound character is obvious, the surface being divided into a large number of corneal facets arranged in a regular manner. The facial suture is distinctly proparian, cutting the lateral cephalic border some way in front of the genal angle and making a right-angled bend at the eye; in front it runs round the front margin of the glabella

from one side to the other, not cutting the front margin of the cephalon, so that the two free cheeks form one inseparable piece. There is no separate rostral plate: the hypostoma is triangular (Fig. 65*d*).

The thorax consists of eleven segments, which differ from those of *Calymene* chiefly in the relative narrowness of the axis (about one-quarter the total width, instead of nearly one-third).

The pygidium consists of twelve or more fused segments, of which only ten have pleurae recognizable: it is longer proportionately than that of *Calymene*, more triangular in outline, its axis narrower, and

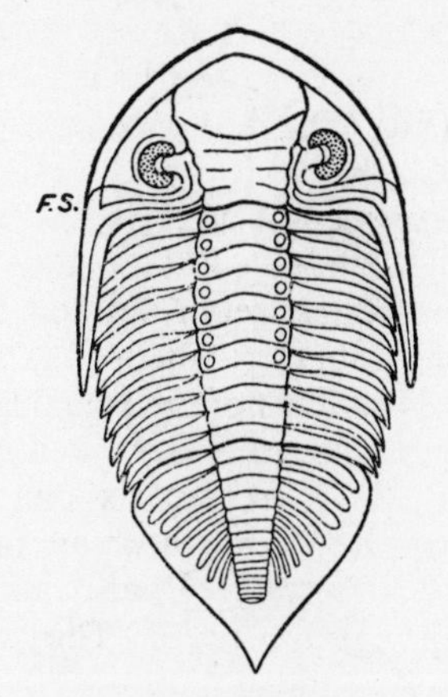

FIG. 62.—DALMANITES VULGARIS (Salter), Wenlock Limestone (Natural size; after Salter).
F.S., Facial suture.

at its posterior end it is drawn out to a sharp point (**mucronate**). The pygidial doublure takes part in the composition of this spine which has a central cavity closed terminally. This tail-spine (much longer on the true *D. caudatus* from the Wenlock Shales) shows, by analogy with living Arthropods, that this trilobite may have obtained its food by thrusting its head forward into the mud, using the tail-spine as its support; alternatively anchored in the mud, the animal with its cephalon flexed into a horizontal position may have awaited its prey. The highly-developed eyes, which control a wide field of vision including a rearwards component, show that it did not live buried in the mud; the eyes could have served to warn it of the approach of enemies. Their position, on the highest part of the cephalon, is clear evidence that this was a bottom-living or **benthic** animal. Nothing is known of the limbs of *Dalmanites*, though allied Phacopacean genera of Devonian age have been found—with a pair of jointed antennae and two-branched leg appendages.

3. **Triarthrus eatoni** formerly wrongly called *becki* (Fig. 63) is a small trilobite of which very beautifully preserved specimens have been found pyritized in the Ordovician Utica Shale of Rome, New York State. In form this is much less tapering than the previous forms, and ends more bluntly behind. The semicircular cephalon has

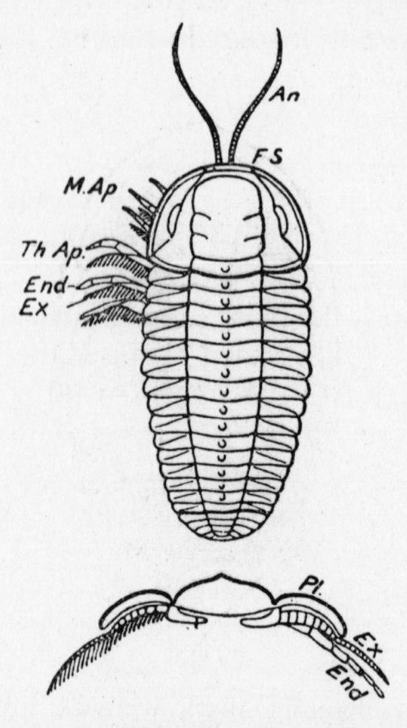

FIG. 63.—TRIARTHRUS EATONI (Hall); Ordovician (Utica Shale), Rome (New York) (× 2; after Beecher, modified)

The lower figure is a diagrammatic cross-section: on the left side the pre-epipodite with its fringe is alone drawn, on the right the pre-epipodite (without the fringe) and telopodite are shown. *An.*, antenna; *Ex.*, pre-epipodite; *End.*, telopodite; *F.S.*, facial suture; *M.Ap.*, mouth appendages; *Pl.*, pleura; *Th.Ap.*, thoracic appendages.

a much less distinct marginal border, the glabella is of almost uniform width, with lateral furrows which are but very slightly inflected from the transverse direction. In all these points the head shows much less difference from the thorax than that of *Calymene* or *Dalmanites*, so that *Triarthrus* is a more generalized form. The cheeks are each little more than half the width of the glabella; a long and narrow eye is in the centre of each; the facial sutures are opisthoparian, but cut the hind margin very close to the genal angle, running obliquely inward and forward (with a bend at the eyes) to cut the anterior margin separately at a distance apart not very much less than the width of the glabella.

The thorax consists of fourteen free segments, of which the first eight are of almost uniform width, after which a slight tapering takes place: each bears a median axial tubercle, as also does the last (occipital) segment of the cephalon and the first of the pygidium. The pygidium is very short (about two fifths as long as broad), and consists of six segments fused dorsally.

The hypostoma is parabolic in outline. From beside it springs a pair of long, many jointed but unbranched antennae, which are the only pair of appendages not conforming to the general type of the rest.

Any one of the largest appendages (from the first eight thoracic segments) shows the following characters: It is attached to the ventral surface of the body by a joint called the **protopodite,** from which a feeble process projects towards the middle line. In the outward direction two branches spring from the protopodite (hence the appendages are said to be biramous): the inner one, or **telopodite,** consists of seven movable joints, not unlike one of the ordinary legs of a crab; the outer, **pre-epipodite,** consists of one long joint followed by a great number of very short joints, the whole fringed with numerous blade-like filaments. Evidently the telopodite is adapted for crawling on the bottom, the pre-epipodite for swimming through the water and to serve as gills for oxygenating the blood.

Thus the appendages of *Triarthrus* perform the functions of crawling, swimming, and perhaps that of seizing food, as do those of such a living Crustacean as *Triops* formerly called *Apus*: to these may be added probably that of respiration, for they present a large surface of thin cuticle to the sea water in a region where it must be constantly agitated by the movements of the limbs. Apart from the antennae which is thought to have a sensory function, there is no evidence of differentiation of function from one limb to another in *Triarthrus*, nor in any trilobite.

Triarthrus was the first Trilobite in which well-preserved appendages were discovered. More recently, appendages have been found in other trilobites, ranging in age from Lower Cambrian to Devonian.

4. **Onnia superba** formerly referred to as *Trinucleus concentricus* (Fig. 64) is a trilobite found in the Upper Ordovician Shales of eastern Shropshire. Complete specimens are rarely found, but the usual (though not maximum) length appears to be about 17 mm., of which 9·5, 4·5 and 3 mm. measure cephalon, thorax and pygidium respectively. The greatest width of the thorax is 16 mm., that of the cephalon 24 mm. This difference is due to the presence of a remarkable flat fringe projecting from the outer margins of the cheek and

glabella, and perforated by a number of pits, arranged in four or five roughly concentric rows. This fringe is composed of two lamellae, one dorsal and one ventral separated by a **marginal suture** which passes on to the dorsal surface near the base of the genal spine. The fringe is about 6 mm. wide at each genal angle, where it contracts into a long, outwardly-curved genal spine, which extends back about 10 mm. beyond the end of the pygidium; at the front end of the cephalon the fringe is about 2 mm. wide, so that the length of the cephalon without the fringe is about equal to that of the thorax and pygidium. Consequently in an enrolled specimen only the cephalon is seen on one surface, and only the thorax and pygidium with the projecting fringe and genal spines on the other: in fact it is rather doubled-up than rolled-up.

The cephalon within the fringe consists of a highly raised glabella widening forwards from about 3 mm. to about 5 mm., and smooth protuberant cheeks. The glabella shows scarcely any sign of segmentation, and the cheeks show no trace of eyes or facial suture. Thus

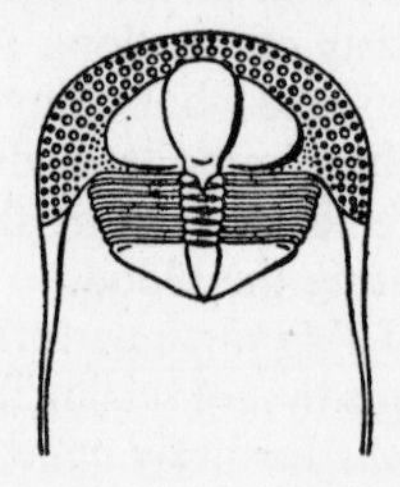

FIG. 64.—ONNIA SUPERBA (Bancroft), Caradoc, Onny River, Shropshire (Natural size, after Geological Survey).

Onnia was a blind trilobite, and had possibly arrived at that condition by degeneration consequent on complete adaptation to the mud-grubbing life for which its shape is so well suited. In certain genera of Trinucleidae, the family to which *Onnia* belongs there are traces of simple eyes in the young, and the late Cambrian genus *Orometopus*, thought by some to be ancestral to the Trinucleidae, has well-developed compound eyes and opisthoparian sutures. With the loss of the eyes devices such as a facial suture for their protection when the exoskeleton was discarded were unnecessary and moulting could be effected by means of a suture placed along the cephalic margin. The disappearance of glabellar furrows is also a specialization, and traces of them can be seen in some species of *Trinucleus*, e.g., *T. fimbriatus*.

The thorax consists of six segments; the axis is only 2 mm. wide, or one-sixth of the total width. The pygidium is short, obtusely triangular in shape, the axis tapering, the number of axial rings being about four.

What may be the most natural broad lines of classification of Trilobites is still imperfectly understood because many essential structures are unknown. The course of the facial sutures, the shape and lobation of the glabella, nature of the ventral sutures, the hypostoma, the number and character of the thoracic segments and the shape and size of the pygidium must all be considered, but it is not easy to say which should have priority.

In respect of the last point, the term **micropygous** denotes a relatively small pygidium following a large number of free segments; **heteropygous,** a larger pygidium and fewer free segments; **isopygous,** a pygidium equal in size to the cephalon.

It suffices here to list some of the principal superfamilies, in order of appearance and time-range.

A. SUPERFAMILIES MAINLY CONFINED TO CAMBRIAN

Individual trilobites medium sized unless otherwise stated.

1. **Olenellacea** [Mesonacida]. Elongated body. Without facial sutures. Micropygous. Surface never granulose. Cephalon semicircular with prominent genal spines. Glabella long, reaching border, with

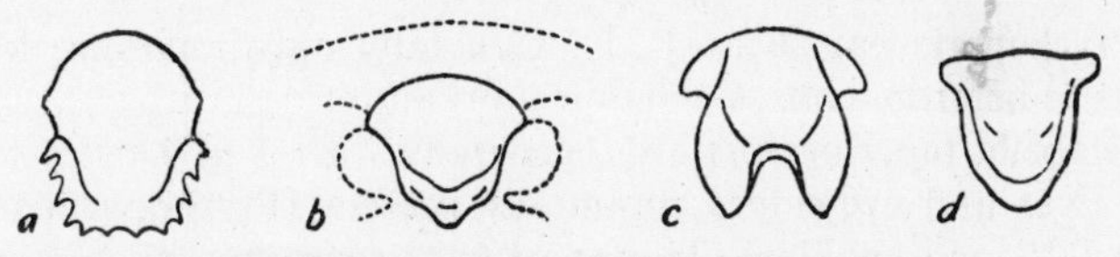

FIG. 65.—HYPOSTOMATA OF VARIOUS TRIBOLITES

a, *Elliptocephala asaphoides* Emmons, Lower Cambrian (× 3; after Walcott); *b*, *Ogygiocarella debuchii* (Brongniart), Llandeilo; dotted lines indicate outline of doublure (× $\frac{2}{3}$; after Salter); *c*, *Basilicus tyrannus* (Murchison), Llandeilo (× $\frac{3}{4}$; after Salter); *d*, *Dalmanites vulgaris* (Salter), Salopian (× $\frac{3}{4}$; after Salter).

distinct furrows some of which may be transverse. Large crescentic eyes, palpebral lobes merging into glabella in front. Rostral plate transverse and sickle-shaped. Thorax of 14–25 segments, tapering back; pleurae furrowed, commonly ending in spines. Pygidium narrow, of few segments. Lower Cambrian. *Olenellus* (Fig. 67*a*), *Callavia*, *Elliptocephala* (Fig. 65*a*). Northern Hemisphere.

2. **Redlichiacea.** Opisthoparian; otherwise resemble Olenellacea in some respects but glabella generally tapers forwards and palpebral lobes join glabella further forward. Thorax of 11–17 segments. L. Cam.–M. Cam. *Redlichia* (L. Cam.) Korea, China, W. Pakistan, Iran, Australia.

3. **Ellipsocephalacea.** Opisthoparian. Micropygous. Like Redlichiacea but eyes more distant from glabella with which usually connected by eye-ridges. Free cheeks generally narrow. Thorax of 12–14 segments. L. Cam.–M. Cam. *Strenuella* (L. Cam.) tapering glabella, *Protolenus* (L. Cam.) tapering glabella with 3 pairs of furrows, *Ellipsocephalus* (L.–M.Cam.) waisted glabella, glabellar furrows obsolete.

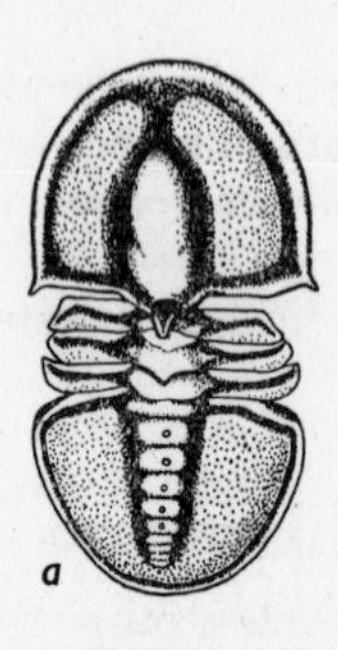

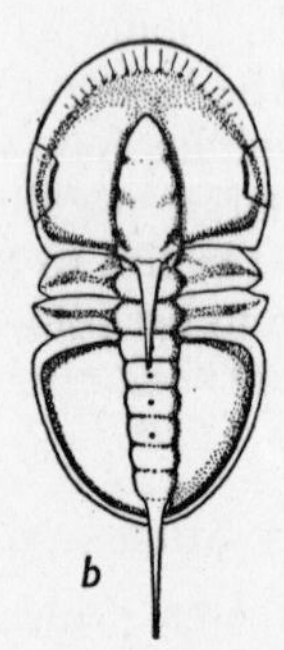

FIG. 66.—EODISCIDACEAN TRILOBITES
a, Eodiscus punctatus (Salter), Newfoundland; *b, Pagetia bootes* Walcott, British Columbia, Middle Cambrian (× 5; after Rasetti).

4. **Ptychopariacea.** Ovate body. Generally opisthoparian. Micropygous to heteropygous. Cephalon short and broad with convex border. Glabella tapering forwards, generally with 3 to 4 pairs of furrows. Eyes and eye-ridges commonly present (Ptychopariidae), or absent (Conocoryphidae). Thorax of 6–25 segments, pleurae nearly flat, furrowed. L.–M. Cam. *Ptychoparia* (M. Cam.), *Atops* (L. Cam.), *Conocoryphe* (M. Cam.) thought by some to be a degenerate Ptychopariid in which because of the loss of eyes, its dorsal cephalic sutures became nearly marginal.

5. **Eodiscidacea.** Small ovate body. Proparian (Pagetiidae) or without facial sutures (Eodiscidae). Isopygous. Cephalon border may be crenulated, tuberculated or smooth. Glabella usually well defined and tapered. Hypostoma present (Pagetiidae). Thorax of 2–3 segments, pleurae furrowed. Pygidial axis of 4–12 rings, pleural regions may or may not be furrowed. L.–M. Cam. *Eodiscus* (M. Cam., Fig. 66*a*), *Pagetia* (L.–M. Cam., Fig. 66*b*) narrow free cheeks,

6. **Agnostacea.** Small ovate body. Without facial sutures. Isopygous. Cephalon long, without eyes; genal spines exceptionally present. Glabella commonly not reaching border; with anterior transverse furrow and two basal lobes. Hypostoma unknown. Thorax of 2 segments which are nodose. Pygidium with 3 or less rings on axis, pleural regions commonly unsegmented, border evenly rounded or bearing a pair of spines posteriorly. L. Cam.–Ord. *Agnostus* (U. Cam.) *Tomagnostus* (M. Cam., Fig. 67*j*).

7. **Paradoxididacea.** Elongated medium to large body. Opisthoparian. Micropygous to heteropygous. Cephalon semicircular with long genal spines. Glabella expanding in front and generally reaching border; commonly with 3 to 4 pairs of furrows some of which may be transverse. Eyes usually long and well separated from glabella. Rostral plate may be separate or fused with hypostoma. Thorax of 13–22 segments, with flat furrowed pleurae, spinose laterally. M. Cam. Pygidium small in *Paradoxides*, wide in *Centropleura.*

8. **Olenacea.** Usually small. Opisthoparian, exceptionally Proparian. Micropygous or rarely heteropygous. Cephalon short and broad with rounded or spinose genal angles; usually much wider than thorax. Cephalic border narrow and thread-like. Glabella tapering forwards with simple or bifurcate glabellar furrows. Eyes small to medium; in most forms connected to glabella by eye-ridges. Thorax of 9–24 segments, furrowed and spinose pleurae. Pygidium short and broad with or without marginal spines. U. Cam.–U. Ord. *Olenus* (Fig. 67*c*), *Peltura* (proparian in early developmental stages, opisthoparian in adult), *Ctenopyge*, *Sphaerophthalmus*, *Angelina* (U. Cam.), *Triarthrus* (Ord., Fig. 63).

B. SUPERFAMILIES APPEARING IN THE TREMADOC

The Tremadoc is the uppermost Cambrian of British authors, lowest Ordovician of others. The first three superfamilies do not survive the Ordovician.

9. **Shumardiacea.** Small. Without facial sutures. Heteropygous. Glabella expanded in front into a transverse roll. Eyes unknown. Hypostoma unknown. Thorax of 6 segments, pleurae ridged and spinose terminally. *Shumardia* (Trem.–U. Ord.).

10. **Asaphacea.** Ovate. Opisthoparian. Isopygous. Surface smooth or with fine striae, never granulose. Cephalon commonly with only a faintly defined border or none. Glabella rarely well defined. Eyes crescentic, stalked in one Asaphid genus. Free cheeks separated anteriorly by a median suture on doublure (Asaphidae) or fused

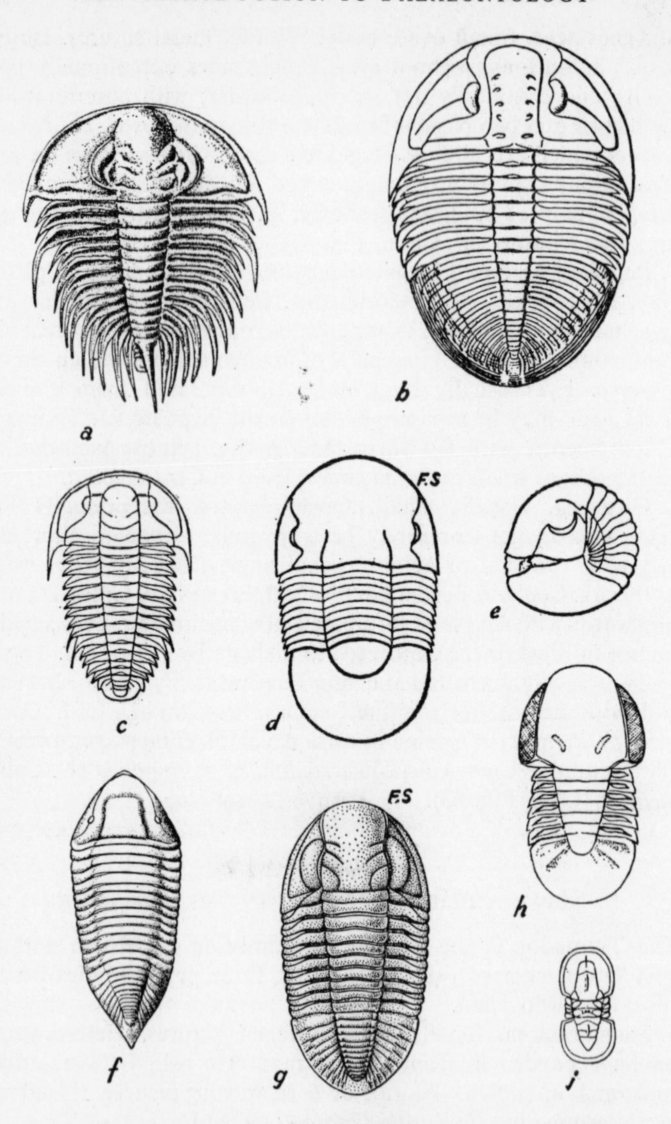

FIG. 67.—TRILOBITES

a, Olenellus thompsoni (Hall), Lower Cambrian, Parker's Quarry, Georgia, Vermont ($\times \frac{1}{2}$). *b, Ogygiocarella debuchii* (Brongniart), Llandeilo ($\times \frac{1}{2}$). *c, Olenus cataractes* Salter, Upper Cambrian ($\times$ 1). *d, Illaenus davisii* Salter, Ordovician ($\times \frac{1}{2}$). *e, Bumastus barriensis* Murchison, Silurian; side view of an enrolled specimen ($\times \frac{1}{2}$). *f, Trimerus* [*Homalonotus*] *delphinocephalus* Green, Silurian ($\times \frac{1}{4}$). *g, Cummingella jonesi* (Portlock) Lower Carboni-

(Nileidae). No rostral plate. Thorax of 9 (rarely 7) segments furrowed. Pygidium rounded or in some Asaphidae with terminal spine; pleural lobes commonly feebly segmented. ASAPHIDAE: *Niobe*, *Asaphellus* (Trem.), *Asaphus*, *Basilicus* (Fig. 64*c*), *Isotelus*, *Ogygiocarella* (Figs. 65*b*, 67*b*), (Ord.). Trends in the family towards loss of segmentation in cephalon and pygidium, also of axial furrows and to development of deep notching of hind margin of hypostoma. Upper Cambrian (pre-Tremadoc) genera known in Sweden. NILEIDAE: *Nileus* (Trem.–U. Ord.) large eyes adjoining glabella.

11. **Cyclopygacea.** Gibbous, mostly small, rarely large. Opisthoparian. Heteropygous. Thin-shelled. Surface smooth or with fine striae. Cephalon without border. Glabella large; occipital furrow absent. Fixed cheeks minute. Eyes enormous, continuing on to ventral surface; uniting in some genera; lenses quadrate and biconvex, up to 3,500 in one eye. No rostral plate or connective sutures. Thorax of 5–6 segments, subparallel or diverging backwards, furrowed, pleurae, Pygidium axis short, of few rings. *Cyclopyge* (Trem.–U. Ord.)., Fig. 67*h*), probably pelagic in habit and coming to surface at night.

12. **Trinucleacea.** Ovate. No facial sutures, or opisthoparian in some families. Heteropygous. Cephalon semicircular (Trinucleidae) or triangular (Raphiophoridae). Glabella inflated, subquadrate or expanded in front. Cheeks inflated. Eyes present in Orometopidae, eye tubercles or none in Trinucleidae, absent in Raphiophoridae. Genal spines usual. Thorax of 5–8 segments. Pygidium triangular or subtriangular. *Orometopus* (Trem.) with eyes, facial sutures, 8 thorax segments. TRINUCLEIDAE: *Trinucleus*, *Cryptolithus*, *Onnia* (Fig. 64), all Ordovician, with pitted wide cephalic fringe and marginal cephalic suture. RAPHIOPHORIDAE: *Ampyx* (Ord.) blind, with median spine in front of glabella, 6 thorax segments. *Raphiophorus* (Ord.–Sil.) like *Ampyx* but 5 thorax segments of which first is longer.

13. **Cheiruracea.** Proparian, exceptionally opisthoparian. Heteropygous. Cephalon with glabella extending to border, expanding forward or tapering, in whole or in part inflated; with as many as 4 lateral or trans-glabellar furrows. Fixigenae with or without spines. Eyes small or absent. Rostral plate small. Thorax of 9–19, typically 11 segments with smooth, ridged or furrowed pleurae. Pygidium with spinose pleurae. Trem.–M. Dev. *Cheirurus* (U. Ord.–Mid. Dev., Fig.

FIG. 67.—TRILOBITES (*continued*).

ferous (× 1). *h*, *Cyclopyge rediviva* (Barrande) Ordovician (Natural size). *i*, *Tomagnostus fissus* (Linnarsson), Middle Cambrian (× $\frac{5}{2}$). *a*, from Poulsen; *b*, *c*, *i*, from Geological Survey; *d*, *e*, after Salter; *f*, from Sdzuy; *g*, from Weller; *h*, from R. and E. Richter. *F.S.*, is anterior part of facial suture.

68*a*), pre-occipital lobes, 11 thor. segments; *Sphaerexochus* (M. Ord.–Sil), gibbous pre-occipital lobes, 10 thor. segments; *Sphaerocoryphe* (Ord.), *Deiphon* (Sil.), glabella bulbous in front of pre-occipital furrow probably pelagic. *Placoparia* (Ord.), opisthoparian, eyes absent.

14. **Harpacea.** Ovate. Without facial suture. Micropygous. Cephalon with enormous horseshoe-shaped pitted fringe extending full length of body; with marginal cephalic suture. Glabella tapering forward, inflated. Eyes small with few lenses, joined to glabella by eye-ridges. Thorax of 12–29 segments, pleurae flat and furrowed. Trem.–U. Dev. *Harpes* (Dev.).

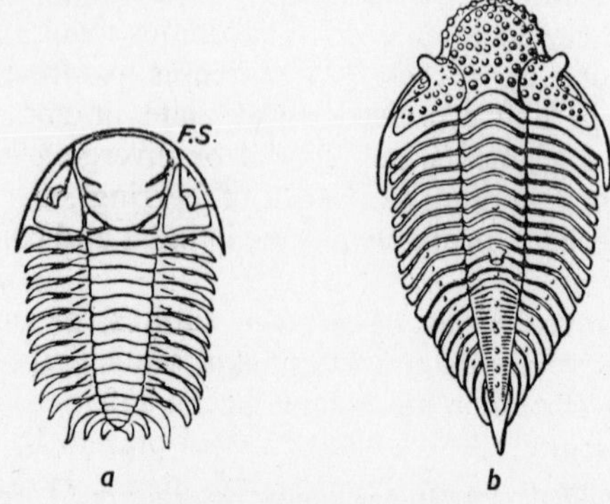

FIG. 68.—TRILOBITES WITH PROPARIAN FACIAL SUTURES

a, *Cheirurus bimucronatus* (Murchison) (Original). *b*, *Encrinurus punctatus* (Wahlenberg) (After Geological Survey). Both Wenlock Limestone, natural size.

15. **Lichacea.** Medium to very large. Opisthoparian. Hetero- and iso-pygous. Tuberculated, in some genera spinose. Glabella lobate. Eyes stalked in some genera. Free cheeks broad. Rostral plate present. Thorax of 10–11 segments. Pygidium with three pairs of pleurae with leaf- or thorn-like spines. Trem.–U. Dev. *Lichas* (Ord.–Sil.), *Conolichas* (Ord.), *Ceratolichas* (M. Dev.), *Uralichas* (Ord.) provides the largest trilobites estimated to be 70 cm. long.

C. FAMILIES NOT SURVIVING SILURIAN, OF SUPERFAMILIES APPEARING IN ORDOVICIAN

16. **Illaenacea**—ILLAENIDAE: Ovate. Commonly gibbous. Opisthoparian. Isopygous. Surface without granulose or tuberculate markings. Axial furrows on cephalon and pygidium not well marked. Glabella expanding forwards. Glabellar and occipital furrows absent.

Rostral plate present. Thorax of 8–10 segments, pleurae unfurrowed. Pygidium with axis and pleural fields smooth. *Illaenus* (Ord., Fig. 67*d*), *Bumastus* (Ord.–Sil., Fig. 67*e*).

17. **Encrinuracea.**—ENCRINURIDAE: Ovate. Proparian. Isopygous. Glabella with subparallel sides or expanding forward. Eyes small, in some genera stalked. Rostral plate wide to very narrow. Hypostoma with elongated middle portion. Thorax of 10–12 segments; pleurae furrowed or not. Pygidium tending to have many more annulations than pleurae which may or may not be spinose. *Encrinurus* (Fig. 68*b*), glabella tuberculate, expanding forward. Pygidium subtriangular. *Dindymene* (Ord.), eyes absent. *Staurocephalus* (Sil.), anterior lobe of glabella inflated almost to a sphere, rostral plate partly formed of ventral surface of glabella.

D. FAMILIES SURVIVING INTO DEVONIAN, OF SUPERFAMILIES APPEARING IN ORDOVICIAN

18. **Illaenacea.**—BRONTEIDAE: Ovate. Opisthoparian. Isopygous. Surface punctate or tuberculate. Glabella expanding forward. Eyes crescentic. Rostral plate present. Genal angles spined. Thorax of 10 segments. Axis of pygidium very short, with no axial rings, with pleurae radiating from it (fantailed trilobite). *Scutellum* [*Bronteus*] (Sil.–U. Dev.).

19. **Calymenacea.** Elongated. Transitional between opistho- and proparian. Hetero- to iso-pygous. Glabella narrowing forwards with 4 or less pairs of lateral furrows. Genal spines mostly lacking. Thorax of 13 segments. Pygidium with lateral and hind portions turned downwards. CALYMENIDAE: *Calymene* (Sil.–M. Dev., Fig. 61) semicircular cephalon with well-developed border; glabellar furrows, commonly well marked; rostral plate transverse and ventral; pygidium semicircular with 5–8 axial rings. Surface granulose. HOMALONOTIDAE: *Homalonotus* (Sil.), *Trimerus* (Sil.–M. Dev., Fig. 67*f*) tend to have convex triangular cephalon and pygidium; glabellar and axial furrows ill defined; rostral plate sub-triangular; in later genera partly on dorsal surface. Thoracic pleurae tend to be unfurrowed. Surface smooth.

20. **Odontopleuracea***. Ovate. Opisthoparian. Heteropygous. Tuberculated and spinose. Cephalon convex, with genal spines. Glabella deeply lobate. Eyes small, commonly united to front of glabella by eye-ridges, stalked in some genera, in others obsolete. Rostral plate present. Thorax 8–10 segments; pleurae broad, with

* Recent finds in Russia suggest that members of this superfamily occur in the Middle Cambrian.

or without furrows, terminating in one or more spines. Pygidium with 2–3 axial rings and spinose margin. Probably pelagic. *Odontopleura* (M. Sil.), *Acidaspis* (U. Ord.–M. Dev.).

E. Superfamilies Appearing in Ordovician and Surviving Beyond Devonian

21. **Phacopacea.** Ovate to elongate. Proparian. Hetero- to isopygous. Glabella expanded forwards. Eyes generally large and crescentic. No rostral plate. Genal angles rounded or spinose. Thorax of 11 segments, pleurae furrowed. Pygidium semicircular or triangular. Tendency to loss of function of facial sutures in some later genera and to absence of eyes. L. Ord.–Up. U. Dev.

PHACOPIDAE: *Phacops* (Sil.–Dev.), pre-occipital lobes fused and separated from rest of glabella by a furrow; genal angles rounded.

DALMANITIDAE: Genal angles spinose. *Dalmanitina* (Ord.), glabella reaches front of cephalon. *Dalmanites* (Sil.–L. Dev., Figs. 62, 65*d*), border in front of glabella.

CALMONIIDAE: Posterior glabellar furrow is deepest. *Acaste* (Sil.). Frontal and anterior glabellar lobes enlarged.

PTERYGOMETOPIDAE: *Chasmops* (Ord.). Two posterior glabellar lobes much reduced, pygidium large and well segmented axially and laterally.

22. **Proetacea.** Opisthoparian. Hetero- to iso-pygous. Cephalon semicircular with border mostly well developed. Glabella tapering forwards in most genera. Thorax of 6–17 segments (typically 10), pleurae furrowed. Ord. to Perm.

PROETIDAE: Elongate, elliptical; eyes crescentic close to glabella; thorax of 8–10 segments. M. Ord.–Carb. *Proetus* (Ord.–M. Dev.).

PHILLIPSIIDAE: Isopygous, thorax of 9 segments—surviving to Permian of the Tethys region; in Britain to Upper Carboniferous. *Phillipsia* (L. Carb.), *Cummingella* (L. Carb., Fig. 67*g*), glabella waisted; *Ditomopyge* (U. Carb.), bispinose pygidium when young.

OTARIONIDAE: [Cyphaspidae], heteropygous; glabella inflated, in some genera short, commonly with pair of basal lobes; thorax of 11–17 segments, outer ends of pleurae rounded. Ord.–L. Carb. *Otarion* [*Cyphaspis*] (M. Ord.–U. Dev.).

BRACHYMETOPIDAE: Isopygous; tuberculated short glabella with basal lobes; pygidium with more axial rings than pleurae. L. Dev.–U. Carb. *Brachymetopus*.

The Trilobites appear suddenly in the Lower Cambrian, but the character of the fauna is not primitive: the supposed common ancestor of *Olenellus* and *Agnostus* must have lived far earlier, and has left no trace. A possible explanation is that Pre-Cambrian trilobites had an uncalcified cuticle.

Something concerning the later history of the group may be learned from the above list of families. The last-known British Trilobites are found in the later marine bands in the Middle Coal Measures of South Wales (Cefn Coed) and South Staffordshire.

The ordinary trilobites seem to have lived on the sea-bottom (*benthic* fauna), but to have been able both to crawl and to swim; but particular genera or families took, some to a mud-grubbing, others to a swimming life (*nektic* or *pelagic*), and became adapted to those special conditions. Adaptation to life in the mud is shown by a shovel-like head; the eyes as far from the margin as possible (to be out of the mud) or in some cases lifted up on long stalks, or else the eyes are lost altogether; the pygidium in these mud-grubbers may end in a spine. Adaptation to a pelagic life is shown by a more or less globular head with eyes near the margin, sometimes by very thin shell, or a flattened pygidium, with a great development of spines to assist buoyancy.

The benthic trilobites, which form the majority, were more restricted than pelagic forms geographically. They may therefore assist in the delimitation of marine zoological provinces in Palaeozoic times, and so help towards the reconstruction of past geographies. Such provinces are regarded by some geologists as established for the Cambrian, Ordovician, and Devonian periods, though the Silurian marine faunas differ little over the whole world.

Ontogeny of Trilobites.—Owing to the inability of the animal to grow otherwise than by shedding its exoskeleton and growing a new one, the adult trilobite exoskeleton, unlike that of a brachiopod or mollusc, retains no trace of its early form. The developmental history of a trilobite is traced by collecting remains of a sufficient number of individuals of all stages of development (whether cases of juvenile mortality, or merely the skins moulted in healthy development) to be sure they belong to the same species. In a number of cases the general course of development has been established. The earliest stage known has been termed the **protaspis** larva, and is probably the form in which most trilobites were hatched. At this stage, the whole dorsal shield is less than 1 mm. long, and consists principally of cephalon, the thorax and pygidium being generally very rudimentary. The trilobation is distinct, and the glabella shows five well-marked segments. The eyes are marginal in many genera, probably as an adaptation to the free-swimming life usual in marine arthropod larvae. By repeated moults the larva passes through stages in which the postcephalic part of the body increases in length and in number of segments, the thorax gradually attaining the adult number; these are formed at the front margin of the hindmost pygidial segment and an freed one by one from the front of the larval pygidium and when the adult member is reached, in some instances (e.g., *Shumardia*) the pygidium itself increases its own segments to the adult content.

The remaining divisions of the Arthropoda must be dealt with much more briefly.

PHYLUM: **ARTHROPODA**

Metamerically segmented animals with jointed appendages, the anterior segments fused into a head.

CLASS: **TRILOBITA**

Classification already given.

CLASS: **CRUSTACEA**

Usually aquatic and breathing by gills borne on the appendages, with one or two pairs of antennae. Typically with a Nauplius larva with only three pairs of appendages.

SUBCLASS: **Branchiopoda.**—Generally with a carapace and thin biramous thoracic appendages of phyllopodan type. Cam.–Rec.

SUBCLASS: **Ostracoda.**—Small bivalved carapace with not more than two pairs of recognizable trunk links and these not phyllopodan. Ord.–Rec.

SUBCLASS: **Copepoda.**—Minute, free or parasitic. No carapace. Typically with 6 pairs of trunk limbs of which first is uniramous. Mio.-Rec.

SUBCLASS: **Cirripedia.**—Barnacles. Fixed. With carapace composed of several plates. Typically 6 pairs of biramous thoracic appendages. Jur.–Rec.

SUBCLASS: **Malacostraca.**—With 19 segments (or in Phyllocarida a few more) the head of 5 (bearing two pairs of antennae, 1 of mandibles, 2 of maxillae), the thorax of 8 with walking limbs (of which the first 1, 2, or 3 may be fused with the head). Chief orders known fossil—Phyllocarida, Isopoda, Decapoda (with suborders Macrura and Brachyura).

CLASS: **CHELICERATA**

Without antennae. Segments 18 or fewer.

SUBCLASS: **Merostomata.**—Aquatic, breathing by gills. Orders—Xiphosura, Eurypterida.

SUBCLASS: **Arachnida.**—Terrestrial, except for some early forms which may be aquatic. Breathing by lung-books or tracheae. Chief orders—Scorpionida, Araneida (Spiders) and Acarida (Ticks). So far only known fossil in Europe and North America.

CLASS: **ONYCHOPHORA**

Thin-cuticled caterpillar-like Arthropoda, with antennae, retaining some features of Annelida. Terrestrial, represented today only by the genus *Peripatus*, confined to the Southern Hemisphere (and Sumatra), and only known fossil from the marine Middle Cambrian of western Canada (*Aysheaia*).

CLASS: **MYRIAPODA (Centipedes and Millipedes)**

With numerous similar segments. Terrestrial, breathing by tracheae.

CLASS: **INSECTA (Hexapoda)**

With head bearing antennae, mandibles and two pairs of maxillae, thorax with 3 pairs of appendages and 2 pairs of wings, and abdomen without appendages. Breathing by tracheae.

Notes on the Chief Classes and Orders of Arthropoda (other than Trilobites) found Fossil

The **Branchiopoda** are mostly non-marine, but as common in excessively salt as in fresh waters. The form of the body and the appendages have some resemblance to those of a generalized trilobite; but (1) there is no trilobation; (2) no fusion of posterior segments into a pygidium; and (3) the exoskeleton of thorax and abdomen is thin, and to protect the hinder regions the head-shield is extended back as a loose covering called a **carapace.** In some forms, especially in the fossil "*Estheria*"*, this carapace is divided into two by a median suture so as to form practically a bivalve shell: this being marked by concrescent ridges closely resembling the growth-lines of a mollusc, it is very easy to mistake "*Estheria*" for such a lamellibranch as *Posidonia*. The following points should be looked for to distinguish "*Estheria*": (1) There is no hinge-structure as in lamellibranchs; (2) the material of the shell is not calcareous but chitinous; (3) with a lens there may be seen between the apparent growth-lines a network of striae which is never seen in lamellibranchs. "*Estheria*" is locally common from Devonian to Lower Cretaceous, in brackish-water deposits of arid regions. *Protocaris* is a Cambrian

* This name familiarly used by palaeontologists, is the homonym (see Chap. XIII, p. 267) of a genus of insects. It is not possible to determine whether the fossils belong to either of the recent genera, *Cyzicus* or *Leptestheria*, or to some extinct allied genus. The familiar but inapplicable name is therefore given between quotation marks.

marine form scarcely distinct from the modern freshwater *Triops* [*Apus*] which itself is recorded from the Triassic.

The **Ostracoda** are more thoroughly bivalved than "*Estheria*," as the two valves enclose the whole body, and they are calcareous. They never show any "growth-lines," and are nearly all of too small a size to be mistaken for lamellibranchs. The Silurian *Leperditia* is a giant among ostracods, being nearly 20 mm. in length: a glance at the hinge will show that it is not a lamellibranch: there is also a tubercle on each valve, anterodorsally, which does not occur in a molluscan shell. Although the majority of ostracods are marine, it is in freshwater deposits (such as the Purbeck Marls and Wealden Shales) that they occur in such abundance as to attract attention in spite of their small size. Ostracods range from Ordovician to Recent. They may be of zonal value. Certain Cambrian fossils, once placed in the Branchiopoda (Conchostraca) are now considered to belong to an early group of Ostracoda.

The **Cirripedia** (barnacles) are the only arthropod group in which the adults are all fixed. They are all marine, and occur sparsely in formations of Upper Jurassic age and later. They secrete a shell of at least five (commonly many more) pieces, which shows little analogy with that of any other arthropod; and in the works of early palaeontologists such as Cuvier and the Sowerbys they will be found figured among Mollusca. Supposed Palaeozoic Cirripedes (*Turrilepas*, etc.) are now referred to the Echinoderma (see p. 196).

The **Phyllocarida** are an almost extinct group, mainly Palaeozoic, differing from phyllopods in little else than the great restriction in the number of segments and the lesser backward extension of the head-shield. Some forms, e.g., *Hymenocaris* (Cambrian) and *Echinocaris* (Devonian), are fairly common fossils.

The **Isopoda** (wood-lice) are freshwater and terrestrial forms, sharing with trilobites the faculty of enrolling. Unlike trilobites, however, their number of segments is fixed. They are rare as fossils, though found occasionally from the Devonian onwards: one species, *Archaeoniscus brodiei*, occurs in enormous numbers in certain of the Lower and Middle Purbeck Beds of the south of England.

The **Decapoda Macrura** (lobsters, etc.) occur in various formations from the Triassic onwards. Well-preserved specimens are abundant in the Lithographic Stone of Solnhofen, Bavaria.

The **Decapoda Brachyura** (crabs) first appear in the Middle Jurassic. A species, *Palaeocorystes stokesi*, abounds in one bed in the Gault and (as a derived fossil) in the Cambridge Greensand. Other forms are found in phosphatic nodules in the London Clay.

The **Eurypterida** had the body divided into three regions; the prosoma or cephalothorax, the mesosoma or pre-abdomen of 7 segments and the metasoma or post-abdomen of 5 segments and a telson. The prosomal appendages all bear serrated edges around the mouth. None has the character of antennae (a fundamental feature distinguishing Chelicerata from Crustacea and Trilobita); the skeletal surface shows a scale-like pattern.

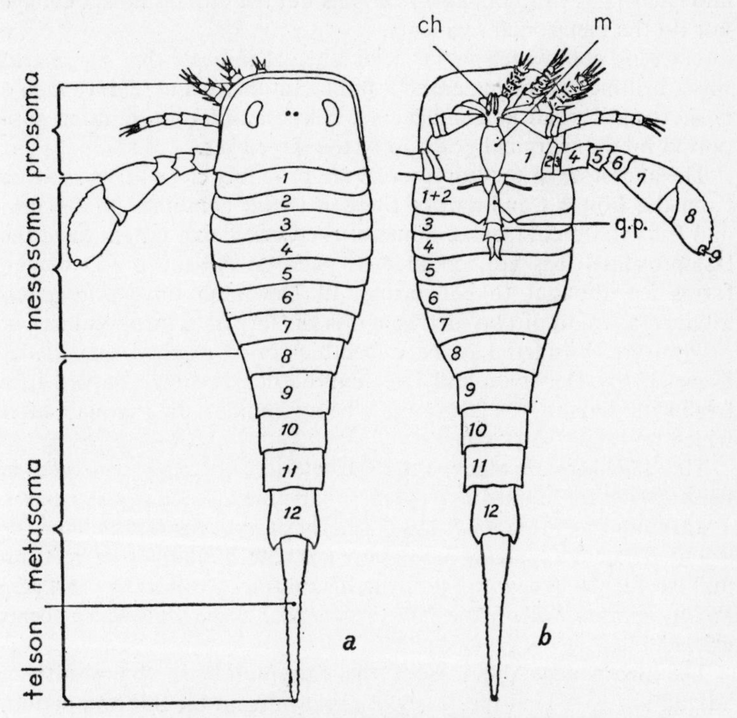

FIG. 69.—EURYPTERUS REMIPES (Dekay), Silurian, New York State
a, dorsal view; *b*, ventral view; *ch.*, chelicera; *g.p.*, male genital appendage; *m.*, metastoma (After Clarke and Reudemann).

Eurypterus (Fig. 69) has a long, tapering body, with a quadrate prosoma and a long pointed telson. The prosoma bears a pair of large reniform compound eyes on the upper surface, and a pair of much smaller simple eyes near the middle line. In the centre of the undersurface is the mouth, with the six pairs of prosomal appendages radiating from it. The first pair (Fig. 69) is 3-jointed, very small and backwardly directed, ending in pincers called the **chelicera.** The next four are 7- or 8-jointed walking-legs, protruding beyond the margin of the prosoma, and the last is a pair of large 8-jointed swimming paddles. Behind the mouth is a large median plate (**metastoma**).

The first segment of the mesosoma is called the pre-genital; on the ventral side it is rarely seen, being covered by the proximal joint of the last prosomal leg but the metastoma is said to belong to it. The second mesosomal segment carries the dimorphic genital appendage and gills. The 3rd to 6th also bear gills but the 7th has no appendage nor do the metasomal segments.

The long pointed telson has been taken to indicate that *Eurypterus* was a benthic burrowing animal; but in another genus, *Pterygotus*, it is short and broad and indicates a nektic animal, an opinion supported by the marginal position of the large eyes.

The ancestors of the Eurypterids are probably close to *Paleomerus* from the Lower Cambrian of Sweden which combines Xiphosurid and Eurypterid characters. Some of the Ordovician genera have the last prosomal legs not modified as paddles. All the pre-Devonian forms are thought to be marine. In Devonian times *Pterygotus* attained a length of two metres and is the largest arthropod known; *Stylonurus*, considered to be a benthic form, reached one metre. Some of the Devonian and later eurypterids became adapted to a freshwater habitat, the last of these being found in the Permian, after which the order was extinct.

The **Xiphosura** are allies of the Eurypterids, differing from them in the fusion of the segments of the meso- and meta-soma into one piece. Transitional forms (*Limuloides* formerly called *Hemiaspis*) have the six mesosomal segments free, with a reduced number of metasomal segments. They range from Cambrian to Recent, and are mainly marine, but in the later Palaeozoic some freshwater forms existed.

The **Scorpionida** differ from the Eurypterids in the results of adaptation to a land-life. They are known from the Silurian period, and the scorpions of today can be traced mainly to a single group found in the Carboniferous.

The **Araneida** (spiders) are derived from scorpions by a shortening of the abdominal region. They are known from the Coal Measures, but are very rare as fossils until the Oligocene, where they occur in the amber (fossil resin) of the Baltic area.

The **Acarida** (ticks) commonly have a body showing no external segmentation. A single genus is known from the Devonian (Rhynie Chert) otherwise they are unknown until the Oligocene amber of the Baltic region.

The **Myriapoda** (millipedes and centipedes) are very rare as fossils, but occur as early as the Devonian.

Insecta.—In the Middle Devonian chert of Rhynie (Aberdeenshire) occur the earliest unquestionable remains: a primitive wingless insect, *Rhyniella praecursor*, of the still surviving group Collembola (spring-tails). Several orders of winged insects are known from the Carboniferous. Insects are sporadic in occurrence as fossils; they are abundant in occasional beds in the Coal Measures (particularly in Commentry in France), in some beds of the British Lias, in the Oligocene amber of the Baltic coast, and in a Miocene lake-deposit at Florissant, Colorado; but generally very rare.

Short Bibliography

TRILOBITA

BEECHER, C. E.—(1) "Outline of a natural classification of Trilobites," *Amer. Journ. Sci.* (4), iii (1897). (2) A series of articles on appendages of *Triarthrus* and *Trinucleus*, *Amer. Journ. Sci.* and *Amer. Geologist*, 1893–6. (3) "Larval stages of Trilobites," *Amer. Geologist*, xvi (1895).

COBBOLD, E. S.—Papers on the Cambrian Trilobites of Comley, *Quart. Journ. Geol. Soc.*, 1910–36.

DOLLO, L.—"La Paléontologie Ethologique," *Bull Soc. Belge Géol. Pal. Hydr.*, xxiii (1909).

HARRINGTON, H. J. and 17 other authors.—*Treatise on Invertebrate Paleontology. Part O, Arthropoda* 1 *Trilobitomorpha*, 560 pp. (1959) Kansas Univ. Press.

HUPÉ, P.—"Classification des Trilobites," *Ann. Pal.*, xxxix; xli (1953–5).

LAKE, P.—"Cambrian Trilobites," *Palaeont. Soc.* (1906–46).

RAW, F.—(1) "The Development of *Leptoplastus*. . . and other Trilobites . . . ," *Quart. Journ. Geol. Soc.*, lxxxi, 223 (1925). (2) "Ontogenies of Trilobites and their significance," *Amer. Journ. Sci.* (5), xiv, 7 (1927). (3) "Mesonacidae of Comley . . . ," *Quart. Journ. Geol. Soc.*, xcii, 36 (1936).

RAYMOND, P. E.—(1) "Beecher's Classification," *Amer. Journ. Sci.*, xliii, p. 208 (1917). (2) "The Appendages, Anatomy and Relationships of Trilobites," *Mem. Connecticut Acad. Arts and Sci.*, vii, 1 (1920).

REED, F. R. C.—Numerous papers in *Quart. Journ. Geol. Soc.* (from 1896) and *Geol. Mag.* (from 1894).

SHIRLEY, J.—"Some British Trilobites of the family Calymenidae," *Quart. Journ. Geol. Soc.*, xcii, 384 (1936).

STØRMER, L.—(1) "Are the Trilobites related to the Arachnids?" *Amer. Journ. Sci.* (5) xxvi, 147 (1933). (2) "Studies on Trilobite morphology" (in three parts), *Norsk. geol. Tidssk.*, xix, xxi, xxix (1939–42–51).

STUBBLEFIELD, C. J.—(1) "Notes on the development of a Trilobite, *Shumardia pusilla* (Sars)," *Journ. Linn. Soc.* (*Zool.*), xxxvi, 345 (1926). (2) "Cephalic Sutures and their bearing on current classifications of Trilobites," *Biol. Rev. Cambridge Phil. Soc.*, xi, 407 (1936). (3) "Evolution in Trilobites," *Quart. J. Geol. Soc.*, cxv, 145 (1959).

SWINNERTON, H. H.—"The facial suture of Trilobites," *Geol. Mag.* (6), vi, 103 (1919).

VOGDES, A. W.—(1) "Classed and annotated Bibliography of Palaeozoic

Crustacea," *California Acad. Sci. Occ. Papers*, iv (1893), with Supplement in *Proc. Calif. Acad.*, v (1895). (2) "Palaeozoic Crustacea: publications and notes on the genera and species during . . . 1895–1917," *Trans. San Diego Soc. Nat. Hist.*, iii, 1–141 (1917). (3) Palaeozoic Crustacea. Part II. A list of the genera and subgenera of the Trilobita. *Ibid.*, 87–115 (1925).

Whittard, W. F.—"The Ordovician trilobites of the Shelve inlier, West Shropshire," *Palaeont. Soc.* (1955 continuing).

Whittington, H. B. and Evitt, W. R.—"Silicified Middle Ordovician trilobites," *Mem. Geol. Soc. Amer.*, no. 59 (1954).

Crustacea

Peach, B. N.—"On the higher Crustacea of the Carboniferous rocks of Scotland," *Mem. Geol. Surv. Great Britain: Palaeont.* i (1908).

Ulrich, E. O., and Bassler, R. S.—"Cambrian bivalved Crustacea of the order Conchostraca," *Proc. U.S. Mat. Nus.*, lxxviii, Art. 4 (1931).

The following monographs of the Palaeontographical Society deal with fossil Crustacea:

Bell, T.—Malacostraca (1858–63).

Brady, Crosskey and Robertson.—Post-Tertiary Entomostraca (1874).

Darwin, C. R.—Cirripedia (1851–5).

Jones, T. R.—(1) Cretaceous Entomostraca (1850, Suppl. 1890). (2) Tertiary Entomostraca (1857, Suppl. 1889). (3) Estheriae (1863).

Jones, T. R. and Woodward, H.—Palaeozoic Phyllopoda (1888–99).

Woods, H.—Macrura (1924–31).

Chelicerata

Holm. G.—"Ueber die Organisation des *Eurypterus fischeri* Eichw.," *Mém. Acad. Imp. Sci. St. Pétersbourg*, viii, no. 2 (1898). Figures first pair of appendages.

Moore, P. F.—"On gill-like structures in Eurypterida," *Geol. Mag.*, lxviii, 62 (1941).

Pocock, R. I.—"Terrestrial Carboniferous Arachnida of Great Britain," *Palaeont. Soc.*, lxiv (1911).

Resser, C. E.—"A new Middle Cambrian Merostome Crustacean," *Proc. U.S. Nat. Mus.*, lxxix, no. 33 (1931).

Størmer, L.—"On the relationships and phylogeny of fossil and recent Arachnomorphia," *Skr. Vid. Akad. Oslo*, I M.–N.Kl. no. 5 (1944).

Størmer, L., Petrunkevitch, A. and Hedgpeth, J. W.—*Treatise on Invertebrate Paleontology Part P, Arthropoda* 2; *Chelicerata*. 181 pp. (1955) Kansas University Press.

Woodward, H.—"Fossil Merostomata," *Palaeont. Soc.* (1866–78).

Insecta

Bolton, H.—(1) "Fossil Insects of British Coal Measures," *Palaeont. Soc.* (1921–2). (2) "Fossil Insects of the South Wales Coalfield," *Quart. Journ. Geol. Soc.*, lxxxvi, 9 (1930). (3) "New Forms from the Insect Fauna of the British Coal Measures," *Quart. Journ. Geol. Soc.*, xc, 277 (1934).

HANDLIRSCH, A.—"Neue Untersuchungen über die fossilen Insekten," *Naturhist. Mus. Wien* [Vienna], *Ann. Bd.*, xlviii (1937), xlix (1939).

SCOURFIELD, D. J.—"The oldest known fossil Insect," *Nature*, cxlv, 799 (25th May, 1940).

TILLYARD, R. J.—(1) "Evolution of the Class Insecta," *Amer. Journ. Sci.* (5) xxiii, 529–39 (1932). (2) *Tracing the dawn of life further backwards*, Cawthorn Inst., Nelson, N.Z. (1936). (3) "Ancestors of the Diptera," *Nature*, cxxxix, no. 3506, p. 66 (9th January, 1937).

ZEUNER, F. E.—"Biology and Evolution of Fossil Insects," *Proc. Geol. Assoc.*, li, 44 (1940).

VI

THE VERTEBRATA

THE Vertebrata are commonly accepted as the dominant group of the Animal Kingdom. That position might be disputed by the Arthropoda, for the latter are certainly more all-pervading and have insinuated themselves everywhere where life is found, even into many nooks and crannies which the Vertebrata have left alone. But it is to the credit of the Vertebrata that they have almost completely avoided the dirty corners of parasitism and degeneracy in which arthropods abound, and that their dominance is mainly a matter of power, of high organization, and finally of intelligence.

In the most primitive Vertebrata the chief hard parts are external (**exoskeleton**), and though an internal or **endoskeleton** is always

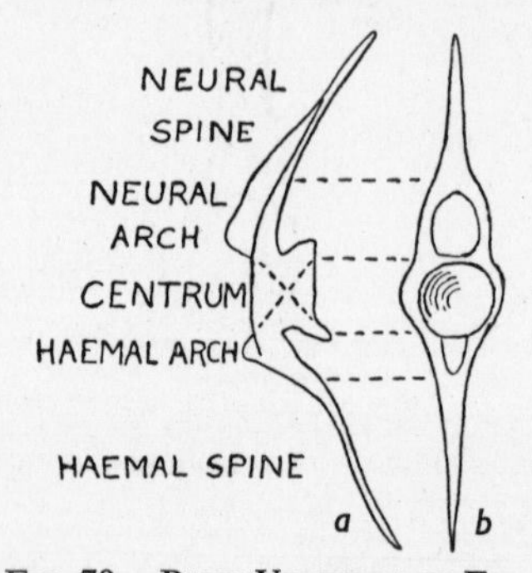

FIG. 70.—BONY VERTEBRA OF FISH

a, side view; *b*, end view. The dotted lines, crossing the centrum in *a*, denote the depth of the bi-concavity (After Gunther, modified).

present it was at first of organic cartilage-like material and rarely preserved fossil; subsequent evolution is marked in general by a steady increase in importance of the endoskeleton, which becomes a series of articulated bones hardened by calcium phosphate, and by a decrease in that of the exoskeleton.

The name "Vertebrata" seems to imply the possession of a back-bone composed of many separate vertebrae, but in practice it includes many primitive fishes in which the place of a back-bone is taken by a stiff cylindrical cellular rod, the **notochord,** underlying the spinal cord. (For this reason many zoologists prefer the term **Craniata,** as the rudimentary cartilaginous brain-case already exists in these "Vertebrates without vertebrae.") In the higher fishes and all land-vertebrates, the notochord is first surrounded and eventually nipped out by a series of short cylinders of cartilage (in some fishes) or of bone (in all other cases), the bodies or **centra** of the vertebrae. There also appears a corresponding series of arches (of cartilage or bone) which covers and protects the spinal cord—the **neural arches**—and each of these finally becomes united with a centrum to form the complete **vertebra.**

In most fishes with vertebrae capable of fossilization the **centrum** is

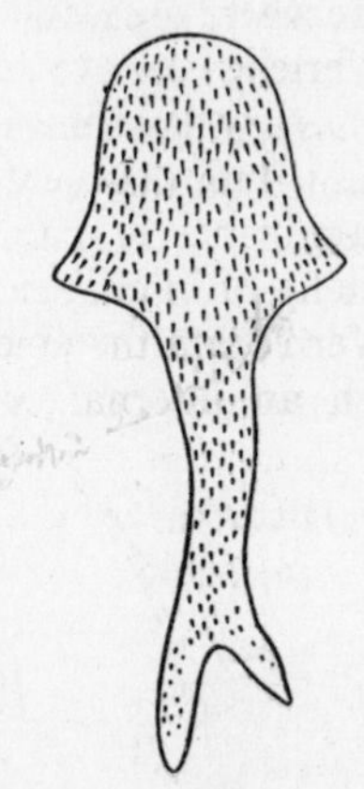

FIG. 71.—LANARKIA SPINOSA Traquair, Downtonian, Lesmahagow (× ½; after Traquair)

This differs from its contemporary, *Thelodus*, in having spiny skin-teeth. The head and trunk have been flattened down from above, but the tail is turned over on its left side.

biconcave (**amphicoelous**), having a deep conical hollow at each end (Fig. 70), occupied by the remains of the notochord. These hollows may be connected by a narrow opening, through which the notochord is continuous, so that an internal mould of the vertebral column suggests a connected series of hour-glasses. The neural arch may be united with the centrum or detached. In the tail-region, there is a similar **haemal arch** on the under side, which protects the main caudal blood-vessels; but in the trunk region this is replaced by the ribs.

In many amphibia and reptiles the amphicoelous form of centrum persists, though the hour-glass neck disappears. In the higher reptiles the form changes, either to **procoelous** concave in front, protuberant behind) as in Crocodiles, or **opisthocoelous** (reverse of procoelous). These changes allow of greater flexibility, which attains its highest form in the saddle-shaped articular surfaces of Birds. In Mammals the centra are flatfaced, and at each end there is a thin plate, separately ossified (**epiphysis**).

The structure of the **cranium** or skull, which lies at the head of the vertebral column, is too complex to be described in this elementary account of the Vertebrata, though it is of supreme importance in tracing the history of the phylum.

The first form taken by the exoskeleton is that still found in the sharks, constituting the "shagreen": every point on this rasping surface marks a single "skin-tooth" or **placoid scale,** the structure and development of which are essentially those of teeth (Fig. 74). Our teeth are in fact the greatly-modified survivors of what was once a general body-covering, retained on the jaws for special purposes but no longer needed on the skin.

Water-breathing Vertebrata

The first British Vertebrata are found as rarities in the Lower Ludlow but their remains abound at the base of the Devonian System, in the Ludlow "Bone Bed." Here occur in quantity the little skin-teeth of *Thelodus* (Fig. 73*a*), which, as shown by a complete body

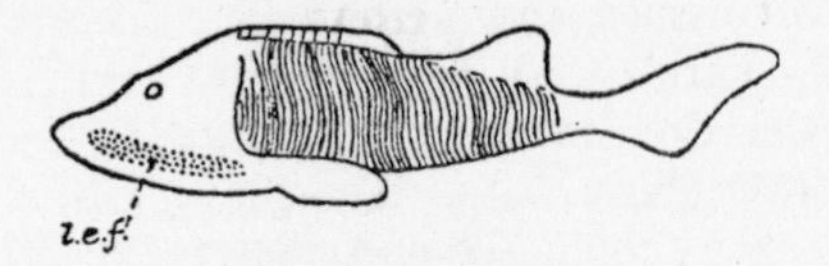

FIG. 72.—RESTORATION OF A CEPHALASPID
Side-view; *l.e.f.*, lateral electric field (× $\frac{1}{8}$; after Stensiö).

(only a few inches long) found in Scotland, is a primitive (or possibly degenerate) member of a group, Ostracodermi (Fig. 71), which, like the modern lamprey, were without the biting jaws found in nearly all Vertebrata. In North America, remains of similar forms have been found in the Ordovician. In more typical members of the same class, found also in the Silurian, but more abundantly in the Devonian (e.g., Cephalaspids, Fig. 72), the skin-teeth are fused into large armour-plates, and (by adaptation to a similar life) many of them came to mimic, to some extent, the contemporary Eurypterids (ante, p. 147). The microscopic structure of the armour is quite different in the two cases, and though much ingenuity has been wasted in trying to prove an arthropod ancestry for Vertebrata, we may feel confident that the resemblance is only a very striking case of convergence. The preservation of some of these Ostracoderms is so perfect that internal moulds have been taken, revealing the details of the nervous system and other soft structures, and it has been inferred that electrical organs were present (Fig. 72, *l.e.f.*). The Ostracoderms died out at the end of the Devonian.

In the lower of the two classes of typical fishes (Chondrichthyes, p. 166) the internal skeleton is composed of cartilage, and is therefore rarely preserved fossil. These are sharks and rays, known principally to fossil-collectors by their teeth (Fig. 73), and the bony spines of "ichthyodorulites" which support the front margin of the fins in some cases. Ichthyodorulites (*Onchus*) accompany skin-teeth in the Ludlow Bone Bed. Sharks' teeth are common in some beds in most geological systems. They are usually sharp lacerating teeth, pointing inwards on the jaws, so as to prevent prey from escaping; they are arranged in rows, new rows developing on the inside of the jaw throughout life, and moving forwards as the older teeth fall out. Thus loose teeth are commoner than complete series, and it is difficult to know whether slight differences among isolated teeth are due to their belonging to different species of sharks or to different parts of the jaw of the same species.

Like other vertebrate teeth, those of the sharks consist of a **root,** buried in the soft tissues, and a **crown** which is exposed. Up to the Jurassic period, shark's teeth have shallow, undivided roots (Figs. 73*b*, *c*); from the Cretaceous period onwards, forms with deep, divided roots occur (Figs. 73*d*, *e*, *f*).

The tooth is mainly composed of **dentine,** a calcareous tissue full of fine tubules into which pass processes from the soft tissues of the central **pulp-cavity** (Fig. 74), the opening into which is called the **nutritive foramen.** The crown has its surface covered with a thin layer of **enamel,** a much denser material than the dentine. In the fossil state, teeth are usually highly phosphatized, and may acquire a black colour.

Sharks' teeth are among the most indestructible of fossils, and worn examples are common among derived fossils.

Some sharks and the rays, instead of swallowing their prey whole, grind it in their mouths: in these, the teeth have lost their sharp points and compressed shape, and form a mosaic with a rough surface, extending over the palate and part of the floor of the mouth as well as the jaws (Fig. 73*j*).

No attempt can be made here to describe the details of the endoskeleton of the Chondrichthyes, but a few words may be given to the general form of the body. This is elongated, tapering towards both

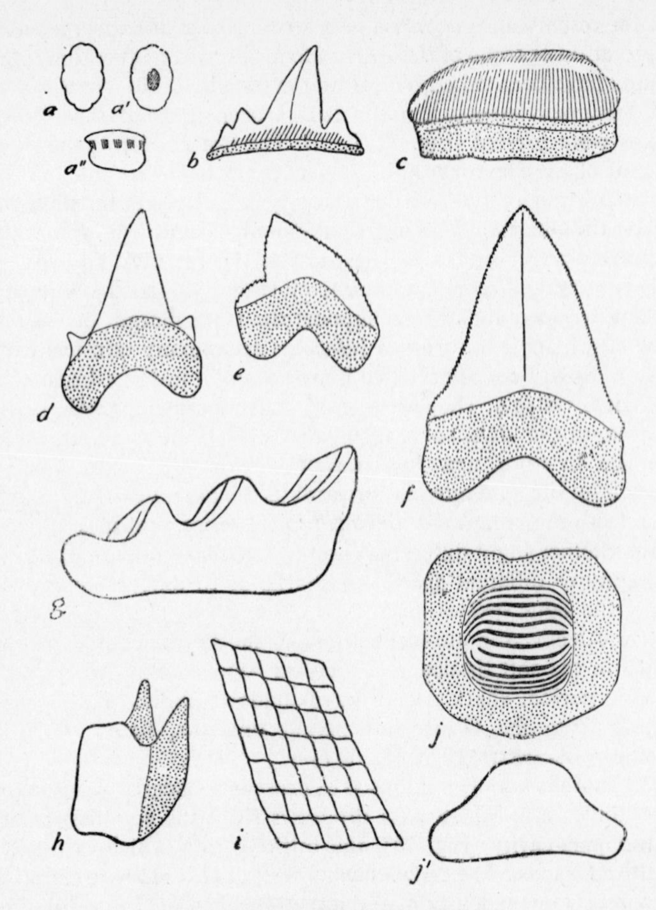

FIG. 73.—FISH TEETH AND SCALES

a–g, *j*, teeth; *h*, *i*, scales. The dotted part in *b–f* is the root; in *h* it is the part overlapped by other scales. *a*, *Thelodus pagei* (Powrie), Downtonian, skin-tooth from above; *a'*, from below, showing nutritive foramen; *a''*, from side; *b*, *Hybodus grossiconus* Agassiz, Bathonian; tooth (natural size); *c*, *Acrodus nobilis* Agassiz, Lower Jurassic; tooth (natural size); *d*, *Lamna appendiculata* Agassiz, Upper Cretaceous; tooth (natural size); *e*, *Corax falcatus* Agassiz, Upper Cretaceous; tooth (natural size); *f*, *Carcharodon megalodon* Agassiz, Miocene-Pliocene; tooth ($\times \frac{3}{8}$); *g*, *Ceratodus altus* Agassiz, Rhaetic; tooth ($\times \frac{3}{4}$); *h*, *Acrolepis semigranulosus* Traquair, Lower Carboniferous; a single ganoid scale ($\times \frac{3}{4}$); surface pattern omitted; *i*, *Lepidotes minor* Agassiz, Purbeck Beds (Upper Jurassic); part of scaly armour of trunk ($\times \frac{3}{4}$); *j*, *Ptychodus mammillaris* Agassiz, Upper Cretaceous; tooth, surface view; *j'*, profile ($\times \frac{3}{4}$). The area shown dotted is part of the crown and should be shown with fine, more or less concentric lines. *a*, *h*, after Traquair; *b*, *c*, *f*, *g*, after Agassiz; *d*, *e*, *i*, *j*, after Smith Woodward.

ends, laterally compressed and streamlined in the nektic (swimming) sharks, depressed in the benthic (bottom-living) skates and rays. Besides the two pairs of fins (answering to the limbs of land-vertebrates) there are a variable number of median fins, spaced out along the whole of the dorsal edge, but only on that part of the ventral edge that lies behind the paired fins. The end of the tail is bent up and combines with a median ventral fin to form the "heterocercal caudal fin," characteristic not only of all cartilaginous fishes and some bony fishes, but also of the Ostracoderms (Figs. 71, 72, 75*a*).

The higher grade of fishes (Osteichthyes, bony fishes) have an endoskeleton of bone, but they retain an exoskeleton of scales of some sort. They vary greatly in shape from greatly compressed forms, whose height is nearly equal to their length, to long cylindrical eel-like forms. Most fishes have a fusiform shape, lying between these two extremes.

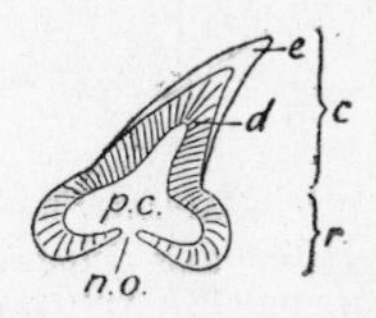

FIG. 74.—TOOTH STRUCTURE (Diagrammatic vertical section)
c, crown; *d*, dentine; *e*, enamel; *p.c.*, pulp-cavity; *n.o.*, nutritive opening; *r*, root.

The bony fishes fall into three divisions:

1. The **Crossopterygii** had a covering of thick enamel-covered scales (ganoid scales) forming a mail-coat, two dorsal fins, and paired fins of rounded shape with large bony supporting elements and a fringe of many finger-like rays. In some the structure of the bony elements foreshadows the limb-structure of the land-vertebrates. The Palaeozoic members of this subclass were freshwater, but in the Mesozoic they took to a marine life. Though formerly held to be extinct after the Cretaceous period, living genera including *Latimeria* are now known off the east coast of South Africa.

2. The **Actinopterygii** are the most numerous of the bony fishes. In early forms the scales are enamel-covered but later they become thin and flexible; there is only one dorsal fin as a rule; and the bony elements of the paired fins are very short, the bony rays long and sharp. They fall into three orders: CHONDROSTEI, the Sturgeons (*Acrolepis*, Fig. 73*h*); HOLOSTEI, the freshwater bony pikes (*Lepidotes*, Fig. 73*i*), and TELEOSTEI, the great majority of bony fishes. (For ranges see p. 167.)

The two first of these were formerly united with the Crossopterygii under the name of Ganoids (in reference to their enamel-covered scales).

3. The **Dipnoi** (Lung-fish) are interesting as being fishes which show how a transition from water-breathing to air-breathing life may have taken place, though they cannot have included the actual ancestors of the land-vertebrates, which must have been Crossopterygians. The Dipnoi appear in the Devonian and the three surviving genera are confined to rivers in the Southern Hemisphere, which are liable to drought, when the lung is used for breathing instead of the gills. The best-known of these modern genera is the Australian *Neoceratodus*, which has paired fins more like the crosso- than the actino-pterygian type (Fig. 76*a*), and a pair of large ridged teeth in each jaw, formed

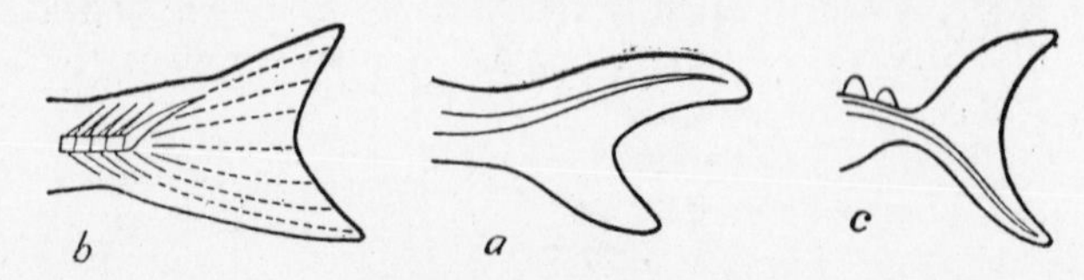

FIG. 75.—CAUDAL FINS

a, Heterocercal fin of a shark, vertebral axis rising, second lobe of fin developed below it; *b*, homocercal fin of bony fish, vertebral axis as in *a*, but the whole fin externally symmetrical; *c*, hypocercal fin of *Ichthyosaurus*, externally like *b*, but vertebral axis running into lower lobe.

by the fusion of many rows of small teeth. Similar teeth (Fig. 73*g*) called *Ceratodus* were known as fossils in the Rhaetic beds before the living *Neoceratodus* was discovered.

Air-breathing Vertebrata (Tetrapoda)

These are usually divided into Amphibia, Reptiles, Birds, and Mammals, but palaeontological discoveries have so extended the idea of a reptile that it is no longer logical to separate birds from that category, and only three classes will be recognized here.

The **Amphibia** are distinguished in that they retain signs of their aquatic ancestry by passing through a larval stage in which they have gills and (with a few exceptions) live like fishes: the tadpole-stage of the frog is a familiar example. At the present day they are represented only by the primitive and somewhat degenerate newts and salamanders, the highly-specialized frogs and toads, and the equally specialized snake-like Coecilians of the tropics (not known fossil). The first two orders are unknown before the Upper Jurassic or Lower Cretaceous epoch; and between them and the other fossil Amphibia there is at present a great gap in time, no certain remains of Amphibia being known from anywhere in the Triassic or Lower Jurassic strata. From Upper Devonian to Permian there are found Amphibia with complex bony skulls, which resemble those of crossopterygian fishes on the one hand and those of reptiles on the other far more than those of modern newts and frogs. That they were really Amphibia is indicated by their retention of gill-arches in the throat, and by the fact that their ribs do not meet ventrally in a breast-bone, so that they must have breathed like a frog by swallowing air, not sucking it in. The limbs of these Stegocephalia (as the three extinct orders of Amphibia are collectively called) were essentially those of a land-animal (Fig. 76*c*). The fin of the fish, adapted to serve as a flexible

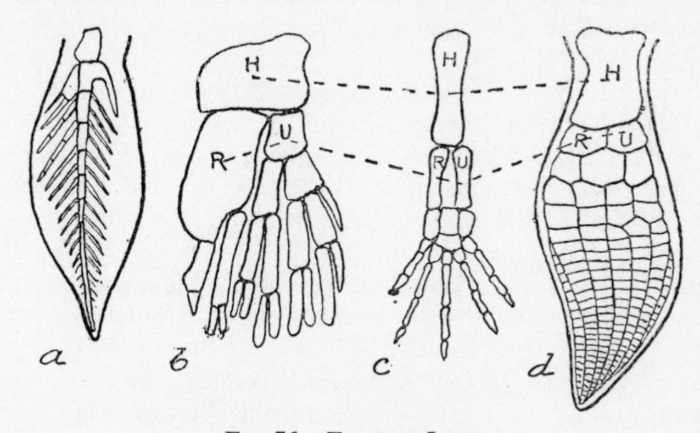

FIG. 76.—FINS AND LIMBS
a, *Neoceratodus*; *b*, *Sauripteris*; *c*, a newt; *d*, *Ichthyosaurus*. The dotted lines connect homologous structures.

oar, had gradually been transformed into a limb capable of performing the much more complex movements required by a land-living animal. The type of fin which is most easily comparable with the terrestrial limb is that of the crossopterygian *Sauripteris* (Fig. 76*b*). If we compare this with the leg of a newt, which shows the simplest form of terrestrial limb, we see the essential changes in the latter are—(1) the lengthening of the basal element (**humerus** or **femur**); (2) the lengthening and equalizing of the next two elements (**radius** and **ulna,** or **tibia** and **fibula**), which make possible a movement of rotation impossible in the fin; and (3) the great reduction in number of the rays and their separation to form five fingers or toes, capable of independent movement (Figs. 76*b*, *c*).

Sauropsida (Reptiles and Birds). The Reptiles of the present day consist of three familiar orders—the lizards and snakes, tortoises and crocodiles. To these must be added, as representing a separate order, Rhynchocephalia, the New Zealand lizard *Sphenodon*, sole survivor of an important Permian and Triassic series. The birds (though commonly dignified as a class **Aves**) should also be counted as another order of Sauropsida. These are the few scattered survivors of a series of forms so vast and varied that the Mesozoic era in which they flourished has been called the Age of Reptiles. They spread over the surface of the continents, hitherto almost devoid of animal life, and exhibited the same "adaptative radiation" to all possible methods of life that the Mammals showed afterwards in the Cainozoic era. There were herbivorous reptiles and carnivorous reptiles, aquatic, terrestrial, and aerial reptiles. Some of the latest and most specialized of Cretaceous Reptiles, belonging to the group Dinosauria, are shown reconstructed in Fig. 78.

We cannot here deal with the many instructive facts shown by the study of fossil reptiles, but must confine ourselves to the case of those orders of reptiles, which took to an aquatic or an aerial life. One of the former, whose remains are fairly common in some of the Jurassic strata, especially the Lias, is the Ichthyopterygia (the Ichthyosaurs). These occupy the same position among Reptiles that the whales do among Mammals. They underwent complete adaptation to a marine life, and approximated to fishes in shape and general structure, but in accordance with the principle of irreversibility in evolution they could not lay aside all the characters they had gained during their ancestral land-life, nor could they recover certain fish-structures which their land ancestors had lost—the gills in particular. If we compare the limbs (or paddles) of an Ichthyosaur with the fin of *Sauripterus* or *Ceratodus* on the one hand and the limb of the newt on the other, we see that while in outline, absence of separate fingers, and slight possibility of movement between the numerous small, closely-packed bones it agrees with the fish, its fundamental structure is that of a land-animal (Fig. 76*d*). So, too, with the tail-fin; externally it resembles the homocercal tail of a bony fish, but the end of the vertebral column bends down into the lower lobe (Fig. 75*c*). In these and other ways Ichthyosaurs show convergence towards fishes, but the resemblances are not exact and are mainly external.

The vertebrae of *Ichthyosaurus* (Fig. 77) are common in Jurassic clays, and easily recognized by their extreme shortness—the length of a vertebral body being only about one quarter of its height or breadth; they are biconcave, and the arches are loose.

Another Mesozoic order of marine reptiles is the Sauropterygia (e.g., *Plesiosaurus*), distinguished by a long neck and small head (in contrast to the large head and very short neck of Ichthyosaurs), by the vertebrae being about as long as high, with very shallow concavities at the ends, and the limbs rather less modified from the land-type. Another group of lizards (Mosasauria) took to marine life in the Cretaceous period, but died out before they had become greatly specialized. The earliest crocodiles were also marine, but the Chelonia (tortoises and turtles) were mainly of land and freshwater habitat and only took to the sea late in the Mesozoic era.

Along two distinct lines Mesozoic reptiles adapted themselves to flight (Fig. 79). The Pterosaurs (or Pterodactyls) had a membranous wing, for the support of which the fifth finger was enormously lengthened, the leg and tail also helping to keep it stretched. They lived through the Jurassic and Cretaceous periods. In the birds, on the other hand, there are only a short thumb and two fingers, the

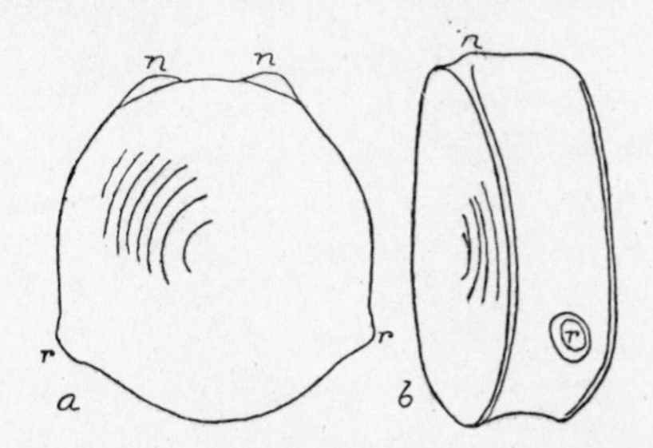

FIG. 77.—VERTEBRA OF ICHTHYOSAURUS
a, front view; *b*, oblique side view; *n*, *n*, position of neural arch attachment; *r*, *r*, rib articulation (× ½; original).

latter combining to support the wing but not being greatly lengthened. (In flying Mammals—bats—yet another type of membranous wing is developed, all four fingers being lengthened to support the wing, like the ribs of an umbrella.) The earliest birds (*Archaeopteryx* and *Archaeornis*) were found in the Upper Jurassic Lithographic Stone of Bavaria. The few known Mesozoic birds all have teeth, and *Archaeopteryx* and *Archaeornis* alone of all birds have a long, many-jointed tail.

Of the more purely terrestrial subclasses of reptiles, the Synapsida (from which the Mammalia may have sprung) and especially the Order Therapsida, had their headquarters in the Southern Hemisphere (South Africa), while the Dinosaurs (Saurischia and Ornithischia) originated in the Northern Hemisphere: but both extended their range to the antipodal region. There are late Triassic forms from South Africa, Yunnan and North Carolina, as to which opinions differ whether they are Therapsid reptiles or multituberculate mammals. By the end of the Cretaceous period all reptilian orders had died out, except the few that still survive today.

The **Mammalia** are represented in a few Mesozoic rocks by lower jaws and teeth of very small forms, which are not closely akin to any existing mammals. Why scarcely any other remains than lower jaws

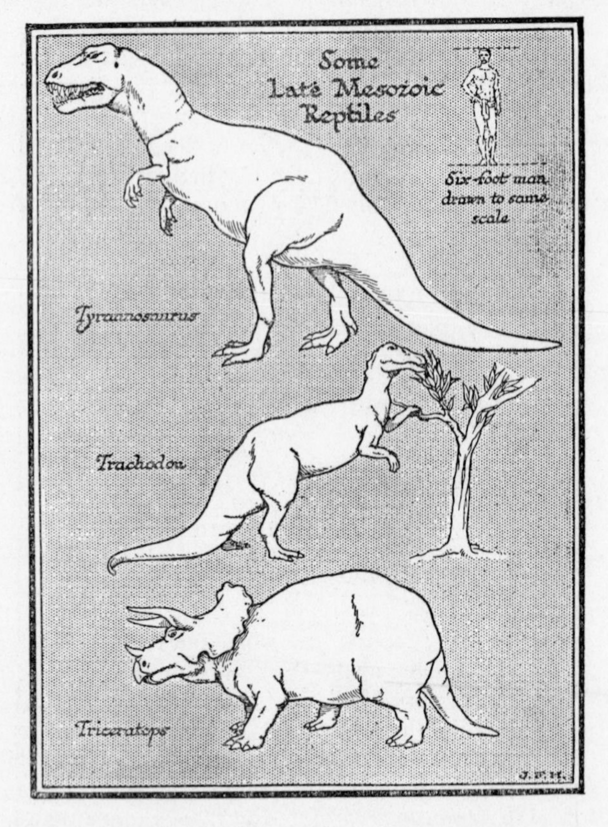

FIG. 78.—LATE MESOZOIC DINOSAURS
Tyrannosaurus is a Saurischian Dinosaur, *Trachodon* and *Triceratops* are Ornithischian Dinosaurs, all from the Upper Cretaceous of North America.

should be found has never been satisfactorily explained. While the Reptiles dominated the world, the Mammals seem to have been a very lowly and insignificant group, but on the extinction of most of the former at the end of the Cretaceous period the Mammals soon became the dominant land-animals.

The Paleocene mammals are all of small size, with the neck not flexible and the trunk passing gradually into the tail. Their limbs are not greatly removed in structure from the primitive five-toed (**pentadactyle**) type, except for the presence of a projecting elbow in the fore-limb and heel in the hind-limb: the two bones in the second limb-segment are free to move on one another so that the limbs can be

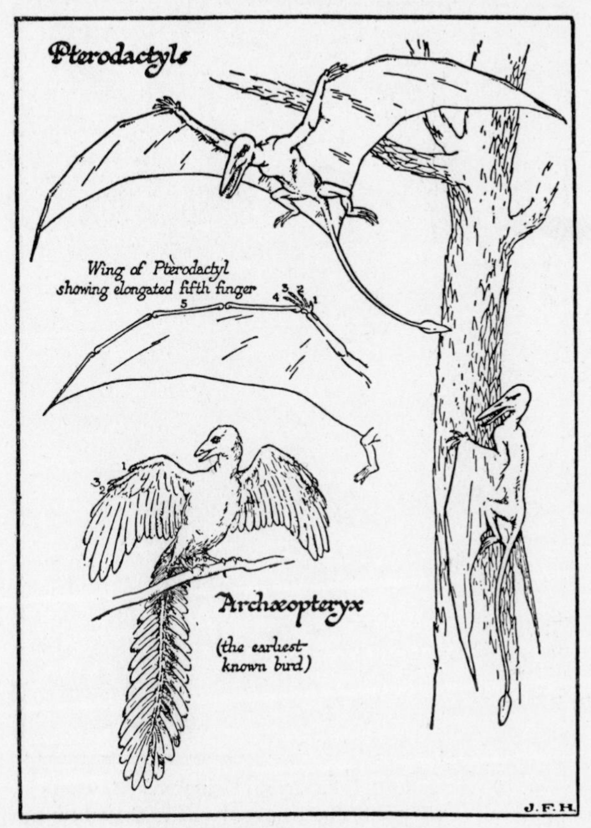

FIG. 79.—PTERODACTYL AND BIRD

rotated. There are forty-four teeth in all, in continuous series, with low crowns (**brachydont**) and a simple arrangement of conical tubercles on the surface. The brain (as shown by internal moulds of the skull) is small like that of a reptile.

From these primitive forms higher forms are soon developed in a series of adaptative radiations, like those shown by the Mesozoic reptiles. All the orders of Mammalia are thus separated in the course of Eocene times. Except for the Insectivora and Rodentia, which remain small, they all show a gradual increase in size, the largest forms being chiefly found in Pliocene or Pleistocene strata, after which glacial (or

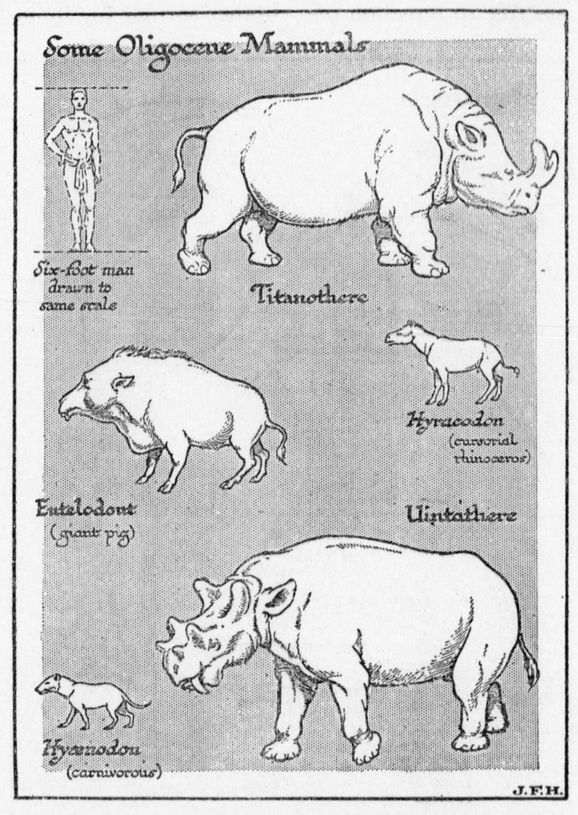

FIG. 80.—SOME LATE EOCENE AND OLIGOCENE MAMMALS

Uintatheres died out at the end of the Eocene. None of the others survived the Oligocene period, except the Entelodonts which died out in the Miocene. All are Ungulates except *Hyaenodon*.

other adverse) conditions exterminated the majority of these gigantic forms, leaving a reduced fauna of small and medium-sized mammals. The brain also increased greatly in many branches, and limbs and teeth underwent various specializations in accordance with the food and habitat.

The most interesting developments are perhaps shown by the Ungulata (hooved mammals). In these the grinding teeth became high-crowned (**hypsodont**) to stand long wear, and developed varying patterns so that as the crowns wear down they show a surface partly

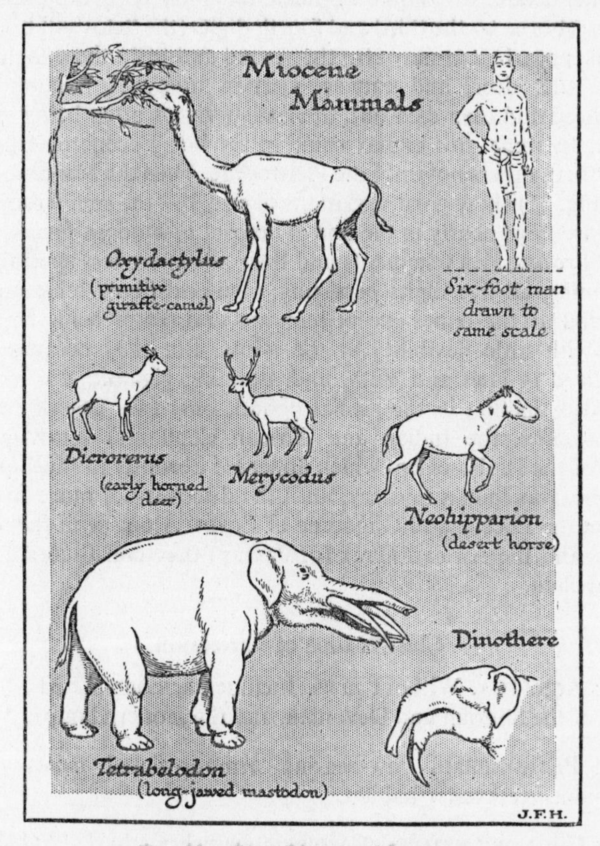

FIG. 81.—SOME MIOCENE MAMMALS

Merycodus is ancestral to the North American prongbuck, intermediate between horned and antlered Ruminants. The typical Mastodons and Elephants derive from a source like *Tetrabelodon* by shortening of the lower jaw. *Dinotherium* is a peculiar offshoot of the Proboscidean line.

of hard enamel, and partly of softer dentine, and therefore never wear smooth. The limbs became longer and vertical in position, lifting the body high off the ground, while at the same time they lost the power of rotation (useless to a running animal) by the fusion of the two bones in the second limb-segment. In one branch of the Ungulata (Perissodactyla, odd-toed) the middle digit bears the weight of the body, and the others gradually shorten and finally are nearly lost in the modern horse (*Equus*), the ancestry of which has been worked out in some detail. In another branch (Artiodactyla, even-toed) the weight is borne by the third and fourth digits (the "cloven" hoof), the first disappearing, as may also the second and fifth: these include the pigs, cattle, deer, and camels. Certain branches became highly specialized and then died out, the last survivors often being gigantic in size. In some, this climax came in the late Eocene or Oligocene (Fig. 80), in others not until the Pleistocene. Thus the Miocene Ungulates (Fig. 81) show great variations in size, but the smaller are those which are most nearly in the line of descent of modern forms.

The Proboscidea (elephants and their allies) are less specialized in their limbs, all five digits persisting, ulna and fibula remaining distinct, and all the bones except humerus and femur being short and thick, with little flexibility at the joints. But they became highly specialized in jaws and teeth, and attaining gigantic size narrowly escaped extinction in the glacial period, only two genera (*Elephas* and *Loxodon*, the Indian and African elephants) surviving. The origin of the Proboscidea, which appeared abruptly in the European Miocene, was long a mystery, until the discovery of more primitive allies in the Eocene and Oligocene of Egypt. Along with the aquatic Sirenia and the African hyrax (or daman) they constitute an Order Subungulata.

Classification of Vertebrata

CLASS: **AGNATHA.** Without jaws. Includes armour-plated Ostracoderms of the Silurian and Devonian, and the modern lampreys.

CLASS: **PLACODERMI.** Jaw-bearing armour-plated fishes of the Devonian. Freshwater habitat.

CLASS: **CHONDRICHTHYES.** Cartilaginous fishes (sharks and skates).

ORDER 1. **Acanthodii.** Freshwater fishes, with a very strong spine (ichthyodorulite) in front of each fin; one or two median dorsal fins; ganoid-like scales. Sil.–Perm.

2. **Pleuracanthodii** (ICHTHYOTOMI). Early freshwater sharks, with continuous median fins and paired fins of *Ceratodus* type. Dev.–Perm.

3. **Cladoselachii** (PLEUROPTERYGII). Marine sharks with broad-based paired fins of very simple structure; two median dorsal fins. Dev.–Perm.

4. **Elasmobranchii** (PLAGIOSTOMI). Ordinary marine sharks and skates. Carb.–Rec.

5. **Holocephali.** Chimaeras. Trias.–Rec.

CLASS: **OSTEICHTHYES.** Bony fishes.

SUBCLASS: **Crossopterygii** (see p. 157). Dev.–Cret. (with one Recent survivor).

SUBCLASS: **Actinopterygii** (see p. 157). One median dorsal fin.

ORDER 1. **Chondrostei.** Dev.–Rec. (e.g., *Acrolepis*, Fig. 73*h*).

2. **Holostei.** Perm.–Rec.

3. **Teleostei.** Trias.–Rec.

SUBCLASS: **Dipnoi** (see p. 158). Dev.–Rec.

[The above 4 classes are included in the Superclass **PISCES** or Fishes. All those that follow constitute the Superclass **TETRAPODA.**]

CLASS: **AMPHIBIA.**

ORDER 1. LABYRINTHODONTIA. Carb.–Trias.

2. LEPOSPONDYLI. Carb.–Perm.

3. PHYLLOSPONDYLI. U. Dev.–Perm.

4. URODELA (newts). Lower Cret.–Rec.

5. GYMNOPHIONA or COECILIA (snake-like). Recent.

6. ANURA (frogs). Jur.–Rec.

CLASS: SAUROPSIDA (Reptiles and Birds).

SUBCLASS: **Anapsida.**

ORDER 1. COTYLOSAURIA. U. Carb.–Trias.

2. CHELONIA. Tortoises and Turtles. Perm.–Rec.

SUBCLASS: **Synapsida.** Extinct reptiles related to Mammalia.

ORDER 1. PELYCOSAURIA. U. Carb.–Perm.

2. THERAPSIDA. Perm.–Trias.

SUBCLASS: **Archosauria.** The Dinosaur-Bird line.

ORDER 1. THECODONTIA. Trias.

2. CROCODILIA. Jur.–Rec.

3. PTEROSAURIA (Pterodactyls). Membrane-winged flying reptiles. Jur.–Cret.

4. SAURISCHIA (Dinosaurs in part). Trias.–Cret.

5. ORNITHISCHIA (Dinosaurs in part). Trias.–Cret.

6. AVES. Birds. Jur.–Rec.

Other orders, not grouped in subclasses.

RHYNCHOCEPHALIA. Trias.–Rec.

SQUAMATA. Lizards and Snakes. Cret.–Rec.

ICHTHYOPTERYGIA (Ichthyosaurs). Short-necked swimming reptiles. Trias.–Cret.

SAUROPTERYGIA (Plesiosaurs). Long-necked swimming reptiles. Trias.–Cret.

CLASS: **MAMMALIA.**

SUBCLASS: **Prototheria** (MONOTREMATA). Egg-laying mammals of Australasia. Plio.–Rec.

SUBCLASS: **Allotheria** (MULTITUBERCULATA). Trias.–Eoc.

SUBCLASS: **Trituberculata.** Jurassic.

SUBCLASS: **Metatheria** (MARSUPIALIA). Pouched mammals. U. Cret.–Rec.

SUBCLASS: **Eutheria** (PLACENTALIA).

SUPERORDER 1. UNGUICULATA. Clawed mammals, including Insectivora (U. Cret.–Rec.), Rodentia, Carnivora, and the two highly specialized orders, Edentata (South American ant-eaters, sloths, etc.), and Chiroptera (bats). Paleoc.–Rec.

2. UNGULATA. Hooved mammals, divided into many orders, of which the most important are Artiodactyla and Perissodactyla (which, with some extinct orders, are of Northern origin), Proboscidea, Hyracoidea and Sirenia (Subungulata, African), and several extinct orders (South American). Eoc.–Rec., except extinct orders which died out mainly in either late Eoc., Olig., or Pleistocene.

3. CETACEA (whales and dolphins, derived from primitive carnivores). Eoc.–Rec.

4. PRIMATES (lemurs, monkeys, and man). Paleoc.–Rec.

Short Bibliography

GENERAL

COLBERT, E. H.—*Evolution of the Vertebrates* (1955) New York.

ROMER, A. S.—(1) *Vertebrate Paleontology* 2nd Ed. (1945) Chicago. The standard textbook with comprehensive bibliographies and generic lists. (2) *Man and the Vertebrates* (1954) London.

SIMPSON, G. G.—(1) *Life of the Past: an introduction to Paleontology* (1953) Yale. (2) *The Major Features of Evolution* (1953) New York.

WATSON, D. M. S.—*Palaeontology and modern biology* (1951) London.

AGNATHA AND FISHES

GOODRICH, E. S.—"Scales of fish, living and extinct, and their importance in classification," *Proc. Zool. Soc. London* (1907).

MOY-THOMAS, J. A.—(1) *Palaeozoic Fishes* London (Methuen) 1939. (2) "Early Evolution and relationships of Elasmobranchs," *Biol. Rev.*, xiv (1939).

ROMER, A. S.—"Early Evolution of Fishes," *Quart. Rev. Biol.*, xxi (1946).

SCHAEFFER, B.—"Triassic Coelacanth *Diplurus*, with observations on the evolution of Coelacanthini," *Bull. Amer. Mus. Nat. Hist.*, xcix (1952).

STENSIÖ, E. A.—(1) *The Downtonian and Devonian Vertebrates of Spitsbergen: I. Cephalaspidae* (1927) Oslo. (2) *The Cephalaspids of Great Britain* (1932) publ., Brit. Mus. (Nat. Hist.) (3) "On the Placodermi of the Upper Devonian of East Greenland II." *Medd. om Grønland*, cxxxix (1948).

WATSON, D. M. S.—"The Acanthodian Fishes" *Phil. Trans. Roy. Soc.* (B) (1937).

WHITE, E. I.—"The Vertebrate Faunas of the Lower Old Red Sandstone of the Welsh Borders," *Bull. Brit. Mus.* (*Nat. Hist.*), i, No. 3 (1950).

AMPHIBIA

SÄVE-SÖDERBERGH, G.—"Preliminary note on Devonian Stegocephalians from East Greenland," *Med. om Grønland*, xciv (1932).

WATSON, D. M. S.—(1) "The Structure, Evolution and Origin of the Amphibia," *Phil. Trans. Roy. Soc.* (B) (1919 and 1926). (2) "The Origin of Frogs," *Trans. Roy. Soc. Edin.*, lx (1940).

WESTOLL, T. S.—"The Origin of the Tetrapods," *Biol. Rev.*, xviii (1943).

REPTILIA

BROOM, R.—*The Mammal-like reptiles of South Africa and the Origin of mammals* (1932) London.

COLBERT, E. H.—*The Dinosaur Book* (1951) New York.

MOOK, C. C.—"The Evolution and Classification of the Crocodilia," *Journ. Geol.*, xlii (1934).

SWINTON, W. E.—*Fossil Amphibians and Reptiles* (1958) Brit. Mus. (Nat. Hist.).

WATSON, D. M. S.—(1) "On *Seymouria*, the most primitive known reptile," *Proc. Zool. Soc.* (1918). (2) "On *Bolosaurus* and the Origin and Classification of Reptiles," *Bull. Mus. Comp. Zool.* (1954) Harvard.

WILLISTON, S. W.—*Osteology of Reptiles* (1925) Cambridge.

AVES

EDINGER, Tilly.—"The Brains of Fossil Odontognathae," *Evol.*, v (1951).

HEILMAN, G.—*The Origin of Birds* (1926) London.

SWINTON, W. E.—*Fossil Birds* (1958) Brit. Mus. (Nat. Hist.).

MAMMALIA

EDINGER, Tilly.—"Evolution of the Horse brain," *Mem. Geol. Soc. Amer.*, xlv (1948).

FLOWER, W. H. and LYDEKKER, R.—*Mammals, living and extinct* (1891) London.

LE GROS CLARK, W. E.—(1) *History of the Primates* 6th Edn. (1958), Brit. Mus. (Nat. Hist.). (2) *The Fossil Evidence for Human Evolution* (1955).

MATTHEW, W. D.—*Climate and Evolution* (1939) New York.

OSBORN, H. F.—*The Age of Mammals* (1910) New York.

SCOTT, W. B.—*History of Land Mammals in the Western Hemisphere* (1913) New York.

SIMPSON, G. G.—"The Beginning of the Age of Mammals," *Biol. Rev.* (1937).

SIMPSON, G. G.—(1) "Principles of Classification and Classification of the Mammals," *Bull. Amer. Mus.* (*Nat. Hist.*), lxxxv (1945). (2) *Horses* (1951) New York.

WATSON, D. M. S.—"The Evolution of Proboscidea," *Biol. Rev.*, xxi (1946).

VII
THE GRAPTOLITES

THE Graptolites are an extinct group, found only in the Palaeozoic rocks, and most abundantly preserved in black shales, which contain few other fossils except horny brachiopods.* Their habit of growth is more suggestive of plants than animals, consisting as they do of numerous similar parts borne on a "stem" which may be branched. Such a colonial habit is, however, known in many living aquatic forms which are undoubtedly animals: it is easily arrived at when a relatively simple organism takes up a fixed habit. So far as this feature goes the graptolites might be assigned to either of several animal phyla, such as the Hydrozoa or Bryozoa; but they show methods of growth unknown in those phyla, and it has recently been suggested that their affinity may be sought in certain lowly living relatives of the Vertebrata.

The Vertebrata (Craniata) constitute by far the largest division of the phylum **Chordata,** and neither in them nor in the division Cephalochorda (*Amphioxus*) is there any possibility of budding to form a compound or colonial organism. But in the two other divisions—Urochorda (Tunicates) and Hemichordata, budding does take place. It is a general feature of the Tunicates, and of one of the orders of the Hemichordata (Pterobranchia), and it is the latter which shows resemblances to the Graptolites, particularly in its stolonal mode of branching. That this is the true affinity of the Graptolites can hardly be proved, but on evolutionary grounds it is likely that at the time when Vertebrata were at the outset of their career, their humbler relatives should have been at the height of their development, and if capable of fossilization at all might form an important part of the fossil fauna.†

The skeleton or **periderm** of graptolites is composed of some black (brown when seen magnified in transmitted light) horny chitin-like organic material. When preserved in shales it is usually crushed flat,

* For two different views on the origin of these graptolitic shales, see the papers quoted in the Bibliography (pp. 182–3), by Marr, J. E., and Elles, G. L. (2).

† This surmise finds some justification in the recent description of two problematical fossils, both from the Silurian beds of south Lanarkshire: *Ainiktozoon* (Scourfield, D. J., *Proc. Roy. Soc. London*, February, 1937) and *Jamoytius* (White, E. I., *Geol. Mag.*, March–April, 1946).

and appears as a thin film, sometimes of a whitish siliceous material, more often of graphite. More rarely the material was replaced by pyrite before being crushed; still more rarely it has been preserved uncrushed and unaltered in limestone or shale from which it can be extracted by chemical treatment or studied by cutting serial sections. (See Chap. XII, p. 259.)

1. **Didymograptus murchisoni** (Fig. 82) is an abundant graptolite preserved in white siliceous powdery material in the black shales of Ordovician (Llanvirn) age of Abereiddy Bay, Pembrokeshire (and elsewhere). Its shape is well described by the name "tuning fork graptolite." Corresponding to the stem of the fork is a hollow, acutely conical body (crushed flat in the shale) called the **sicula,** its apex drawn out into a short thread-like tube called the **nema.** There is

FIG. 82.—DIDYMOGRAPTUS MURCHISONI (Beck), Upper Llanvirn. (Natural size; after Elles and Wood)
S, Sicula

reason to believe that by the nema the graptolite was attached to floating sea-weed, from which it hung downwards, and we will describe the fossil as if in this position. In this description many statements cannot be fully verified on ordinary specimens, the facts having been demonstrated on Swedish specimens of pendent Didymograptids preserved uncrushed in limestone.

The sicula probably lodged the first-formed individual, from which others were afterwards produced by a process of budding (Fig. 83). These are arranged along two branches, or **stipes,** hanging down almost vertically but slightly divergent. Each stipe consists of a series of somewhat cylindrical cups, **thecae,** which lodged an individual each, and a **common canal** uniting their inner ends. The sicula and stipes constitute the **rhabdosome,** or complete skeleton, of the whole colony. The end at which is the sicula is called the **proximal** end, the opposite end the **distal.**

The two stipes are not, however, absolutely symmetrical. The first theca of one stipe arises directly from the sicula, and successive thecae, with the common canal pertaining thereto, are budded off in turn; but the first theca of the other stipe is budded off from the first theca of the first stipe, so that its portion of common canal has to cross one side of the sicula to the other. This is called the **crossing canal,** and the side on which it lies is called the **reverse,** while the side on which the sicula is in front is the **obverse.** After initial thecae have thus been formed for each stipe, further thecae are developed alike on each. Each theca is situated at an acute angle to the direction of the stipe and overlaps the adjacent theca considerably. If tangents are drawn to the outer face of the first theca on each side, the angle enclosed by

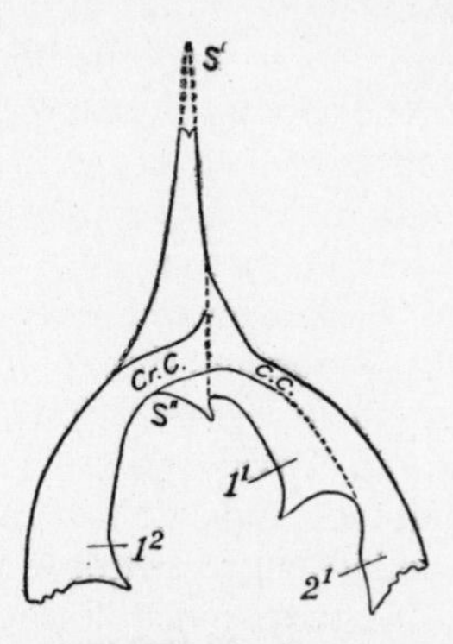

FIG. 83.—DIDYMOGRAPTUS MINUTUS Törnquist, Lower Ordovician (*Orthoceras* Limestone), Oeland

Initial portion, reverse side: *S'*, Apical part of sicula; *S''*, aperture of sicula; *Cr. C.*, crossing canal; *c.c.*, common canal; 1^1, 2^1, first and second thecae of first stipe; 1^2, first theca of second stipe (× 19; after Holm).

them is the **angle of divergence,** which in this species is less than 45°. This angle is not maintained by the stipes, as lower down they become subparallel: a rhabdosome so disposed is described as **pendent.** Eight or nine British species of *Didymograptus* have this pendent form: they are distinguished from one another by differences of detail in the thecae.

There are other species which begin their growth like *D. murchisoni* and form two stipes, but with a different angle of divergence or different final direction.

Thus if the angle of divergence is about 90° and the stipes continue at that angle, the rhabdosome is said to be **declined**; if starting at less than 90° they gradually spread out almost horizontally, **deflexed**; if with an angle of divergence approximating 180°, they are termed

horizontal; if curved upwards, **reclined**; if curved upwards and then downwards, **reflexed** (Fig. 84). Even in these last two cases (known only in one species each) the original angle of divergence is less than 180°. The constancy of these angles and curvatures, as well as the frequent abrupt bends in stipes that have been snapped by too great strain (as in Fig. 88*c*) shows that the rhabdosome was a fairly stiff structure, with about the same elasticity as a bird's feather.

All these forms, covering about 30 British species alone, are habitually included in the one genus *Didymograptus*. But the modern conception of a genus is that of a number of species connected together by common descent. That *Didymograptus*, in its accepted extent, is a true genus has been rendered very doubtful by the following observations.

The genus *Didymograptus* is found in the Lower Ordovician and has a surviving species in the basal Caradoc. In the Lower Ordovician are found graptolites which start growth like *Didymograptus*, but branch again, forming four or eight stipes, and in the Tremadoc Beds (transitional between Cambrian and Ordovician) are still more complexly branched forms. Nicholson and Marr pointed out in 1895 that similarities of stipe-disposition and thecal form can be traced through the morphological genera *Bryograptus* (or *Clonograptus* or *Loganograptus*), *Tetragraptus* and *Didymograptus* in these successive periods, and they suggested that these indicate the true lines of descent, while the number of stipes underwent reduction from many to four, and at last to two, in at least five independent lineages (Fig. 84).

Phyllograptus (Fig. 84) consists of species closely allied to the reclined Tetragrapti, but with the four stipes confluent (as indicated in the lower figure, which represents a cross-section). This type of growth is no longer reclined, but **scandent** (climbing).

However the extent of the "genera" so far mentioned may have to be altered, they appear to form a natural family—the Dichograptidae, of which the five series mentioned would be subfamilies. The family is distinguished by the simple structure of its thecae, and by the fact that the first-formed thecae always grow downwards from the sicula. It ranges from Tremadoc to early Caradoc.

2. **Climacograptus wilsoni** (Fig. 85) is a graptolite which occurs in such abundance at the base of the Hartfell Shales in the Southern Uplands of Scotland that it has been chosen as a zone-fossil. It has an unbranched stipe with a double column of thecae (**biserial** rhabdosome), and may reach a length of six centimetres or more. When complete, the proximal end shows the sicula, having at its base a very large vesicle (a specific character, not found in other Climacograpti), and passing at its apex into a rod, the **virgula.** This virgula is compar-

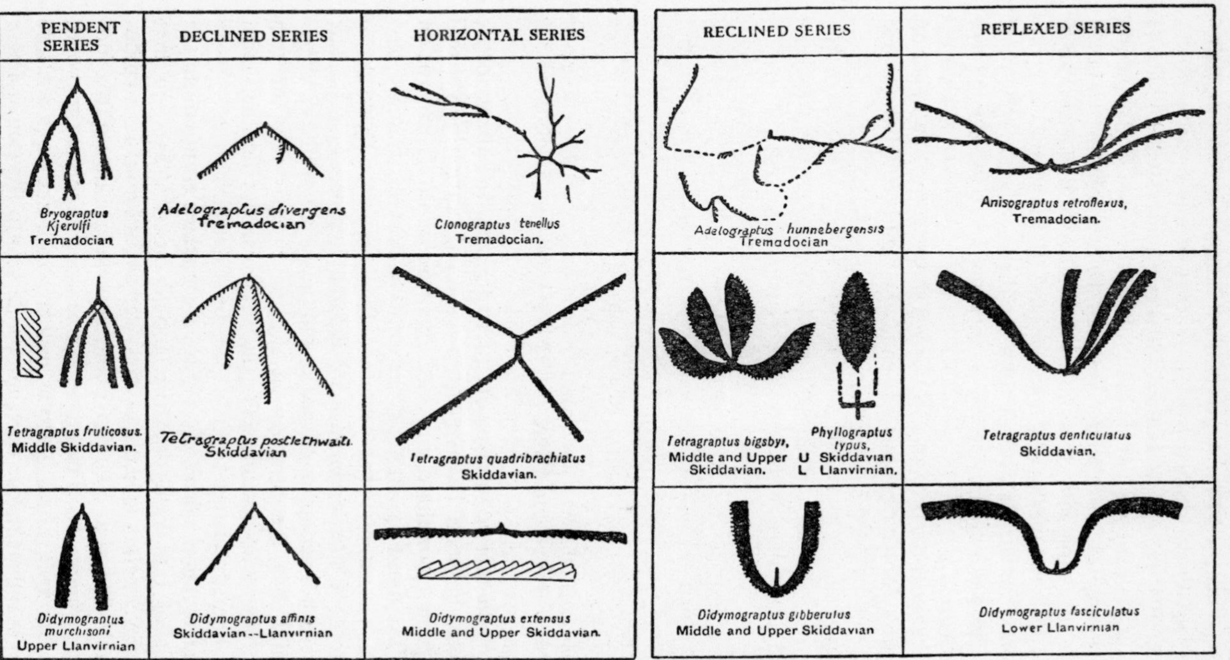

FIG. 84.—EVOLUTION OF THE DICHOGRAPTIDAE

Five parallel series, of which two are incomplete. It will be noted that the evolution is most rapid in the case of the horizontal forms and slowest in that of the pendent.

The insertions of *Adelograptus hunnebergebsis* and *Tetragraptus denticulatus* in this table are only tentative.

able with the nema of *Didymograptus*, but serves an additional function. The first theca budded from the sicula begins to grow downwards as in *Didymograptus*, but soon makes a sharp bend and grows upwards, and all subsequent thecae grow in this way. Consequently the virgula comes to be imbedded between the two columns of thecae and serves as a support to them. This is another example of a scandent form. In many species of *Climacograptus* (though apparently not in *C. wilsoni*) the virgula extends for some distance beyond the most distal thecae, and probably served to suspend the rhabdosome from floating seaweed.

The thecae near the proximal end have a simple form, but higher up they acquire a double right-angled bend, which gives the rhabdosome

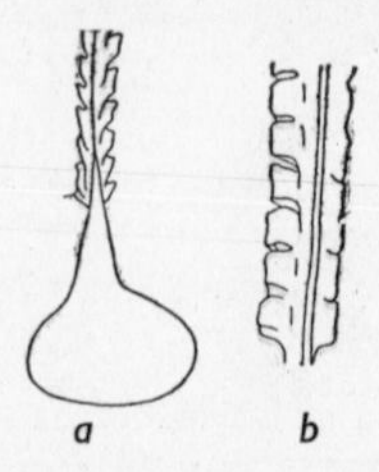

FIG. 85.—CLIMACOGRAPTUS WILSONI Lapworth, Lower Caradoc
a, Initial part of rhabdosome ($\times \frac{5}{6}$), showing sicula with large vesicle, and six thecae of each column; *b*, distal part of rhabdosome ($\times \frac{5}{2}$), showing form of thecae and "excavations" clearly on the left side (After Elles and Wood).

the appearance of a double row of square bodies with narrow spaces (**excavations**) between. Each square is the distal portion of a theca, with its aperture horizontal and opening upwards into the excavation above it. The middle portion of each theca is bent horizontally, and the proximal portion is again vertical.

Specimens of *Climacograptus* are usually flattened in the plane which bisects both columns of thecae, and thus give a **profile** view: but occasionally they are flattened in the plane separating the two columns: they then show one set of thecae only in full face, and from its ladder-like appearance this is called the **scalariform** view.

3. **Monograptus priodon** (Fig. 86) is an abundant graptolite in the lower part of the Wenlock Shale, though not restricted to that horizon; nevertheless it had a world-wide distribution. Its rhabdosome is unbranched and attains a considerable length—up to a foot (25 cm.) and more. The sicula lies on one side of the proximal end, and from its apex arises a stout virgula which supports a single column of scandent thecae on the opposite side to the sicula. These thecae lie obliquely and overlap one another for about half their

length; the free portion is bent first outwards and then downwards, so as to have a hook-like appearance. The curvature of the theca and the direction of the aperture are thus the opposite from those of *Climacograptus*. The first theca grows upwards from the sicula: the first stage of downward growth, corresponding to the permanent direction of *Didymograptus*, and still preserved in *Climacograptus*, is skipped altogether in *Monograptus*.

The graptolites (**Graptolithina**) are divided into Dendroidea and Graptoloidea.

Dendroidea.—This group, so-called because of its tree-like many branched growth, has not yet been well classified, as the structure of its members has only recently been thoroughly studied. The most familiar species, *Dictyonema flabelliforme*, has a sicula, the bud from

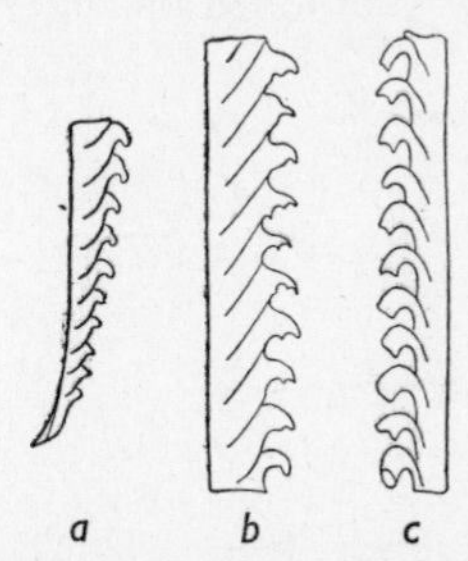

FIG. 86.—MONOGRAPTUS PRIODON (Bronn), Wenlock Shale

a, Initial part of rhabdosome, showing sicula with virgula arising from its apex; *b*, distal thecae, reverse aspect, apertures partly embedded in the shale; *c*, distal thecae in low-relief, obverse aspect, showing apertural margins ($\times \frac{5}{2}$; after Elles and Wood).

which quickly divides into three primary stipes: these further divide and spread out into a cone- or bell-shaped net, the stipes being connected by fine hollow cross-threads (**dissepiments**) (Figs. 87, 88*a*). The thecae open on the inner face of the cone, and are of three kinds—**autothecae** (like the ordinary thecae of Dichograptids), **bithecae** (smaller and more cup-like), and **stolothecae**, containing the **stolon**, from which are budded both other types. When found fossil in shales, the cone-shaped net has been compressed into a fan-shaped structure lying on a bedding place; the outer surface however on occasions retains a weak convexity. As bithecae have been found in the simplest of the Dichograptids—*Bryograptus* and *Clonograptus*—it is probable that *D. flabelliforme* was ancestral to the Graptoloidea, especially as it seems to be the only Dendroid that lived suspended from floating seaweed. All other species of *Dictyonema* (Cam.–Carb.), and all other Dendroidea (collectively with the same long range) either lack a nema or it is thickened so that the rhabdosome was rooted in the sea-floor and thus became **benthic.**

Graptoloidea.—These appear all to have had the same habitat as *Dictyonema flabelliforme* (pseudo-planktonic or epi-planktonic, i.e., attached to floating organisms or **plankton**). They are confined to the Older Palaeozoic and are of great value as stratigraphical indices because their diversification of shape took place rapidly and the rhabdosomes were widely distributed. The following classification is that of the monograph by Elles, Wood and Lapworth.

Family DICHOGRAPTIDAE.—Uniserial Graptoloidea, with bilateral rhabdosomes, being simple subcylindrical thecae. Branching usually dichotomous, but occasionally irregular. Primary angle of divergence generally 180° or less. Tremadoc to Lower Caradoc.

Chief genera are those mentioned on Fig. 84, characterized by the

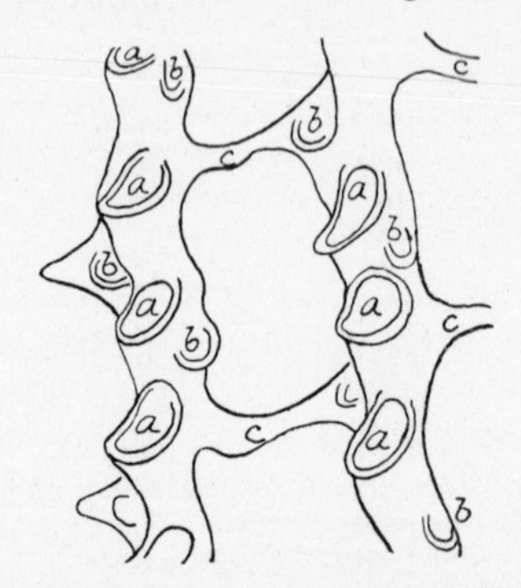

FIG. 87.—DICTYONEMA FLABELLIFORME (Eichwald)
Small part of rhabdosome (× 17·5; after Bulman). *a, a,* autothecae; *b, b,* bithecae; *c, c,* dissepiments.

number of stipes. Of these only *Didymograptus* ranges above the Lower Llanvirn (*Didymograptus bifidus* Zone). *Bryograptus* and *Adelograptus* have imperfectly symmetrical and somewhat irregularly branching rhabdosomes, the former with three primary stipes, the latter with two. It should be noted that Professor Bulman has lately assigned these two genera to the Dendroidea along with *Clonograptus* because some of their early species possessed bithecae.

Family LEPTOGRAPTIDAE.—Uniserial, with slender flexuous rhabdosomes; primary angle of divergence approximately 180°; branching usually lateral; thecae elongated, with slight sigmoid curvature, apertures inclined, situated partly within depressions (excavations), somewhat **introverted** (turned inwards), but not **introtorted** (twisted inwards). Range in Europe: Middle Llanvirn to Middle Ashgill.

Chief genera: *Leptograptus* (Fig. 88*b*), without lateral branching (range of family); *Nemagraptus* (Fig. 88*c*), S-shaped rhabdosome with numerous lateral branches (world-wide in Glenkiln or Lower Caradoc); *Pleurograptus*, with compound lateral branches (one zone in Hartfell Shales, Upper Caradoc).

Family DICRANOGRAPTIDAE.—Uni- or uni-bi-serial, angle of divergence always exceeding 180°; thecae with strong sigmoid curvature; apertures horizontal or inclined, situated within well-defined "excavations" and commonly introverted. Llanvirn to Ashgill.

There are two genera: *Dicellograptus* (Fig. 88*d*), in which the two branches rise up somewhat as in the reclined dichograptids, and *Dicranograptus* (Fig. 88*e*), in which they are united at first (as in *Phyllograptus* or *Climacograptus*), but afterwards separate. This family appears to be transitional between Leptograptidae and Diplograptidae. *Dicellograptus* has the range of the family, while *Dicranograptus* is Caradoc only.

Family DIPLOGRAPTIDAE.—Biserial, with straight unbranched stipes. Thecae tubular, usually in contact for a large part of their length. Range: Arenig to Upper Valentian, with acme from Caradoc to Middle Valentian.

Two genera are usually recognized. *Climacograptus* (Fig. 85) is characterized by the strongly sigmoid curvature of the thecae, and its "excavations"; *Diplograptus* (Figs. 88*f*, *g*) has straight or nearly straight thecae, overlapping and with aperture everted (turned outwards). *Diplograptus* has been subdivided into several subgenera such as *Glyptograptus* (Arenig–Valentian) *Orthograptus* (Caradoc–Val.) and *Petalograptus* (Val.); a modern tendency is to regard these as genera. Dr. Ruedemann discovered in New York State specimens showing that *Diplograptus* lived in clusters hanging from a float, and suggesting that numerous siculae were developed in reproductive vesicles under the float and grew out from these as a second generation of rhabdosomes (Fig. 88*g*); this supercolony is termed a **synrhabdosome.**

The families GLOSSOGRAPTIDAE and RETIOLITIDAE are allied to the last, but show peculiar developments of the periderm—firstly, in a change from a membrane of uniform thickness to one with strengthening ribs, at last becoming a complex network, in the meshes of which is left a membrane so delicate that it is generally lost in the fossil state; secondly, in a great development of projecting spines. The latter form the special feature of the Glossograptidae (extending through nearly the whole Ordovician), the network being most highly developed in Retiolitidae (almost confined to Silurian).

These three biserial families, with scandent thecae, are believed to

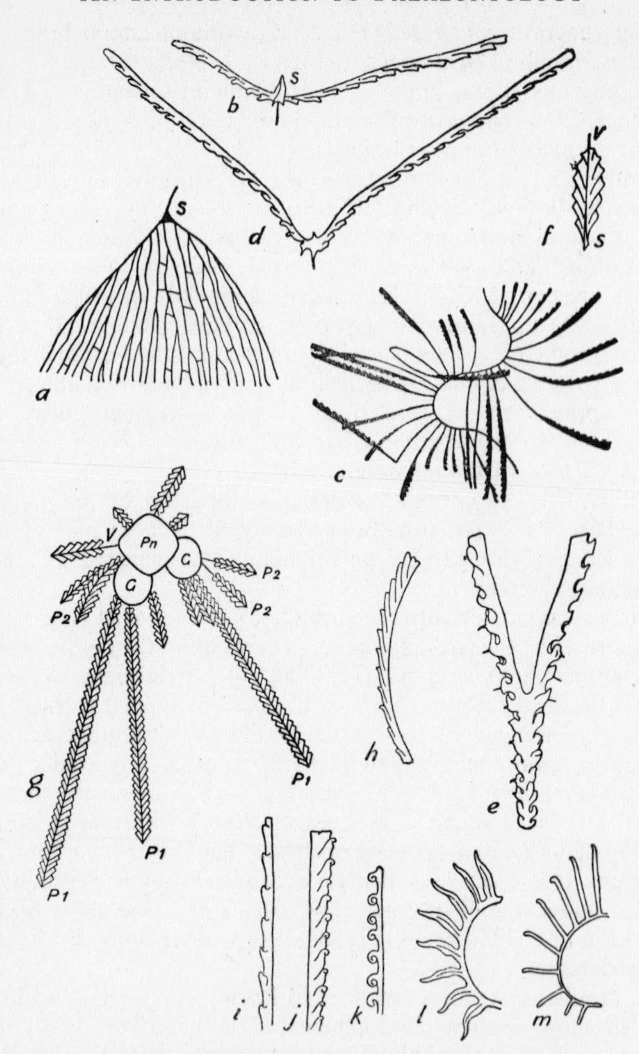

FIG. 88.—DICTYONEMA AND VARIOUS GRAPTOLITES

a, *Dictyonema flabelliforme* (Eichwald), Tremadoc; part of rhabdosome, with sicula and nema (Natural size; full length over 80 mm., maximum width 75 mm.). *b*, *Leptograptus flaccidus* (Hall), Caradoc ($\times \frac{5}{2}$); *S*, Sicula; *c*, *Nemagraptus gracilis* (Hall) , Caradoc ($\times \frac{1}{2}$); *d*, *Dicellograptus gurleyi* Lapworth, Llandeilo, New York ($\times \frac{5}{2}$); *e*, *Dicranograptus nicholsoni* Hopkinson, Caradoc ($\times \frac{5}{2}$); *f*. *Diplograptus* (*Petalograptus*) *palmeus tenuis* (Barrande), Valentian; *S*, Sicula; *V*, virgula; *g*, *Diplograptus* (*Orthograptus*) sp. M. Ordovician, New York; Floating synnhabdosome (natural size; after Ruedemann); *Pn*., Pneumatophore (float); *G*, gonangia (reproductive

have originated from more than one source in the earlier families, and to have passed through stages represented by the reclined Dichograptids, *Phyllograptus* and *Dicranograptus*. It is noteworthy that *Dicranograptus* affords an example of **caenogenesis** (ante, p. 118), since it starts with the more highly evolved biserial form and later reverts to the simpler uniserial form; but the term caenogenesis strictly speaking belongs to individual rather than to colonial development.

Family DIMORPHOGRAPTIDAE is transitional between Diplograptidae and Monograptidae, being uniserial only in the early part of the rhabdosome and biserial afterwards: thus they give another example of caenogenesis. The only genus, *Dimorphograptus*, is confined to the Lower Valentian, being contemporaneous with the earliest Monograptids, so that it cannot be regarded as a degeneration from the latter.

Family MONOGRAPTIDAE.—Uniserial and unilateral. The characteristic Silurian family, ranging from the bottom almost to the top of that system.

The chief genus, *Monograptus* (Figs. 86, 88*h–m*) is divided by Elles and Wood into the following seven groups, characterized by differences in the thecae, and each named after a typical species.

1. *M. cyphus:* thecae simple, straight, overlapping tubes, with even apertural margins (Middle Valentian to top Salopian; Fig. 88*h*).

2. *M. sandersoni:* thecae with flowing sigmoid curvature and oblique apertural margins (Lower and Middle Valentian; Fig. 88*i*).

3. *M. vomerinus:* thecae short tubes with abrupt sigmoid curvature and some torsion of axis (Middle Valentian to Upper Salopian; Fig. 88*j*).

4. *M. priodon:* thecae with apertural region more or less isolated and retroverted (Middle Valentian to Upper Salopian; Fig. 86).

5. *M. lobiferus:* thecae with apertural region coiled into a definite lobe (Middle Valentian to Lower Salopian; Fig. 88*k*).

6. *M. convolutus:* thecae triangulate or conical, with reflexed apertural margins (Middle and Upper Valentian; Fig. 88*l*).

FIG. 88.—DICTYONEMA AND VARIOUS GRAPTOLITES (*continued*).

vesicles); P_1, first generation of rhabdosomes; P_2, second generation; *V*, virgula; *h*, *Monograptus cyphus* Lapworth, Lower Valentian; thecae near proximal end; *i*, *M. sandersoni* Lapworth, Valentian, distal thecae; *j*, *M. vomerinus* (Nicholson), Salopian; thecae near proximal end; in full relief, showing torsion of thecal axis; *k*, *M. lobiferus* (M'Coy), Valentian; distal thecae; *l*, *M. convolutus* (Hisinger), Valentian; *m*, *Rastrites longispinus* (Perner), Valentian. All × $\frac{5}{2}$, unless otherwise stated. *a*, *e*, *g*, after Ruedemann; *b*, *d*, from Ruedemann, after Lapworth; *c*, from Ruedemann, after Hall; remainder after Elles and Wood.

7. Usually taken as a distinct genus, *Rastrites:* thecae more or less linear and isolate, with reflexed terminations (Middle and Upper Valentian; Fig. 88*m*).

The rhabdosome of *Monograptus* is commonly straight or slightly curved, but increased curvature resulting finally in a spiral is found in all groups except the second and third, and is the invariable case in the last two.

The name *Cyrtograptus* is applied to Monograptids in which young individuals (**cladia**) remain attached to the parent rhabdosome, giving a false appearance of branching; the main stipe is more or less spiral. Whether this is justifiably taken as a generic character may be doubted; but the species with this character all occur within a narrow time-range (practically Lower Salopian), and although some of them are almost identical in other characters with particular species of Monograptus, they are not contemporary with them.

The graptolites have proved of the greatest value as zone-fossils. Thanks to them alone, Charles Lapworth was able to unravel the complexities of structure of the Southern Uplands of Scotland, which had been entirely misinterpreted before. Apart from detailed zoning they provide any geologist with the means of recognizing broad divisions: the complex Dichograptidae are easily recognized as marking the lowest Ordovician, the Monograptidae as distinctive of the Silurian, and so on. Their zonal distribution appears in general to be the same in the Ordovician in Europe and North America, though in the latter region Silurian graptolites (other than Dendroidea) are rare.

Short Bibliography

BULMAN, O. M. B.—(1) "British Dendroid Graptolites," *Palaeont. Soc*. (1927–). (2) "On the Graptolites prepared by Holm," *Arkiv. för Zoologi* (Stockholm, 1932–6). (3) "Programme Evolution in the Graptolites," *Biol. Rev.*, viii, 311 (1933). (4) "The Caradoc (Balclatchie) Graptolites from limestones in Laggan Burn, Ayrshire," *Palaeont. Soc.* (1944–7). (5) *Treatise on Invertebrate Paleontology*, *Part V*, *Graptolithina* 101 pp., (1955) Kansas University Press.

COX, I.—"On *Climacograptus inuiti* sp. nov. and its development," *Geol. Mag.*, lxx, 1 (1933).

ELLES, Gertrude L., WOOD, Ethel M. R., and LAPWORTH, C.—"British Graptolites," *Palaeont. Soc.* (1901–18). This contains a detailed historical account and bibliography.

ELLES, Gertrude L.—(1) "The Graptolite Faunas of the British Isles: a study in Evolution," *Proc. Geol. Assoc.*, xxxiii, 168–200 (1922). (2) "Factors controlling Graptolite Successions and Assemblages," *Geol. Mag.*, lxxvi, 181 (April, 1939).

KOZLOWSKI, R.—(1) "Informations préliminaires sur les Graptolithes du Tremadoc de la Pologne et sur leur portée théorique," *Ann. Mus. Zool. Polonici*, xiii, 183–96 (1938). (2) "Les graptolites et quelques nouveaux groupes d'animaux du Tremadoc de la Pologne," *Palaeont. polonica*, iii (1948).

MARR, J. E.—"The Stockdale Shales of the Lake District", *Quart. Journ. Geol. Soc.*, lxxxi, 113 (1925).

RUEDEMANN, R.—"Graptolites of New York", 2 vols., *Mem. New York State Museum*, vii and xi (1904–8).

STUBBLEFIELD, C. J.—"Notes on some early British Graptolites," *Geol. Mag.*, lxvi, 268 (1929).

ULRICH, E. O. and RUEDEMANN, R.—"Are the Graptolites Bryozoa?" *Bull. Geol. Soc. Amer.* xlii, 589 (1931).

VIII

THE ECHINODERMA

WHEN a marine limestone is broken with a hammer it usually shows a smooth or granular fracture, but here and there on the broken surface may often be seen distinct cleavage-surfaces of calcite: these may be few or many, and in particular cases may be large. They are the fractured surfaces of fossil echinoderms. Whereas the shells of most invertebrates show on a broken surface a fibrous or laminated texture, due to the arrangement of an indefinite number of minute crystals of calcite, it is one of the essential features of echinoderms (and of no other group except the calcareous sponges) that the skeleton is composed of a series of units that may attain a large size, but of which every one is, mineralogically, a crystal of calcite—not in external shape, which is determined by organic secretion), but in molecular constitution. These units have different names, according to shape: the commonest are broad, flat structures, called **plates.** In these the normal to the surface corresponds to the vertical crystallographic axis. When the shape is long and cylindrical, as in the radioles (articulated spines) of sea-urchins, it is the long axis which is the vertical crystal-axis.

Solid crystalline calcite would be much too heavy a material for an organic skeleton; accordingly, in the living state the plates are lightened by rounded cavities, arranged so closely as to make the whole substance spongy, with a definite organic pattern. This network of calcite is called the **stereom.** When the skeleton of a recent echinoderm—say, the radiole of a sea-urchin—is broken, the calcite cleavage is interrupted by these abundant cavities, and is easily overlooked, though careful observation will always show it. But the first stage in fossilization is generally the infilling of these cavities with calcite in crystalline continuity with that around them. The pattern of the cavities may or may not be obliterated in the process; when not, it forms a further means of recognition (besides the crystalline character and cleavage) of echinoderm skeletons in microscopic sections of rocks. Calcite in crystalline continuity may also be deposited around the plates (or other units), if there is opportunity, concealing the organic form and replacing it, as far as circumstances permit, by the geometrical form of a crystal. Thus, on breaking open one of the sea-urchins so common in the Upper Chalk, one occasionally finds that each plate projects into the central cavity as a rhombohedron; externally the contact of the chalk has prevented such growth, and if the interior has been filled with chalk (the more usual case) the plates will preserve their original surface internally also.

There is another general, but not universal, feature characteristic of echinoderms—a five-rayed or pentamerous symmetry. Such a symmetry is common among flowers, but is quite unknown among animals except in the echinoderms. There is one other animal phylum with a radial symmetry—the Coelenterata—and in Cuvier's classification, over a century old, the two were united as Radiata. But not only is the symmetry of the Coelenterata a four-, six-, or eight-rayed symmetry, but it is in general a more perfect symmetry than that of Echinoderma. The former include some forms, such as the jelly-fish, which are absolutely radial throughout their bodies; the Echinoderms include none such, though some attain almost perfect radiality in their skeleton only. In some early Echinoderms radial symmetry is seen in process of development; in some relatively late forms it has been modified into a practically bilateral symmetry.

The skeleton of echinoderms is not truly external like that of Brachiopods or most Mollusca; there are soft living tissues outside it, but these are so thin in most cases that it may be roughly spoken of as external. Still, it is important to remember these external tissues in order to form an idea of the life of echinoderms.

Another special feature of the Echinoderms is the presence of a system of tubes containing sea-water, which plays an important part in their physiology (water-vascular system). The only traces of this system in the skeleton are the perforations seen in the plates of many forms.

The Echinoderma fall into two main divisions—Pelmatozoa or fixed forms, and Eleutherozoa or freely-moving forms. To be precise, the former are fixed at some time in their life (and usually throughout adult life), the latter never. The former were abundant in the Palaeozoic era, rare afterwards; with the latter the reverse is the case. They are further divided into classes, thus:

PELMATOZOA

Eocrinoidea.—Extinct, Middle Cambrian to Ordovician.

Paracrinoidea.—Extinct, Middle Ordovician.

Carpoidea.—Extinct, Middle Cambrian to Lower Devonian.

Cystidea.—Extinct, Cambrian to Permian, but mainly Lower Palaeozoic.

Blastoidea.—Extinct, Silurian to Permian, but mainly Upper Palaeozoic.

Edrioasteroidea.—Extinct, Lower Cambrian to Carboniferous.

Crinoidea.—"Sea-lilies," Lower Ordovician to Recent.

ELEUTHEROZOA

Echinoidea.—"Sea-urchins," Ordovician to Recent.

Stelleroidea.—"Starfish," Cambrian to Recent.

Holothuroidea.—"Sea-cucumbers," Cambrian to Recent.

All echinoderms are marine.

It will be convenient to describe first some examples of crinoids.

1. **Cupressocrinites gracilis** (Fig. 89) is one of many crinoids found in the Middle Devonian "crinoid-bed" of Gerolstein in the Eifel (Western Germany). It consists of a **root, stem,** and **crown.** By the first (not often preserved) it was fixed to the sea-bottom, but this "root" had no absorbent function like the root of a plant. The stem lifted it up high above the mud: as preserved fossil it consists of a large number of squarish discs piled up into a column: these are the stem **columnals.** Through each of them runs a central vertical tube (axial canal) and four smaller peripheral canals. Adjacent columnals were united, in life, by organic tissues, and the surfaces of contact were roughened to give greater grip or "key": on the articular surface of an isolated columnal this gives, with the canals, a pattern which is fairly characteristic of the genus *Cupressocrinites*. During life the slight amount of flexibility of the organic tissue of each joint gave a suitable degree of flexibility to the whole stem.

The **crown** contained the essential vital parts of the animal. It is divided into the **theca** or **calyx,** which is the "body" of the animal and directly articulated to the stem, and five **arms** which, when the animal was feeding were spread out widely, but in the fossil are usually found tightly closed up. The part of the calyx which is visible when the arms are closed is called the **dorsal cup**; it is the ventral **tegmen** which is hidden. The dorsal cup consists of two circlets of five plates each and a single pentagonal plate next to the stem. This last is regarded (by analogy with other crinoids) as a fused third circlet, the five plates of which if separate would be called **infrabasals**; the five plates in the circlet next to it are called **basals**; those in the upper circlet, **radials.** The straight lines along which adjacent plates meet are termed **sutures.** Radials and basals alternate in position, and if the infrabasals were separate they would alternate with the basals. Because of the presence of (fused) infrabasals as well as basals, this crinoid is said to be **dicyclic.** In many crinoids the top columnal directly joins the basals: such crinoids are **monocyclic.**

The five radii which can be drawn from the centre of the calyx through the middle of each radial plate are called **perradii,** and organs, such as the arms, which lie symmetrically upon them are said to be **perradial** in position. The radii half-way between these are called **interradii,** and organs which lie upon them are said to be **interradial** (e.g., the basal plates). These alternating positions are of fundamental importance in all echinoderms, except in those forms in which five-rayed symmetry is not fully developed. It is a universal rule that

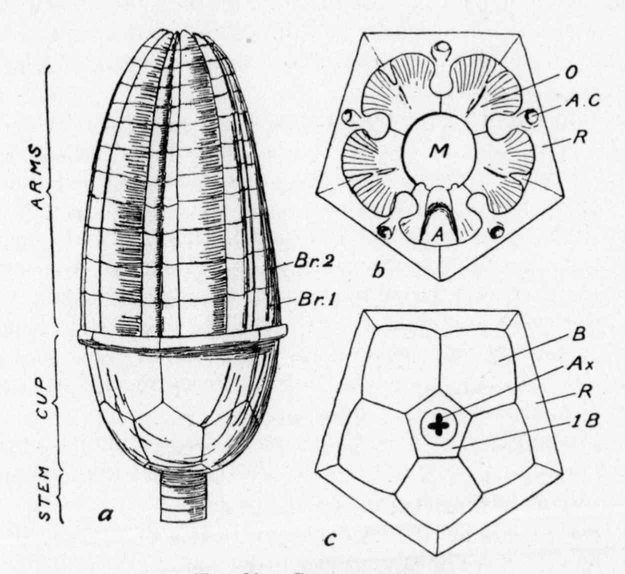

FIG. 89.—CUPRESSOCRINITES

a, *C. gracilis* Goldfuss; *b*, *c*, *C. abbreviatus* Goldfuss; both Middle Devonian, Eifel. *a*, side view of complete crown, with arms closed, and part of stem; *b*, oral view; *c*, basal view of a cup without arms or stem. *A*, Anus; *A.C.*, axial canal; *Ax*, axial canal of stem; *B*, basal plate; *Br. 1*, *2*, first and second brachial plates; *IB*, infrabasal plate; *M*, mouth; *O*, oral plate; *R*, radial plate (All natural size; original).

the mouth and anus, unless they are central, are respectively perradial and interradial in position. With the mouth, however, the central position is the rule, with the anus it is the exception.

In the case of *Cupressocrinites* the mouth is central and appears as a large circular opening between five large plates, interradial in position. Probably the actual mouth was smaller, and other smaller plates surrounded it, now scattered and lost. The five large plates are called **orals** or **deltoids**: one of them differs from the rest in being deeply notched by the anus. This is the only disturbance of radial symmetry in the crown of *Cupressocrinites*, and it marks out the odd oral as **posterior** in position; the two adjacent to it as postero-lateral and the other two as antero-lateral. The margins of the orals that are in contact with their neighbours bear semi-oval notches, which combine into five pear-shaped perforations corresponding to the middle line of each arm. Along these were continued the five food-grooves that occupy the middle ventral line of the arms. As the edges of the five orals meet around the mouth the food-grooves must have finally entered the mouth by tunnels, and the pear-shaped openings themselves may have been converted into tunnels by small overlapping covering-plates (**ambulacrals**).

Each arm consists externally (dorsally) of a single row of large **brachial** plates, tapering to the free end. Internally (ventrally) they have a wide and deep groove, with a row of **pinnules** (like miniature arms) along each side, and a series of small plates (ambulacrals) arching over it and converting it into a tunnel. This groove, in life, was ciliated and constituted the **food-groove** or **ambulacrum,** along which microscopic food was wafted towards the mouth. The pinnules were similarly grooved and ciliated, and served as tributaries. Thus the method of feeding (**microphagous**) is fundamentally the same as in brachiopods and lamellibranchs, but the structures adapted to the same ends are fundamentally different.

Not to be confused with the food-grooves are longitudinal cavities, elliptical in section, in the substance of the brachial and radial plates: these are the **axial canals** and contained a nerve-cord.

2. **Amphoracrinus gilbertsoni** (Fig. 90) is a not uncommon fossil in the Carboniferous Limestone knolls of Clitheroe and elsewhere in the North of England. Usually the calyx is found without arms or stem. The columnals are abundant, though not with certainty distinguishable from those of allied forms. Their articular surfaces show fine radiating lines and a pentagonal axial canal.

The cup is monocyclic (no infrabasals) and the basals are reduced to three by crowding, leading to fusion. The radials are five in number, but when we have identified the basals and these we still have a great many plates in the dorsal cup unaccounted for. Nevertheless, we cannot have mistaken the radials, because there are plates directly above them which lead up to the arms, and that could never be the case with basals. These plates above correspond with the lower brachials of *Cupressocrinites*, but instead of forming part of free arms they are incorporated in the dorsal cup: they are called **fixed** brachials. But holding them fixed are other plates, interradial in position, arranged in bifurcating series: these are called **interbrachials.** The largest of them (Fig. 90*b*, *A*) disturbs the five-rayed symmetry by pushing itself between the radials and meeting two of the basals: it is called the **anal,** not because it carries the anus, but because it forms the base of the interray, high up in which the anus lies.

The ventral surface, instead of being flat and of few plates, as in *Cupressocrinites*, is a lofty dome of many plates, solidly articulated together. Crinoids with this feature are confined to the Palaeozoic, and are known as **Camerata.** This arrangement closes in the mouth completely and shuts it off from all possibility of the excrement from the anus fouling the food. As an additional precaution, the anus is lifted up from the level of the tegmen by a short **anal tube.**

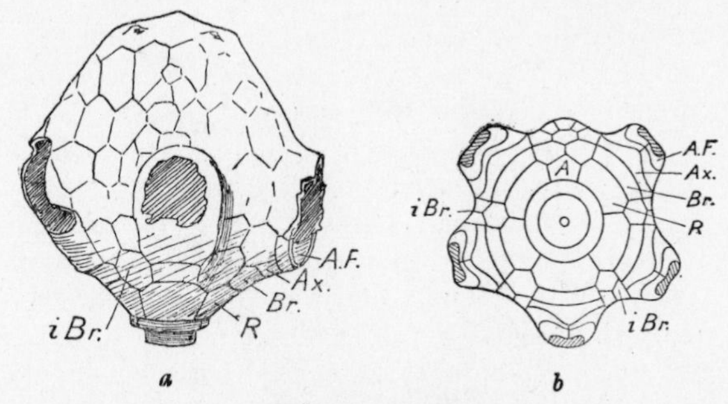

FIG. 90.—AMPHORACRINUS GILBERTSONI (Phillips), Viséan (Carboniferous Limestone), Yorkshire

a, Side view; *b*, base; *A*, Anal plate intercalated between radials; *A.F.*, facet to which arm is articulated; *Ax.*, axillary plate (primaxil); *Br.*, brachial plate primibrach); *i Br.*, interbrachial plate (interprimibrach); *R*, radial plate (Natural size; original).

The arms branch into two immediately they become free from the dorsal cup, and again branch repeatedly.

3. **Marsupites testudinarius** (Fig. 91) is a crinoid characteristic of a definite zone high in the Upper Chalk. It is one of the few crinoids that were not fixed in the adult stage. By analogy with the modern free swimming crinoid *Antedon* we may infer that it was stalked in a larval stage, but of this there is no actual evidence. The top columnal, however, is apparently retained as a large pentagonal plate in the centre of the dorsal cup. Around and above this are five large pentagonal infrabasals, then five large hexagonal basals, and five large pentagonal radials, with facets for the arms. So far as the dorsal cup goes there is perfect five-rayed symmetry, and the dicyclic condition is shown very simply and clearly.

The arms branch repeatedly. The ventral surface is never preserved, but we may assume that, as in other post-Palaeozoic crinoids, it consisted of many small plates forming a flexible surface, on which lay open food-grooves, converging to an exposed mouth. An inter-radial anus would be the only disturbance of symmetry.

Isolated plates of *Marsupites* can easily be recognized by their curious patterns of ridges arranged in sets at right angles to the sutures. An almost identical pattern is found, however, in the Silurian *Crotalocrinites*.

In both monocyclic and dicyclic crinoids certain definite grades of structural complexity are found, which are taken as a basis for classification: (1) Inadunata, in which there is no incorporation of fixed brachials in the dorsal cup; (2) Camerata, with the tegmen in the

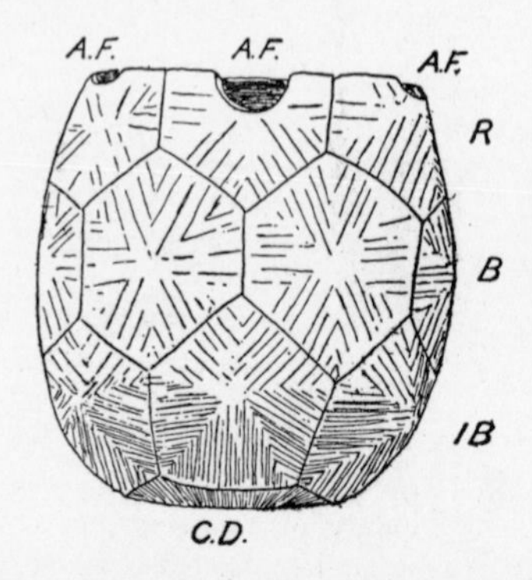

FIG. 91.—MARSUPITES TESTUDINARIUS (Schlotheim), Senonian (Upper Chalk) Cup without arms; *A.F.*, arm-facets; *B*, basal plates; *C.D.*, centro-dorsal plate; *IB*, infrabasals; *R* radials (Natural size; original).

form of a rigid vault covering the mouth and food-grooves (Palaeozoic only); (3) Flexibilia or Articulata, in which the tegmen is composed of small plates loosely articulated, and the mouth and food-grooves exposed (almost exclusively Mesozoic and later).

Among the simplest Inadunata (Larviformia) is the dicyclic *Cupressocrinites*, already described. *Pisocrinus* (Sil., Fig. 93*a*) is a monocyclic of equally simple structure, but much smaller sized cup, though the very slender arms attain as great a length as those of *Cupressocrinites*; the cup-plates have their symmetry disturbed by the subdivision of the right posterior radial obliquely into two plates (*r.p.* and *R'*)—a common feature in many inadunates.

Platycrinites (Carb., Fig. 93*b*) is a monocyclic crinoid, transitional from Inadunata to Camerata. The basals are reduced to three by fusion; the symmetry is almost perfect, there being only a slight difference between the anal interray and the others; the arms branch repeatedly, and have a large crescentic articulation on the radials. The columnals have elliptical surfaces, but the long axis of the ellipse on the lower surface is at an angle to that on the upper surface. Thus the axis of bending of successive joints gradually changes, enabling the stem as a whole to move in any direction in spite of the shape of the columnals. This arrangement gives the stem a curious

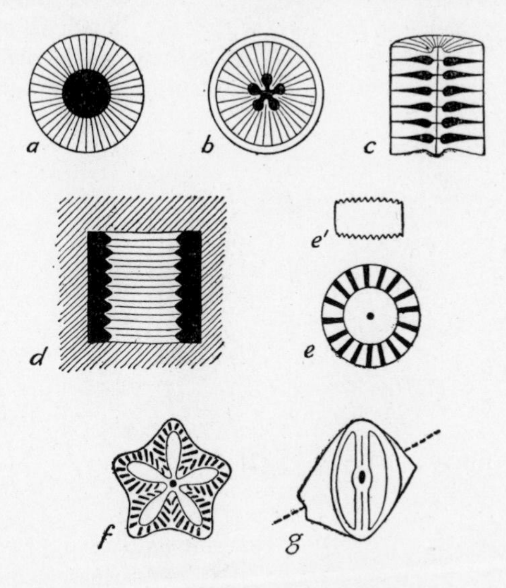

FIG. 92.—CRINOID STEM COLUMNALS

a, b, c, f, g, views of articular surfaces; *c,* vertical section of part of stem; *d,* internal mould in place in rock-matrix; *e'*, side view. The thick black in *a–d* represents cavities (All about natural size; after Goldfuss). *a, Actinocrinus granulatus* (Goldfuss), Devonian; *b, Cyathocrinites rugosus* (Miller), Devonian; *c, d, Cyathocrinites pinnatus* (Goldfuss), Devonian; *e, e' Encrinus moniliformis* (Miller), Triassic; *f, Isocrinus basaltiformis* (Miller), Jurassic; *g, Bourgueticrinus ellipticus* (Miller), Upper Cretaceous (Upper Chalk). Dotted line indicates long axis of articular surface below.

twisted appearance. Good specimens are found in the Carboniferous Limestone of England and Ireland.

Marsupiocrinus (Sil.–L. Dev.) differs in the very rapid branching of each arm into four, the parts below the separation being united by single interbrachial plates, so that many more brachials are incorporated in the cup.

Crotalocrinites (Sil., Fig. 93*c*) is a dicyclic inadunate in which the arms branch with great suddenness, and all the branches of one ray

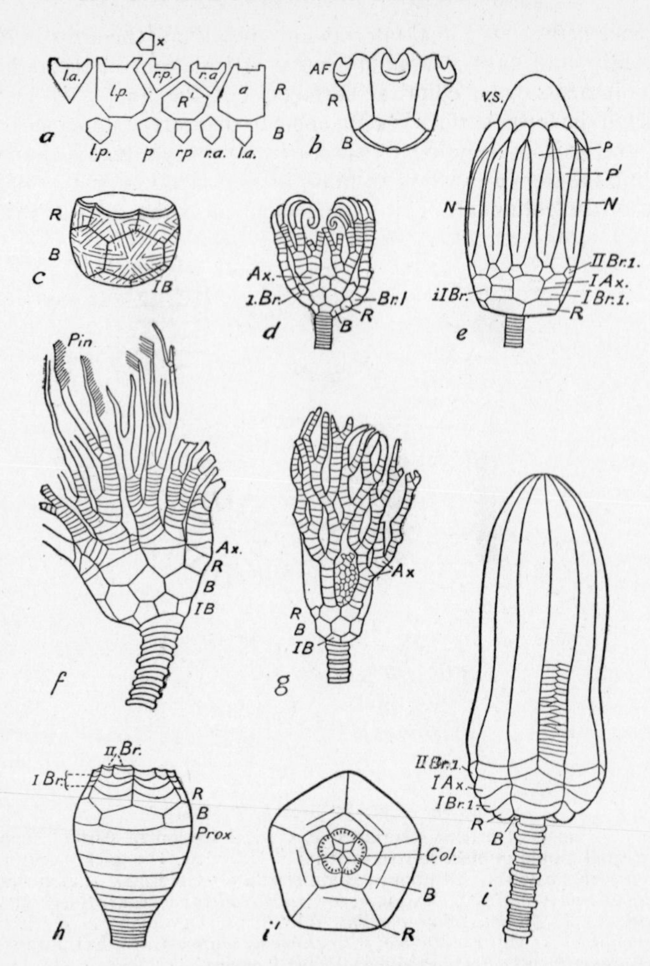

FIG. 93.—VARIOUS CRINOIDS

IB, Infrabasals; *B*, basals; *R*, radials; *Ax*, axillary. *a*, *Pisocrinus*, dissected. *a* Anterior; *p*, posterior; *l*, left; *r*, right; *R'*, radianal; *x*, anal. *b*, *Platycrinites laevis* Miller, Lower Carboniferous; cup only; *A.F.*, Arm-facet. *c*, *Crotalocrinites costatus* (Austin and Austin), Silurian (Wenlock Limestone); cup only. *d*, *Taxocrinus tuberculatus* (Miller), Silurian (Wenlock Limestone); *Br.1*, first brachial; *i Br.*, interbrachial. *e*, *Eucalyptocrinites decorus* (Phillips), Silurian (Wenlock Limestone); *I Br.1*, first primibrach; *I Ax.*, primaxil; *II Br.1*, first secundibrach; *i I Br.*, inter-primibrach; *N*, niche; *V.S.*, visceral sac. *f*, *Woodocrinus expansus* de Koninck (Carboniferous Limestone), Yorkshire; *Pin.*, pinnules. *g*, *Cyathocrinites acinotubus* (Angelin), Silurian (Wenlock Limestone). *h*, *Apiocrinus parkinsoni* (Schlotheim), Bathonian (Bradford Clay); cup and part of stem, and of arms; *Prox.*, proximal columnal; *I Br.*, primibrachs; *II Br.*, secundibrachs. *i*, *Encrinus moniliformis*

are united into a sort of membrane, which was spread out horizontally when the crinoid was feeding at ease and rolled up spirally on an alarm. The stem is as broad as the crown, is composed of very wide and thin columnals, and has a very rough surface.

Cyathocrinites (Sil.–Perm., Fig. 93*g*) and *Gissocrinus* (Sil.–Dev.) are dicyclic inadunates with very simple cup, one anal only intercalated, and much-branched arms. In the former genus the arms branch in a rather lax and irregular manner, and the anal tube is often very long; in the latter the arm-branches are closely packed together, the anal tube is compressed and its plates are wider than high.

Woodocrinus (Fig. 93*f*), of which beautiful examples are found in the Namurian part of the Carboniferous Limestone of the North of England, is a dicyclic inadunate with four-branched short arms and a relatively short stem tapering to a point, without roots.

Among Camerata, *Amphoracrinus* has been described. *Actinocrinus* is similar, but with the superficial area of the tegmen not larger than that of the cup, and a longer and central anal tube. A very remarkable genus is *Eucalyptocrinites* (Sil.–M. Dev., Fig. 93*e*), with the base deeply concave, the visible part of the cup being mainly composed of brachials and interbrachials: the arms bifurcate within the cup, and from the ten sets of interbrachials there grow up vertical pillars which arch out and meet at their upper ends, forming ten niches in which the arms lie when not spread out.

The few Palaeozoic Flexibilia are dicyclic and impinnate (i.e., arms have no pinnules): *Taxocrinus* (Fig. 93*d*) is the commonest. The best-known Triassic crinoid is *Encrinus* (Figs. 92*e*; 93*i*), which seems to be related to Carboniferous inadunates such as *Woodocrinus*. It is dicyclic, but the infrabasals are minute and difficult to find.

Isocrinus (Trias.–Rec., Fig. 92*f*) and *Pentacrinus* (Jur.) not only have minute infrabasals, but the basals are very small, and the radials intervene between them and come into contact with the stem, so that the whole cup, lying between a large stem and large, much-branched and pinnulate arms, is reduced to insignificance. The stem is five-sided, and the columnals show a very striking pattern. *Apiocrinus* (Jur., Fig. 93*h*) has a cylindrical stem, which expands at its upper end, so that its outline passes imperceptibly into that of the crown; the infrabasals are not recognizable (pseudo-monocyclic or cryptodicyclic), the arms branch into two very quickly, and their lower part

---

FIG. 93.—VARIOUS CRINOIDS (*continued*).

(Miller), Ladinian (Muschelkalk); brachials shown on lower part of one arm only; *i′*, base; *Col.*, proximal columnals; *II Br.1*, first secundibrach (All × ½, except *a*) (*a*, *g*, after Bather; *b*, after de Koninck; *c*, *d*, *e*, after Murchison; *f*, after Roberts; *h*, *i*, after Goldfuss).

is incorporated in the cup. Beautiful specimens have been found, and fragments are common, in the Bradford Clay of Bradford-on-Avon. In *Bourgueticrinus* (Cret., Fig. 92*g*) most of the columnals have the same peculiarity as in *Platycrinites* (see p. 191), but are longer and more barrel-shaped; towards the crown they are round and the proximal one is as wide as the cup, which shows five basals and five radials: the arms are rarely preserved. *Antedon* or *Comatula* (Jur.–Rec.) is fixed by a stem in its earliest youth, but afterwards becomes free-swimming. Another free-swimming form is *Saccocoma*, in which the radials form nearly the whole cup; the arms are bifurcated and the ten branches often spirally coiled, the brachials bearing wing-like expansions. This is abundant in the Upper Jurassic lithographic stone of Solnhofen, Bavaria, and plates, of which the calcite has been

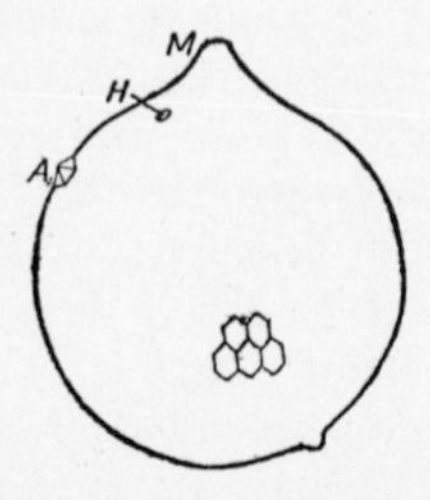

FIG. 94.—ECHINOSPHAERITES AURANTIUM Gyllenhaal, Ordovician, Sweden
Only a few of the plates are indicated. *A*, anal pyramid; *H*, hydropore; *M*, mouth (Natural size; original).

replaced by pyrites, but keeping the coarse meshwork of stereom, are found at the corresponding horizon in the English Kimmeridge Clay.

The **Cystidea** are a somewhat heterogeneous group, consisting of the most primitive forms, in which gradual development of five-rayed symmetry can to some extent be traced.

Echinosphaerites aurantium (Fig. 94) is very abundant in the Cystid Limestone of the Ordovician of Sweden, of which glacial erratics are common on the plains of North Germany. It is an almost spherical body, a slight projection at one point forming a rudimentary stem, while almost opposite it is the slightly-raised rim of a circular opening, presumably the mouth. About one-third the distance from the mouth to the stem is a third projection—a low pyramid of five triangular plates, the **anal pyramid.** Between this and the mouth, but nearer the latter, is a very small round opening, interpreted as a **hydropore,** that is an opening by which water is taken into the system of water-vessels.

The spherical theca is made up of 800 or more polygonal (mostly hexagonal) **thecal plates,** which when weathered show a pattern somewhat resembling that on the infrabasals of *Marsupites*, though due, not to external ornament, but to the internal structure of the plates. The stereom of each plate is thrown into folds perpendicular to the adjacent suture, so that the whole surface appears divided up into a series of pore-rhombs, each belonging half to one plate and half to

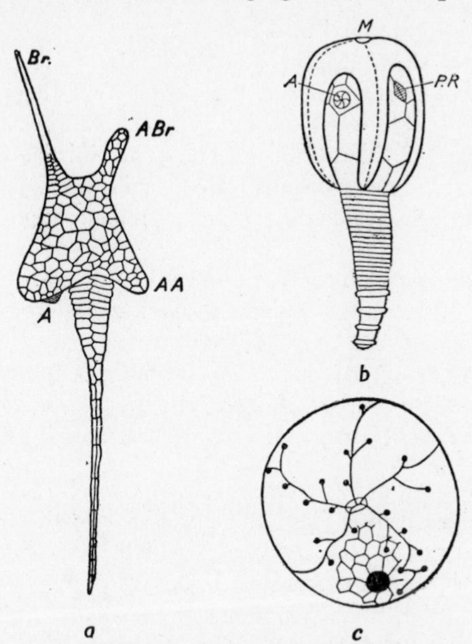

FIG. 95.—VARIOUS EARLY PELMATOZOA

a, *Dendrocystites scotica* (Bather), Ordovician, Girvan; details of plates omitted from brachiole (× ½; after Bather); *b*, *Lepadocrinus quadrifasciatus* (Pearce), Silurian (Wenlock Limestone); slightly enlarged; details of plates on arms omitted (After Forbes); *c*, *Glyptosphaerites leuchtenbergi* (Volborth), Ordovician, Pulkowa (Russia); mouth in centre, with five radiating grooves, irregularly branched, each branch ending in articulation of a brachiole (represented by black spots: where plate-sutures are shown, each of these is seen to be in the centre of a plate); anus below, black (plates of pyramid missing); thecal plates only shown around anus (× ½; after Jaekel). *A*, Anal pyramid; *AA*, anti-anal process; *ABr*, anti-brachial process; *Br*, brachiole; *M*, mouth; *P.R.*, pectinirhomb.

another, the folds in each rhomb forming a parallel series. The Cystids showing this structure form an order Rhombifera.

In specimens much better preserved than usual small areas have been found arising from the mouth, composed of grooved brachial and smaller ambulacral plates. These are usually three in number, but may be two or four. Thus there is nothing in *Echinosphaerites* showing pentamerous symmetry, unless it be the anal pyramid. Certain other cystids, however, give a hint of how the pentamerous condition arose: they have also three food-grooves, but while the anterior one (opposite the anus) remains unbranched, the other two branches soon divide into two, giving five grooves with an imperfect radial symmetry (Fig. 95*c*).

The **Cystidea** are divided into two orders:

1. **Rhombifera,** showing the rhomb-folding of stereom described above. *Echinosphaerites* is about the simplest of these. Others like *Lepadocrinus* (Sil., Fig. 95*b*) show a remarkable specialization of a few (usually three) of the rhombs into large **pectinirhombs** with all the appearance of respiratory organs (gills), while the food-grooves are on the crests of ridges which might be described as arms sessile on the theca.

2. **Diploporita,** in which the stereom cavities are normal to the surface and grouped in pairs (**diplopores**), giving rise to an appearance which must by no means be confused with the very different "pore-pairs" of Echinoidea. There is a fair pentamerous symmetry, and the food-grooves radiate on the surface of the theca. *Sphaeronites* (Ord.) and *Glyptosphaerites* (Ord., Fig. 95*c*) are spheroidal contemporaries of *Echinosphaerites.*

The **Eocrinoidea** have a stem and a sac-shaped theca with several circlets of thecal plates, the top one of which gives rise to arms that bifurcate just above the theca. The structure of the arms resembles that of Cystids rather than Crinoids. The plates are imperforate as in Crinoids. *Macrocystella* (Tremadoc) has 4 circlets of thecal plates and there are thought to be five plates in each circlet.

The **Carpoidea** possess imperforate thecal plates and a tail-like appendage or stem. *Dendrocystites* (Ord., Fig. 95*a*) has one long arm (*Br*), with the mouth at its base, the anus (*A*) at the opposite end near the stem, both on the same side, whilst on the other side are outgrowths (*ABr* and *AA*) which presumably served to balance them. There are more specialized forms like *Placocystites* (Sil.) in which one side of the theca has large plates symmetrically placed along a longitudinal axis; this side, a little more convex than the other, may have lain on the sea-bottom. The other side has many small irregularly arranged plates and an anal vent. The mouth is situated on the edge opposite to that having the stem. Some workers have suggested a resemblance to primitive chordates, others to the group Machaeridia formerly classified with the Cirripedes (p. 146).

The **Edrioasteroidea** have a disc or sac-like flexible theca composed of many plates. The mouth is central on the uppermost surface and five Starfish-like straight or curved ambulacral areas extend towards the periphery. *Edrioaster* (Ord.) and *Lepidodiscus* (Carb.) are genera known in Britain. These are rare fossils but where one example is found, there are usually several as in the case of Starfish,

The **Blastoidea** may be illustrated by *Pentremites pyriformis* (Fig. 96), of the Lower Carboniferous of North America. The shape suggests a flower-bud just opening. There is no true stem, but the base is drawn out into a stem-like stump. It would consist of five lozenge-shaped basals, but two pairs are fused, so that only three basals are counted. The five radials are much longer and would be elongate-hexagonal but for a very deep notch which extends half-way down. Five small orals or deltoids alternate with the radials

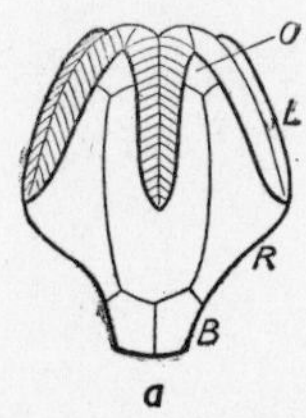

a

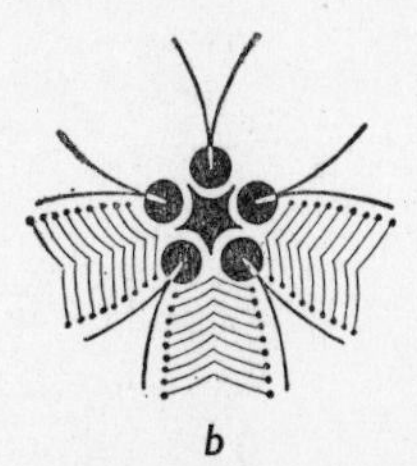

b

FIG. 96.—PENTREMITES PYRIFORMIS Say, Mississippian (Chester Group), Alabama, U.S.A.

a, Side view (Natural size). *b*, Oral region (× 2; original); *B*, Basals; *R*, radials; *L*, lancet-plates; *O*, orals or deltoids. In *b*, the mouth is in the centre, the five spiracles around it, and beyond and alternating with these the five ambulacra.

above. Between these, and in the notches of the radials, lie five **lancet-plates,** each with a central food-groove into which run numerous oblique grooves from right and left alternately. Between the margins of the lancet-plates and the radials are **side-plates,** which in exceptionally preserved specimens bear each a pinnule. From their position in relation to other structures it is obvious that a lancet-plate with its side-plates (constituting an **ambulacrum**) answers to the arm of a crinoid—an arm which is sessile upon and forms part of the calyx. The side-plates bear pores, which open into internal gill-like organs, the **hydrospires.** The mouth is in the centre of the upper surface: around it are five openings, the **spiracles,** interradial in position, divided into two internally: these also communicate with the hydrospires. Each spiracle is roofed by the halves of two lancet-plates and their side-plates. The anus is probably confluent with one of the spiracles.

The **Eleutherozoa** are always freely-moving benthic animals, sometimes carnivorous, sometimes mud- or sand-eaters, but very rarely, if ever, microphagous as are all Pelmatozoa (even the few nektic forms).

Hemicidaris intermedia (Fig. 97) is a sea-urchin of which many very perfect specimens have been found in the Coral Rag of Calne in Wiltshire. The general characters and symmetry of the skeleton suggest at once an affinity to crinoids, but there are no stalk and no arms. The skeleton (or **test**) has somewhat the appearance of a globe with poles flattened, but one (the **oral** pole) so much more so than the other (**aboral**) that the flattened surface is but little way below the equator

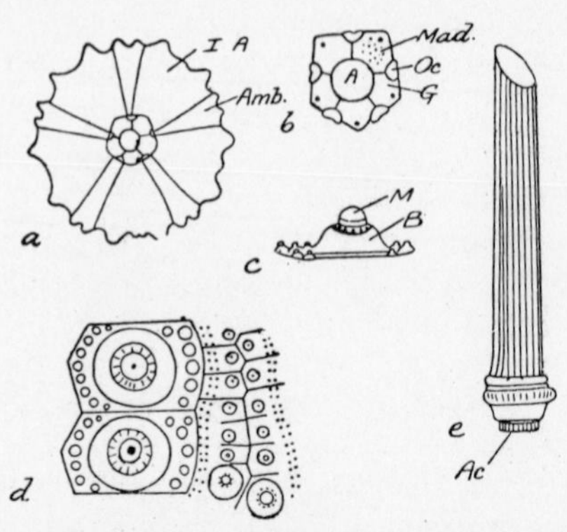

FIG. 97.—HEMICIDARIS INTERMEDIA (Fleming), Oxfordian (Coral Rag), Calne (Wilts)

a, Aboral view (× ½); all details of ambs and interambs omitted; *b*, Apical disc (Natural size). *c*, An interamb plate in profile (× 2); *d*, Two interamb plates with adjacent part of amb (× 2); *e*, Large radiole, truncated by calcite cleavage (× 6/5; after Wright). *A*, Periproct; *Ac.*, acetabulum (articular socket for mamelon); *Amb.*, ambulacral area; *B*, boss; *G*, genital plate; *IA*, interambulacral area; *M*, mamelon; *Mad.*, madreporite; *Oc.*, ocular plate.

(or **ambitus**). A full-grown specimen is about 36 mm. in diameter at the ambitus and 25 mm. high. Sometimes the fossil is found with the movable spines (**radioles**) characteristic of sea-urchins still in their natural position; but as a rule they fell off before burial, and the test only shows the tubercles on which they were articulated. Some of these are large and prominent, and carried thick cylindrical radioles, up to 95 mm. in length. These large (or **primary**) tubercles are arranged in five double rows, between which are narrower areas each with a double row of much smaller tubercles. The difference between these alternate areas is most striking near the aboral pole, and becomes much less so near the oral pole by the increase in width of the narrower areas and in the size of their tubercles. But there is a more important distinction that persists throughout: the narrower areas carry a series of fine pores, always arranged in pairs, and in each area there are two vertical columns of these **pore-pairs.** These pores, in life, transmitted finger-like muscular tubes kept tense by the pressure of water in them from the tubular water-vessels: these are the tube-feet (**podia**), the organs of locomotion of sea-urchins. In life they extend out beyond the long radioles, and their sucker-like ends can adhere firmly to any foreign body.

The five areas containing pores are called the **ambulacral areas,** or, briefly, the **ambs**; the alternate areas are the **interambulacral** areas or interambs. The ambs are perradial, and the interambs interradial. Together they form the **corona,** which is the greater part, but not the whole, of the test (the other parts being the apical system and the peristome).

It is not always easy to see the sutures of the plates which build up the ten areas, especially on the external surface. On the internal surface they are plainer. Each area consists of two vertical columns of plates, the sutures down the middle of an area being zigzag, while those between adjacent areas are nearly straight. In the interambs there is a plate to each large tubercle. In the ambs there is, near the aboral pole, one pore-pair to each plate, and each plate extends across half the width of the amb. Such plates are called **simple primaries.** Lower down **compound** plates appear, each with several pore-pairs. The development of living forms shows that such plates arise by the fusion of simple plates, and sometimes the fused sutures can be traced in a fossil. When this can be done in *Hemicidaris,* it is found that three plates fuse together: the middle one carries most of the large tubercle and may be the only one that occupies the full width of its column (**primary plate**), the other two only covering about two-thirds of that width (**demi-plates**).

The ambs and interambs end orally around a large space, the outline of which has ten notches, which mark the position of external gills. This space in fossils is colloquially termed the mouth, but this is not strictly correct, the real mouth being smaller, and surrounded by a membrane (**peristome**), the plates in which are loose and on its decay fall away, leaving a large space. The mouth was furnished in life with a complex series of biting organs (jaws and teeth) forming the **lantern** (so called from a resemblance noted by Aristotle more than 2,000 years ago). This is rarely preserved, but its existence is shown by the presence of the **perignathic girdle,** a series of internal arches arising from the lowest plates of the corona, and serving as origin for the muscles inserted on the lantern.

Towards the aboral pole the corona is surmounted by a flat series of ten plates called the **apical disc** (or better, **apical system**), in the centre of which is a space, colloquially called the anus, but more correctly the **periproct** (an analogous case to that of the peristome). Of the ten plates, the five smaller are at the ends of the ambs and are called the **ocular** plates; the five larger interradial plates are termed **genital,** because they bear the apertures of the genital ducts. One of the genital plates, in addition to the genital pore, is perforated by a great number of smaller pores, through which water filters into the water-vessels: this plate, called the **madreporite,** constitutes the only imperfection of five-rayed symmetry in the test of *Hemicidaris*. The

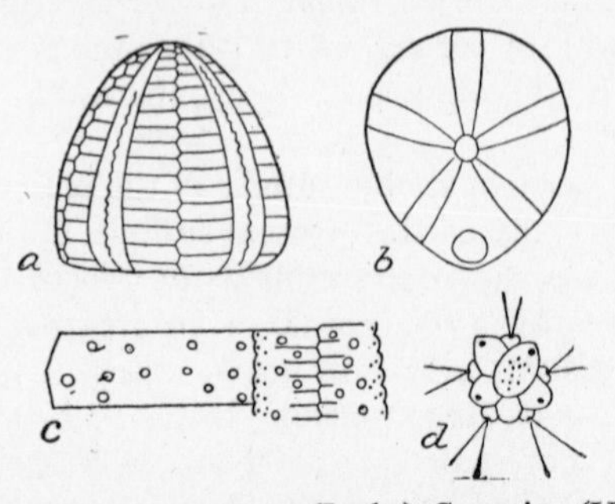

FIG. 98.—CONULUS ALBOGALERUS (Leske), Senonian (Upper Chalk)
a, Side view (× ½); *b*, Oral view (× ½); *c*, One interamb plate and adjacent amb plates (× 2); *d*, Apical disc (× 2; after Wright). Sutures of amb plates omitted in *a*; of all plates in *b*; madreporite dotted in *d*.

ocular plates have been termed radials, and the genital, basals; but they are certainly not strictly homologous with the plates so named in crinoids, which do not encircle the anus.

The radioles of *Hemicidaris* articulate with the tubercles by a ball-and-socket joint. They are moved by muscles in the skin, which lies outside the test. The tubercles, like the radioles which they carried in life, are distinguished according to size into **primary, secondary,** and **miliary.** The primary tubercles consist of a boss and a mamelon (Fig. 97*c*): in *Hemicidaris* the inner part of the boss shows a ring of radial **crenulations,** and the centre of the mamelon has a perforation; so the tubercles are described as **crenulate** and **perforate.**

Conulus albogalerus (Fig. 98) is a common sea-urchin of the Upper Chalk. The first obvious differences from *Hemicidaris* are its more conical shape and much smoother surface. Next we notice that the symmetry is not strictly radial, but is bilateral, the base being oval: near its narrow end is a large opening, the periproct. The peristome, however, is still in the centre of the oral surface, and the very small apical disc is correspondingly central at the aboral apex.

The smoothness of the surface enables the sutures of the plates to be traced easily. The interambs are broad and composed of large, nearly oblong plates, with from one to twelve or more larger flat tubercles and numerous, much smaller tubercles (granules). The ambs are narrow; their plates are much smaller and more irregular, including simple primaries, simple demi-plates, and some compound plates; each has one or sometimes two larger flat tubercles, and a number of granules.

The peristome is small; the lantern and girdle are in a vestigial

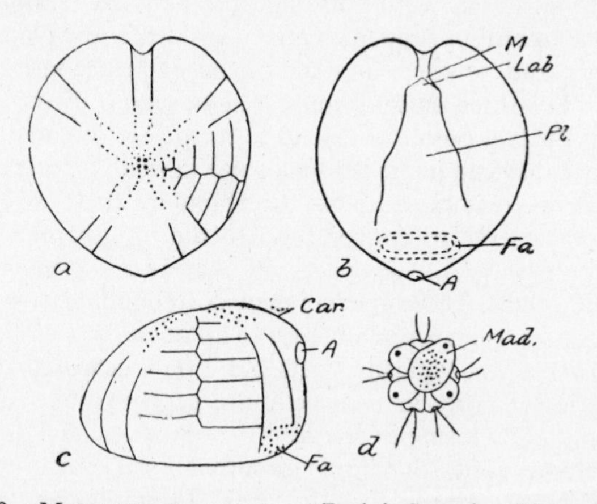

FIG. 99.—MICRASTER CORANGUINUM (Leske), Senonian (Upper Chalk)
a, Aboral view (× ½); *b*, Oral view (× ½); *c*, side view (× ½); *d*, Apical disc (× 4; after Wright). Some of the interamb sutures are indicated in *a* and *c*. *A*, anus; *Car*, carina; *Fa*, subanal fasciole; *Lab*, labrum; *M*, mouth; *Mad*, madreporite; *Pl*, plastron.

condition. The apical disc is not very different from that of *Hemicidaris*, except that the madreporite is larger than the other genital plates and extends into the space cavated by the periproct.

Micraster coranguinum (Fig. 99) is another sea-urchin from the Upper Chalk. Viewed from above or below its outline is heart-shaped, the anterior amb being deeply depressed so as to notch the outline. The other ambs are also sunk in a similar way for some distance from the apical disc, but not far enough to affect the outline. The posterior interamb is raised into an obtuse ridge, the keel or **carina,** which runs almost horizontally back from the apical disc to the hind margin which drops perpendicularly. Sometimes the carina projects slightly above this vertical end, forming a **rostrum.** The periproct lies near the top of this vertical face.

The mouth lies far forward (four-fifths of the whole length) and faces forwards instead of downwards, as a distinct lower lip (**labrum**) projects on its underside: this is the forward termination of the somewhat protuberant posterior interamb, known as the **plastron,** the plates of which are very large (cf. Fig. 106*c*).

In the apical disc the two posterior ocular plates are in contact, the genital plates between them being crowded out; the madreporite is large and extends into the centre of the disc.

The ambs are composed, near the apical disc, of close-set plates, very short and broad, with uniserial pore-pairs. At about half-way towards the ambitus a sudden change takes place: the plates become much larger and squarer, and their pore-pairs become much less noticeable. The three anterior ambs can be traced directly down to the mouth; the two posterior extend to the posterior end of the plastron, whence they can be traced with some difficulty along either side of the plastron. It is evident that the abundant podia of the dorsal surface cannot serve to convey food to the mouth, since they are separated from it by a wide space with very scanty podia; probably they served as gills. These aboral regions of abundant pore-pairs are called the petals, from a resemblance to an open five-petalled flower; but *Micraster* is not typically "petaloid," only **subpetaloid,** because the two series of pores in each petal are nearly parallel, instead of widening out and closing in again.

The interamb plates, having to occupy all the space between the ambs, are for the most part very large. Their regular double-row becomes rather confused on the underside. The posterior interamb on the underside constitutes the plastron.

All plates except those of the petals bear tubercles and granules. On the upper surface the tubercles are small, not very prominent, and scattered, with abundant small granules between. On the under surface they are larger, with more deeply excavated areola, and are much more closely set; the granules, though fewer in number, being also coarser and more close-set. Running round the lower corner of the vertical posterior end there is a narrow ring which at first sight appears smooth, but under magnification is seen to be covered with extremely fine granules. This is a **fasciole,** and from its position is called a subanal fasciole. In life each granule carries a minute radiole, covered with a ciliated epithelium, the agitation of which creates a current of water with a flushing effect.

The Micrasters of the White Chalk, of which *M. coranguinum* is the latest, are famous from the work of Dr. Rowe, who showed in 1899, from the study of large numbers of specimens collected and arranged with care according to their exact zones, that they went through a gradual series of changes which did not bear a close relation to the accepted distinctions of species. These changes (which may be adaptations to gradual increase in depth of the Chalk Sea) run on parallel lines in separate lineages, so that they may serve as zonal indices. The chief changes may be summarized thus:

(1) In general form they tend with the lapse of time to become slightly broader and very distinctly higher, while the ambital outline tends to change from wedge-shape to ovate.

(2) The carina appears and becomes more marked, the highest point which was at first at the apical disc shifting backwards; finally a rostrum appears.

(3) The anterior groove becomes deeper.

(4) The mouth shifts forward and the labrum makes its appearance; at first smooth at the tip, it becomes more and more granulated; the labral plate changes from triangular to oblong, and its tubercles from irregular to linear in arrangement.

(5) The amb-petals become longer and shallower.

(6) The subanal fasciole, from being weak, becomes very strong, and loses a slight tendency towards an **8**-shape which it had at first.

(7) The periplastronal area (the two posterior ambs on the under surface) from smooth passes to granulated and finally to very coarsely granulated (mammillated).

(8) Perhaps the most easily observed change of all is that in the **interporiferous areas** (i.e., the middle strip of the petals, between the pore-pairs). These pass through a series of stages (a) "smooth"; (b) "sutured," i.e., the sutures of the plates are clearly marked; (c) "inflated," i.e., instead of being gently concave, they rise in a double convexity; (d) "subdivided," i.e., the middle line between these convexities becomes a distinct groove; and (e) "divided," i.e., this groove becomes very deep and narrow.

The main division in the classification of Echinoidea is between those with radial and bilateral symmetry, or Regularia and Irregularia. The great differences between these two subclasses may be tabulated thus:

REGULARIA	IRREGULARIA
Radial symmetry, perfect except for madreporite.	Bilateral symmetry, perfect except for apical system.

REGULARIA	IRREGULARIA
Anus within apical system, which is symmetrical and large.	Anus in posterior interamb. Apical system usually asymmetric; small and compact, or elongated.
Mouth central, with well-developed jaws.	Mouth either central (when jaws may be well developed or reduced), or in anterior amb (when jaws are lost).
Corona usually bearing large, prominent tubercles (as well as smaller ones) with long stout radioles.	Corona hardly ever with large, conspicuous tubercles. Radioles small.

The Palaeozoic echinoids are distinguished from all later forms (except one Triassic and one Neocomian genus) by the inconstancy in the number of coronal columns of plates, which may be more or less than twenty, but never that exact number, which is the invariable number among later forms (with the two exceptions named). This has been made the basis of a classification into "Palechinoidea" and "Euechinoidea," but this is unsatisfactory, since it unites a number of divergent Palaeozoic stocks, of which one was apparently more closely related to Mesozoic regular echinoids than the others.

CLASS: **ECHINOIDEA**

SUBCLASS: **Regularia (Endocyclica)**.

ORDER 1. BOTHRIOCIDAROIDA. With a single column of interamb plates. Only genus *Bothriocidaris* (Fig. 100*a*), Ordovician of Estonia, the only known Ordovician echinoid, except for some problematical forms lately found in the Girvan district, Scotland.

2. MELONECHINOIDA. A series in which both amb and interamb columns tend to increase in numbers; no large tubercles, only granules with small radioles. Sil.–Permian. *Palaeechinus* (Sil.–Carb., Fig. 100*b*), *Melonechinus* (Carb., Fig. 100*d*).

3. CIDAROIDA. A series in which the ambs have never more than two columns, but in Palaeozoic genera and in the Neocomian *Tetracidaris* the two interambs have more than two. Even where there are only two interamb columns the interambs are many times wider than the ambs. Large tubercles and radioles on the interambs only. Amb plates simple. Devonian to Recent. *Archaeocidaris* (Carb.–Perm.), *Cidaris* (Trias.–Rec., Figs. 100*c*, *e*).

4. PLESIOCIDAROIDA.—An aberrant Triassic group, with very large

apical disc, and only three plates in the interambs. *Tiarechinus* (Carnian).

5. CENTRECHINOIDA [DIADEMOIDA].—Regular echinoids in which there is a tendency to form compound plates in the ambs, which are not much narrower than the interambs. *Peltastes* (Jur.–Cret.), *Salenia* (Cret.–Rec.), and *Acrosalenia* (Jur.–Cret.) have a large apical disc with an extra suranal plate (Fig. 100*f*). *Hemicidaris* (Jur.) has been described. In *Pseudodiadema* (Jur.–Cret.) three primaries are fused, but the pores remain uniserial except close to the peristome. In *Diplopodia* (Jur.–Cret.), with the same type of compound plate, the pores are **biserial**—i.e., they form two vertical columns within each

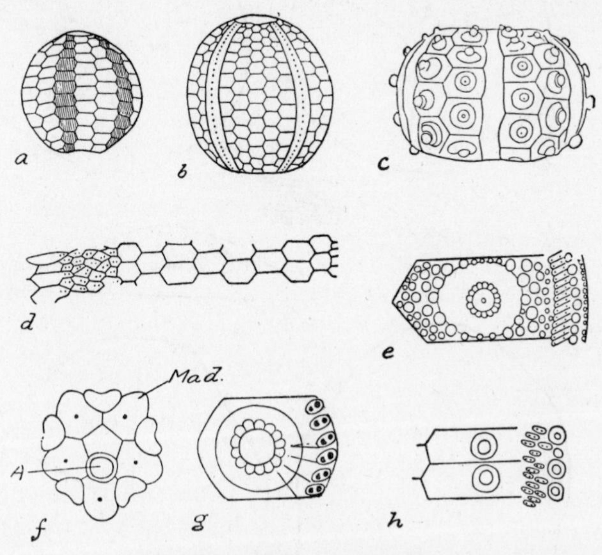

FIG. 100.—REGULAR ECHINOIDS

a, *Bothriocidaris pahleni* (Schmidt), Ordovician, Estonia (natural size); interambs shaded. *b*, *Palaeechinus elegans* (M'Coy), Lower Carboniferous (× ½); details of ambs omitted. *c*, *Cidaris florigemma* (Phillips), Oxfordian (Coral Rag) (× ½); details of ambs omitted. *d*, *Melonechinus liratus* (Jackson), Lower Carboniferous; portion of one amb and one interamb (full breadth) (× ⅔). *e*, *Cidaris smithi* (Wright), Oxfordian (Coral Rag); one interamb plate, with perforate tubercle, and the amb plates of the adjacent single column; slightly enlarged. *f*, *Peltastes wrighti* (Desor), Aptian, Faringdon (Berks); apical disc (Natural size; after Wright). *A*, periproct; *Mad.*, madreporite; between these is the pentagonal sur-anal plate. *g*. *Phymosoma koenigi* (Mantell), Senonian (Upper Chalk); compound amb plate (× 3); traces of the sutures between the original six plates are seen; tubercle imperforate. *h*, *Stomechinus germinans* (Phillips), Lower Bajocian (Pea Grit); two interamb plates and the adjacent amb plates (× 3/2); showing triserial character of ambs, and imperforate tubercles. *a*, *d*, after R. T. Jackson; *b*, after Baily; remainder after Wright.

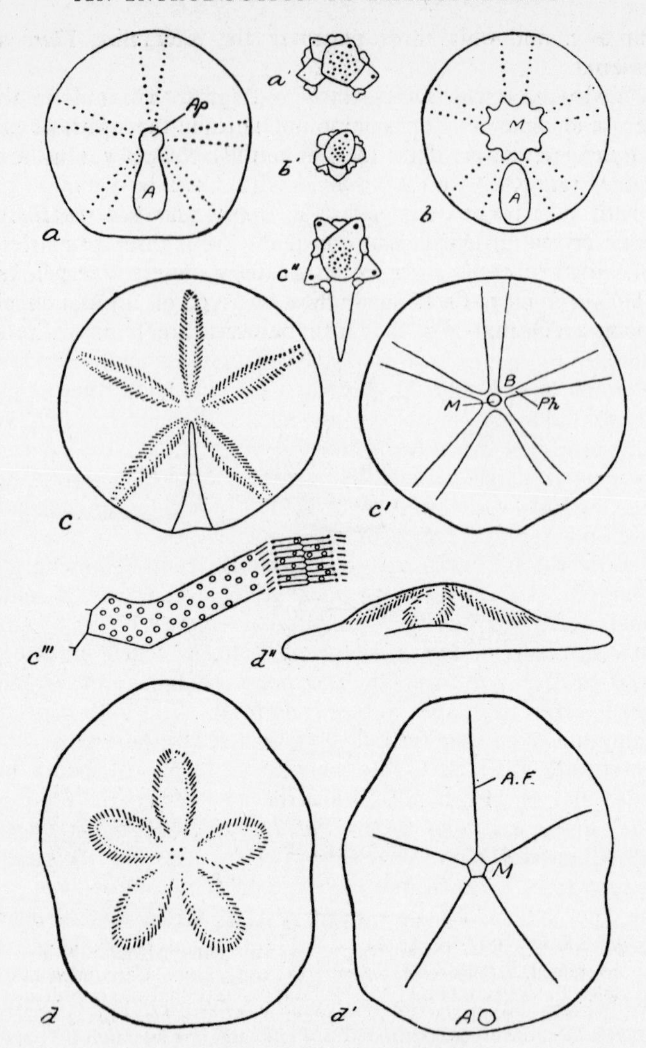

FIG. 101.—IRREGULAR ECHINOIDS

a, *Pygaster morrisi* (Wright), Bathonian; aboral view (× ½); *a′*, Apical disc of *P. umbrella* (Agassiz) (× ¾). *b*, *Holectypus depressus* (Leske), Bathonian; oral view (× ½); *b′*, Apical disc (× ¾). *c*, *Clypeus altus* (M'Coy), Vesulian; aboral view (× ½); *c′*, Oral view (× ½); *c″*, Apical disc (× ⅔); *c‴*, Interamb plate and adjacent part of amb (× 2). *d*, *Clypeaster suffarcinatus* (Duncan and Sladen) aboral view; *d′*, Oral view; *d″*, Side view (All × ½). *a*, *b*, *c*, after Wright; *d*, after Duncan and Sladen. In all, except the apical discs, the sutures between the plates are omitted; in *a* and *b*, pore-pairs are indicated by single dots. In the apical discs the madreporite is dotted; the ocular plates are dis-

column of plates; while in *Stomechinus* (Jur.–Cret., Fig. 100*h*) they are **triserial.**

In *Phymosoma* [*Cyphosoma*] (Jur.–Cret., Fig. 100*g*) six plates, three primary and three demi-, may fuse into a compound plate to carry the big tubercles, which are imperforate (the last three genera having them perforate).

Echinus (Cret.–Rec.), the common sea-urchin, has plates compounded of two primaries and one demi-; pores triserial; poriferous zone narrow; tubercles and radioles numerous and not very large.

SUBCLASS: **Irregularia (Exocyclica).**—Sea-urchins of this subclass appear first near the top of the Lower Jurassic (Lower Bajocian), and the simplest of them, *Pygaster* (Fig. 101*a*), is very little removed from some simple member of the Regular Centrechinoida, the periproct having shifted to a position only just outside the apical system. With this, however, are others with much stronger irregularity (*Hyboclypus*), the mouth having also shifted forwards from its central position and having lost its jaws and teeth. From the beginning we thus find two well-marked and divergent superorders.

SUPERORDER 1: GNATHOSTOMATA.—Mouth central, having girdle and lantern (the latter sometimes vestigial). All ambs similar. Symmetry bilateral, but with strong radial tendency.

There are three orders: (1) Holectypoida, in which the jaws are reduced in size and strength, the pores of the ambs extend in unbroken series from apex to base, and some of the amb-plates may be compound; (2) Nucleolitoida, in which the pore-pairs of the upper surface take on a "dot-and-dash" form, the podia being flattened and serving as gills; while the jaws disappear, the mouth becomes small and with a "floscelle" (see below, under *Clypeus*); and (3) Clypeastroida, in which the jaws are very powerful, and the ambs tend to be petaloid, but there are no compound plates.

The chief genera of (1) are *Pygaster* (Jur.–Cret., Fig. 101*a*), with periproct close behind apical disc; *Holectypus* (Jur.–Cret., Fig. 101*b*), in which it has shifted to the underside; *Discoidea* (Cret.), hemispherical, periproct marginal, with ten radiating ridges internally around mouth, well seen on internal moulds; and *Conulus* (Cret.), already described.

Chief genera of (2) are *Nucleolites* [*Echinobrissus*] (Jur.–Cret.), with oval outline, broader and rather truncated behind, with deep

---

FIG. 101.—IRREGULAR ECHINOIDS (*continued*).

tinguished by smaller size and absence of genital pores; in *b′*, the left anterior ocular plate and pores on four genital plates have been accidentally omitted. *A*, periproct; *A.F.*, actinal furrow; *Ap.*, apical disc; *B*, bourrelet; *M*, peristome; *Ph.*, phyllode.

anal groove on upper surface, ambs subpetaloid, the outer pore of each pair elongated; and *Clypeus* (Jur., Fig. 101*c*), flat, circular, anal groove as in the last, ambs petaloid, elongation of outer pores very great, mouth surrounded by a **floscelle**—five ambulacral grooves of petaloid appearance (**phyllodes**)—with protuberances (**bourrelets**) on the interambs.

A doubtfully allied, rather isolated genus is *Collyrites* (Jur.–Cret.,

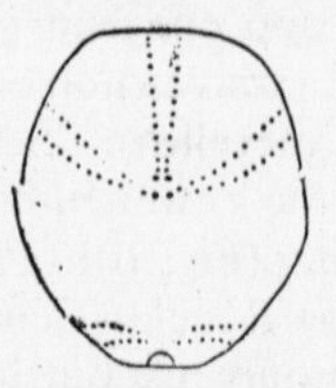

FIG. 102.—COLLYRITES RINGENS (Agassiz), Upper Bajocian
Aboral view. Periproct seen at lower end. The four black spots near centre are the genital pores of the apical disc; from here the trivium radiates, while the bivium is far behind (× ⅘; after Wright).

Fig. 102), an ovoid form in which the two posterior ambs (**bivium**) and the three anterior (**trivium**) are so far separated as to disrupt the apical system, the two posterior oculars being far behind the rest.

In (3) the tendency is for the shape to become very flat (cake-urchins); this diminishes the strength of the corona against vertical pressure, and two methods are adopted to strengthen it. The more usual is for vertical pillars or radiating partitions to be developed internally to support the roof. The other is for the margin to be

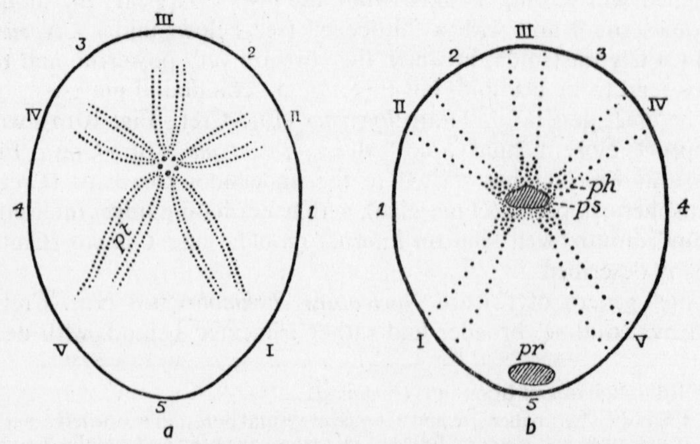

FIG. 103.—ECHINOLAMPAS PROTEUS (Fourtau), Upper Eocene, Libya
a, apical view; *b*, oral view (× ⅘; after Fourtau).

notched in the interambs, the plates in the sides of these notches serving as supports of the roof; further growth may reunite the plates at the margin, converting notches into perforations ("the lunules"). In all but the more primitive genera some of the amb plates widen out and pinch the interambs across until they meet the plates of the next amb (**discontinuous** interambs), and there are furrows (actinal furrows) radiating from the mouth along the ambs.

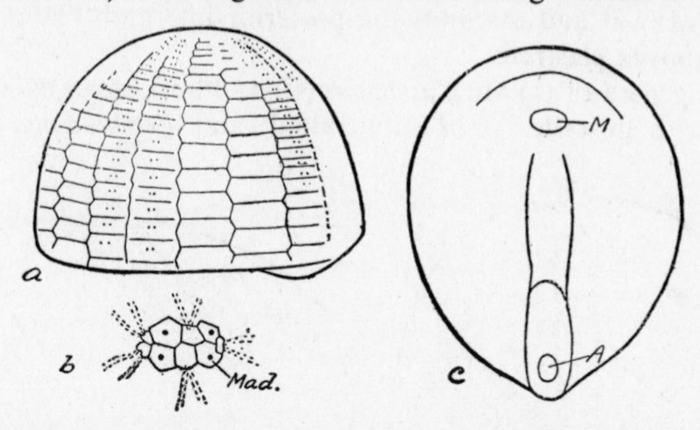

FIG. 104.—ECHINOCORYS SCUTATUS (Leske), Senonian (Upper Chalk)
a, Side view; *b*, apical disc; *c*, oral view (*a* and *c* × ½; after Wright).

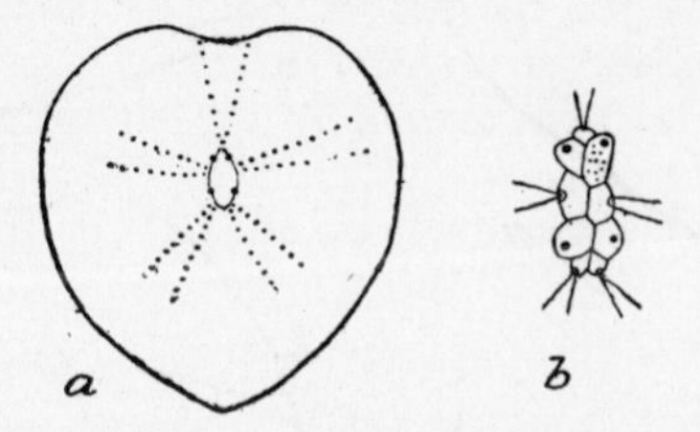

FIG. 105.—HOLASTER SUBGLOBOSUS (Leske), Cenomanian
a, Aboral view (× ½); *b*, Apical disc (× $\frac{3}{2}$; after Wright).

Chief genera: *Echinocyamus* (Cret.–Rec.) and *Fibularia* (Cret.–Rec.) are very small primitive forms with petaloid ambs, but none of the other special features of the suborder. *Scutella* (Eoc.–Rec.) is very flat, circular, with bifurcating actinal furrows; *Mellita* and *Rotula* (Plio.–Rec.) have lunules; *Clypeaster* (Olig.–Rec., Fig. 101*d*) is oval to pentagonal in plan, with straight actinal furrows; in some species advantage is taken of the strengthening pillars to raise the central part of the roof, including the respiratory petals, into a dome. Such species are the largest and most massive of all sea-urchins. All but the most primitive genera of this order are confined to warm seas.

SUPERORDER 2: ATELOSTOMATA.—Mouth shifted forward; lantern and jaws lost. Anterior amb somewhat different from the other ambs. Symmetry strongly bilateral. Tendency to separate the three anterior ambs (trivium) from the two posterior (bivium).

There are two orders—(1) Cassiduloida, in which the mouth, though slightly forward, still opens downwards; and (2) Spatangoida, in which the mouth opens forwards, with more or less of an underlip, and between it and the anus the posterior interamb plates form a strong convex **plastron.**

Chief genera of (1) are *Cassidulus* (Cret.–Plio.), small ovoid; peristome with floscelle, ambs subpetaloid; and *Echinolampas* (Eoc.–

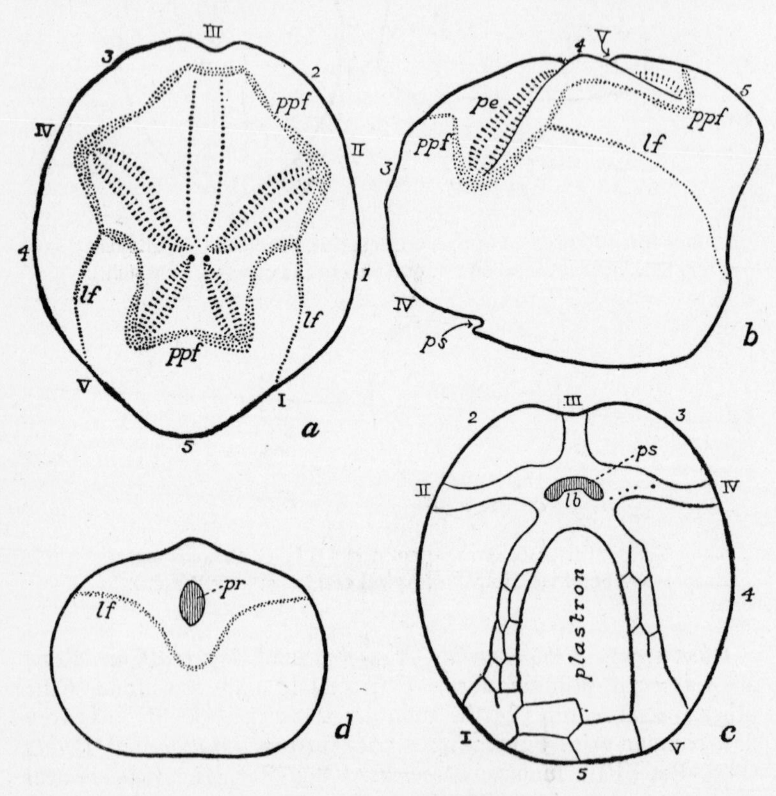

FIG. 106.—SCHIZASTER

a, apical view; *b*, side view, of *S. africanus* de Loriol, Middle Eocene; *c*, oral view; *d*, posterior view, of *S. parkinsoni* (Defrance) Middle Miocene; *lb*, labrum; *lf*, lateral fasciole; *pe*, petal; *ppf*, peripetalous fasciole; *pr*, periproct; *ps*, peristome. Roman numerals denote ambs; Arabic, interambs (× $\frac{3}{4}$; after Fourtau).

Rec., Fig. 103), ovoid, somewhat conical, ambs petaloid, anus submarginal, floscelle present.

Chief genera of (2) are *Echinocorys* [*Ananchytes*] (Cret., Fig. 104), oval, with flat base, periproct submarginal; *Holaster* (Cret.–Mio., Fig. 105), heart-shaped with anterior groove, slight separation of bivium and trivium; *Micraster* (Cret.–Mio., Fig. 99), already described. *Hemiaster* (Cret.–Rec.), and *Schizaster* (Eoc.–Rec., Fig. 106) differ from *Micraster* in not having a subanal fasciole, but have a **peripetalous** fasciole running round the petaloid ambs, which are deeply sunk, especially in *Schizaster*, which has also a **lateral** fasciole branching off from the main fasciole on each side.

The **Echinoidea** are by far the most useful stratigraphically of the Eleutherozoa. In the **Stelleroidea** these are three orders: ASTEROIDEA, ordinary starfish, in which there is no separation of arms and body, and OPHIUROIDEA (brittlestars), in which they are sharply distinguished. These two orders are known from the Ordovician onwards, and where they do occur it is usually in "starfish-beds," where representatives of one or a few species are extremely abundant, yet between such beds there may be great thicknesses of rock destitute of any trace of them. The third order SOMASTEROIDEA (Ord.–U. Dev.) differs essentially from the Asteroidea in that rod-like ossicles branch from the sides of the ambulacral plates.

The **Holothuroidea** have their skeleton reduced to microscopic spicules shaped like anchors, crosses, plates and wheels, so that special search has to be made for them. They are found, here and there, from the Cambrian to the present time.

Short Bibliography

BARRANDE, J.—*Système Silurien de Bohème*, vol. vii, Cystids and Crinoids (1887).

BATHER, F. A.—(1) "British Fossil Crinoids," *Ann. Mag. Nat. Hist.* (6), v–ix (1890–92); (2) review of Wachsmuth and Springer's Crinoid Monograph, *Geol. Mag.* (4), v–vi (1898–9).

BATHER, F. A. and GREGORY, J. W.—"Echinoderma" in E. R. Lankester's *Treatise on Zoology*, part 3 (1900).

DAVIES, A. M.—*Tertiary Faunas* (1935), London (Murby).

D'ORBIGNY, A., continued by COTTEAU, G. H.—*Paléontologie française* [*Echinoidea*] *Terrains jurassiques*, vols. ix, x; *Terrains crétacées*, vols. vi, vii; *Terrains tertiaires*, vols. i, ii (1853–94).

DUNCAN, P. M. and SLADEN, W. P.—"Fossil Echinoidea of Western Sind . . . of Kach and Kattywar," *Palaeont. Indica* (14), i (1882–6).

ETHERIDGE, R. and CARPENTER, P. H.—"Blastoidea," *Brit. Mus. Cat.* (1886).

FORBES, E.—(1) "British Silurian Cystidea," *Mem. Geol. Survey*, vol. ii, part 2 (1848); (2) "Tertiary Echinoderms," *Palaeont. Soc.* (1852).

HAWKINS, H. L.—(1) "The Morphology and Evolution of the Ambulacrum in the Echinoidea Holectypoida," *Phil. Trans. Roy. Soc.* (B), ccix, 377 (1920); (2) "The Lantern and Girdle of some Recent and Fossil Echinoidea," *Phil. Trans. Roy. Soc.* (B), ccxxiii, 617 (1934); (3) numerous papers in *Geol. Mag.* from 1912 on; (4) "Evolution and Habit among the Echinoidea," *Quart. Journ. Geol. Soc.*, xcix, p. lii (1943).

JACKSON, R. T.—"Phylogeny of the Echini," *Mem. Boston Soc. Nat. Hist.*, vii (1912).

MACBRIDE, E. W. and SPENCER, W. K.—Two new Echinoidea, *Aulechinus* and *Ectinechinus*, and an adult plated Holothurian, *Eothuria*, from the Upper Ordovician of Girvan, Scotland, *Phil. Trans. Roy. Soc.* (B), ccxxix, pp. 91–136 (1938).

MOORE, R. C. and LAUDON, L. R.—"Evolution and Classification of Paleozoic Crinoids, *Spec. Papers Geol. Soc. Amer.*, No. 46 (1943).

REGNÈLL, G.—Non-crinoid Pelmatozoa from the Palaeozoic of Sweden, *Med. Lunds Geol. Min. Inst.*, No. 108 (1945).

ROWE, A. W.—"Analysis of Genus *Micraster*," *Quart. Journ. Geol. Soc.*, lv (1899).

SLADEN, W. P. and SPENCER, W. K.—"Cretaceous Echinoderms," vol. ii: "Asteroidea and Ophiuroidea," *Palaeont. Soc.* (1891–1908).

SPENCER, W. K.—"Palaeozoic Asterozoa," *Palaeont. Soc.* (1914–34).

SPENCER, W. K.—"Early Palaeozoic Starfish," *Phil. Trans. Roy. Soc.* B), ccxxxv, 87 (1951).

UBAGHS, G.—(1) "Classe des Crinoides," in J. Piveteau, *Traité de Paléontologie*, pp. 658–773 (1953) Paris (Masson). (2) "Classe des Stelleroides," *Ibid.*, pp. 774–842.

WACHSMUTH, C. and SPRINGER, F.—"North American Crinoidea Camerata," *Mem. Mus. Comp. Zool. Harvard*, xx and xxi (1897).

WITHERS, T. H.—"Machaeridia," *Brit. Mus. Cat.* (1926).

WRIGHT, J.—"The British Carboniferous Crinoidea," Two vols., *Palaeont. Soc.* (1950–60).

WRIGHT, T.—(1) "Oolithic Echinoderms," vol. i: "Echinoidea" (1857–78); vol. ii: "Asteroidea and Ophiuroidea" (1863–80); "Cretaceous Echinoderms," vol. i: "Echinoidea" (1864–82), *Palaeont. Soc.*

IX

CORALS AND OTHER COELENTERATA

THE Coelenterata are the simplest in structure of all animals except the Sponges and Protozoa. One typical form is the hydroid polyp—a cylindrical body, fixed at one end, and having at the other end a mouth, which opens into a digestive cavity occupying the whole interior: around the mouth is a radiating series of tentacles for seizing food. A more complex type of polyp is seen in the sea-anemone, where the cavity is partially divided by radiating **mesenteries** and the mouth is turned in to form a short gullet. Both these are fixed forms, and like other fixed animals have a great tendency to branch and form colonies. There are also free-swimming (nektic) forms, of which the chief are the jelly-fish or medusae. These most delicate of living organisms would appear to be the most hopeless to look for among fossils, yet remains of such have been preserved in such fine-grained sediments as the Jurassic Lithographic Limestone of Bavaria and the Middle Cambrian Shale of Mount Stephen, British Columbia.

The hydroid polyps and medusae belong to the Class **Hydrozoa.** Some of these secrete calcareous skeletons and their fossil remains sometimes form a significant constituent of limestones: they may be spoken of as hydrocorallines—a descriptive term without exact taxonomic value, since it may include skeleton-forming polyparies (polycolonies) of more than one branch of the Hydrozoa. Among fossil hydrocorallines are probably to be counted the extinct Stromatoporoids, which form an important constituent of some of the Ordovician, Silurian, and Devonian limestones, and are also known from the Jurassic and Cretaceous. Their skeleton consists of close-set, wavy, concentric laminae traversed by abundant vertical pillars, and the masses of this structure may attain a considerable diameter and thickness.

Somewhat similar structures are found here and there in later systems—e.g., the spheroidal bodies, about 25 mm. in diameter, found in the Cambridge Greensand, called *Parkeria*. Some fossils of this kind, however, may be calcareous Algae.

The other Coelenterate Class, **Anthozoa,** includes polyps of the more complex, sea-anemone type. These may be solitary or colonial, and with or without a calcareous skeleton.

The term **Coral** is a useful term, especially in palaeontology, though without exact taxonomic value, as it is applicable to any Anthozoan with a calcareous skeleton. Four examples of fossil corals will be described.

1. **Zaphrentis** (Fig. 107) is a genus of which (using the word in its older, wide sense) there are several common species in the Carboniferous Limestone. Though the following description applies essentially to *Zaphrentoides konincki*, much of it is true of other species.

The shape is that of a drinking-horn—that is, a cone with its axis more or less curved. An ordinary specimen is about an inch (25 mm.) in height, and about 12 mm. in diameter at the broad end. The external surface is marked with concentric rings, obviously lines of

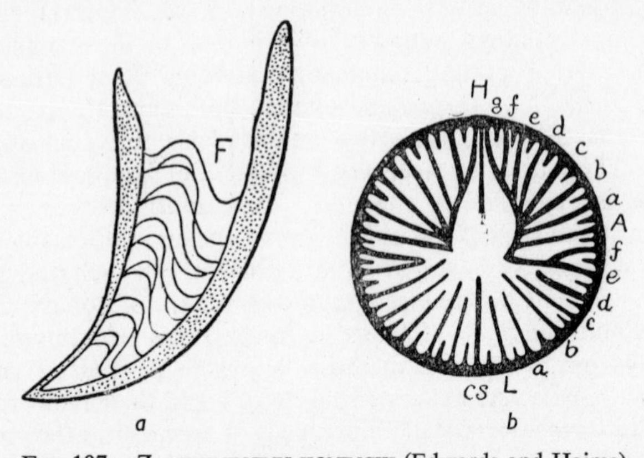

FIG. 107.—ZAPHRENTOIDES KONINCKI (Edwards and Haime)
Diagrams of vertical and transverse sections (After Carruthers, simplified). *a*, Vertical section (× 2); cardinal and counter septa dotted; *F*, fossula. *b*, Transverse section (× 5); *H*, Cardinal; *CS*, counter; *A*, alar; *L*, counter-lateral septa; *a–g*, metasepta, lettered in order of development. The short unlettered septa are minor septa.

growth, and in some species there are also fine vertical ribbings. The curvature of the cone may be in part the effect of the conditions of growth: a cone fixed by its apex and growing larger upwards is in unstable equilibrium. If not firmly fixed it will topple over towards one side, and, fixing itself in that position, will become curved in the effort to grow straight upwards.

At the broad (distal) end the cone is hollowed out into a depression (the **calice**) about 6 mm. deep, with steep sides and a nearly flat floor. A well-marked groove, the **cardinal fossula,** extends from the centre of this floor to that side which corresponds to the convex of the horn (Fig. 107*b*). The sides and floor of the calice bear a radiating series of ridges, the **septa**, which are the most distinctive character of corals in general these are formed of groups of fibres grouped together to form rods or spines called **trabeculae.** In these species the septa are over sixty in number and arranged in an alternation of two series. One series, the **major septa** (or **entosepta**), extend from the outer wall of the calice to near the centre of its floor. The alternate **minor septa** (or **exosepta**) are thinner, less prominent, and confined to the calice-sides. Although the septa are, in a general way, radial, the presence of the fossula makes the whole calice very distinctly bilateral, and there cannot be said to be more than an approximation to radial symmetry. The major septum which corresponds to the middle line of the fossula is much shorter and less prominent than the others: it is called the **cardinal septum.** The next major septa on either side start from the wall nearly parallel, diverging on the floor of the calice, where they form the boundaries of the fossula. It is not easy to make out any further characters of the septa as seen in the calice, but by taking transverse and longitudinal sections through the coral more may be learned.

If a section were taken very near the apex of a well-preserved specimen, it is probable (from what is known in allied corals) that six septa only would be present at the very early stage of growth here preserved. Owing to the destruction of the actual apex in most specimens it is not possible to prove this, but these six primary septa can easily be recognized in slightly later stages, and by means of a series of transverse sections the method by which additional septa are intercalated can be recognized: this method shows that the symmetry of the coral is bilateral, not radial (Fig. 108). One of the six primary septa is the cardinal septum, already noted; another which is and remains exactly opposite to this is called the **counter**-septum. The two on either side of this are the **counter-lateral** septa; the two between them and the cardinal are the **alar** septa.

The six first septa are distinguished as **protosepta,** and the major septa that arise later as **metasepta.** As growth proceeds, additional major septa appear between the cardinal and alar septa, and between the alar and counter-lateral, but none between the counter-lateral and counter-septum (Fig. 109). There are thus four spaces only in which new septa appear, and the newest septum in each space is always that nearest the cardinal septum. (The terms right and left **cardinal-lateral** and **counter-lateral serials** may be used for these four series of metasepta.) The minor septa, on the other hand, appear in rapid succession after the full number of major septa has been formed.

The transverse sections also show that the cone is hollow, consisting of a **lumen** with a comparatively thin wall (the **epitheca**), and divided by the septa into **interseptal loculi.** Near the epitheca there may be seen a few curved calcareous plates crossing some of the loculi: these are called **dissepiments.**

A vertical section shows that the floor of the calice is the last of a series of transverse partitions (**tabulae**) across the lumen, which must be formed, as growth proceeds, in somewhat the same way as the septa are formed in a cephalopod-shell. They do not, however, serve the same function, as all corals are fixed animals; they are much less regular in arrangement, and have no perforation like a siphuncle. Each tabula is convex upwards, and in addition to the cardinal

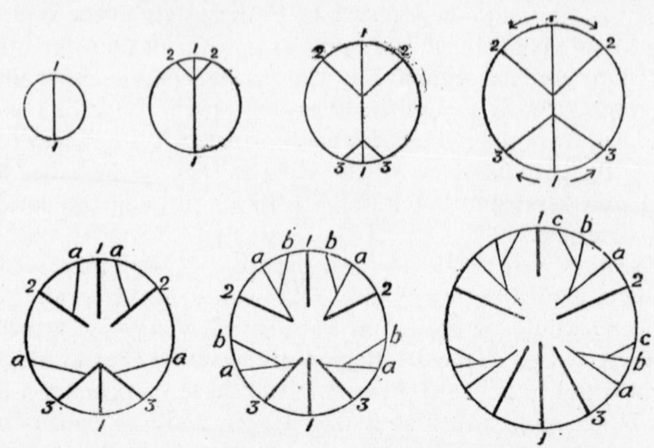

FIG. 108.—DEVELOPMENT OF MAJOR SEPTA IN A RUGOSE CORAL

1, 2, 3, Protosepta; *a*, *b*, *c*, metasepta; 1, cardinal and counter septa; 2, alar septa; 3, counter-lateral septa. Seven successive transverse sections, giving ontogenetic stages, are shown. The numbering and lettering denote the order of development of the septa (After Carruthers and Duerden).

fossula there can be recognized two other similar, but slighter, fossulae, in which lie the two alar septa, and sometimes one enclosing the counter-septum. The counter-lateral septa have no fossulae, consequently some investigators count them as only the first-formed metasepta. The septa are continuous through the tabulae. In those species of "*Zaphrentis*" which have external ribbing, the ribs alternate with the septa.

When a "*Zaphrentis*" grows large it may pass from conical to cylindrical in shape, and then a change takes place in the septa: instead of continuing as vertical plates through the later-formed tabulae, they retain the form seen in the calice, of low ridges on the upper surface of the tabula rising up on the inner face of the wall as low vertical ridges. Such a condition is described as the "amplexoid habit," because it attains its fullest development in the genus *Amplexus*. In this form, the tabulae are so horizontal and far apart, and the septa reduced to such feeble ridges, that J. Sowerby, who first described it more than a century ago, mistook it for a cephalopod-shell, and, thinking that the resemblance to a coral was accidental, called the species he described *Amplexus coralloides*. The same habit is seen, rather less pronounced, in the genus *Caninia*, species of which may attain a length of 40 cm. or more. This is distinguished from "*Zaphrentis*" by the far greater abundance of dissepiments between the septa for some distance in from the theca. These two genera occur, like "*Zaphrentis*," in the Carboniferous Limestone of England and Belgium.

If we now try to form an idea of the animal of which the skeleton

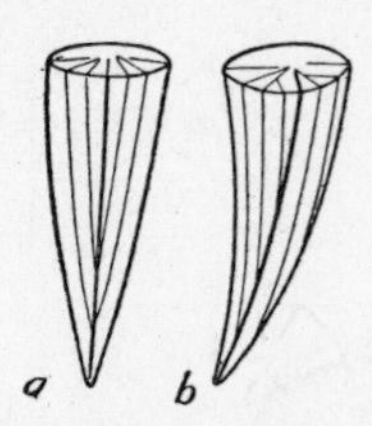

FIG. 109.—SEPTAL DEVELOPMENT IN RUGOSA

Diagrammatic side views of apex. *a*, showing cardinal septum in centre, metasepta arising on both sides of it; *b*, showing alar septum in centre, cardinal on right margin, counter on left, metasepta arising on only one side of alar septum, away from cardinal.

has just been described, we can only do so by analogy with living corals. The main living part of the body of "*Zaphrentis*" must have lain within and beyond the calice, and perhaps it could shrink up within the calice completely when safety required it. In structure it was essentially a polyp of the sea-anemone type. It might easily be supposed that the septa are the skeletons of the mesenteries, but this is not the case: each mesentery occupies the centre of an interseptal loculus, and the mesenteries therefore alternate with the septa. Where the septa come the wall and floor of the polyp are, as it were, indented by them, for the whole skeleton is a secretion by a part of the **external** layer of the polyp.

2. **Lithostrotion pauciradiale** (Fig. 110) is a coral very abundant in some of the uppermost beds of the Carboniferous Limestone (zone D_2). At Wrington, Somerset, for instance, it occurs in great masses, red-stained by iron oxide. It is a **compound** or **colonial** coral, consisting of a mass of vertical cylindrical **corallites** (each secreted by a

single polyp), each about 4 or 5 mm. in diameter, loosely in contact or slightly separated (**dendroid**), and originating one from another by lateral branching. Only near the base of the **corallum** (as the whole compound skeleton or coral colony is termed) is branching abundant, but it occurs occasionally at higher levels. Externally each corallite is marked by slight horizontal wrinkles, closely set. The calice of a full-grown specimen shows twenty-four major septa and as many minor, the former reaching nearly to the centre, the latter scarcely projecting through a narrow peripheral zone of dissepiments (**dissepimentarium**). In the centre of the calice there projects a laterally compressed structure, the **columella.** In a vertical median section of a

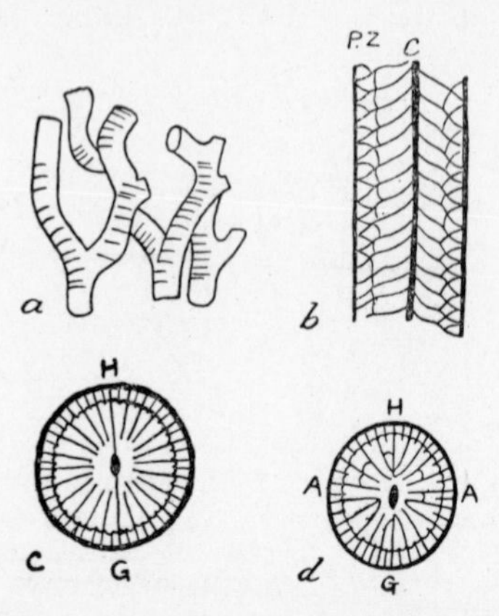

FIG. 110.—LITHOSTROTION PAUCIRADIALE (M'Coy), Viséan (Carboniferous Limestone)

a, Small portion of a young corallum (× ½). *b*, Vertical section of a corallite (× 2). *c*, Columella; *P.Z.*, dissepimentarium. *c*. Transverse section of a full-grown corallite (× 3). *d*, Transverse section of a young corallite (× 5/2); *H*, Cardinal, *G*, counter; *A*, *A*, alar septa. *a*, *b*, *c*, after Edwards and Haime; *d*, original.

corallite this is seen to be a vertical pillar extending the whole length. The lumen is crossed by numerous very regular tabulae, which are very slightly convex upwards except near the columella, where each is lifted up into a cone, and towards the margin, where it falls steeply into the dissepimentarium. The central part of the corallite, where only columella, tabulae and major septa are present is termed the **tabularium.** In the calice of an adult corallite the septa appear almost perfectly radial. The compression of the columella, however, indicates a median plane, and at least one of the septa in line with this (cardinal and counter) joins the columella. On examining a cross-section of the earlier-formed part of a corallite, a distinctly bilateral arrangement can be recognized, and the same protosepta as in "*Zaphrentis*" can be made out (Fig. 110*d*).

In other species of *Lithostrotion* (e.g., *L. martini*, which has twenty-eight major septa) the protosepta are better defined: in these species, however, the much greater abundance of dissepiments tends to confuse the appearance of a cross-section.

Some species of *Lithostrotion* have cylindrical corallites in a loose bundle, forming a **fasciculate** (where the corallites are parallel but not in contact, e.g., *L. junceum*; or a dendroid (irregularly branched) corallum, as in *L. pauciradiale* (Fig. 110*a*); in some other species the corallites are close to each other though not in contact but joined by

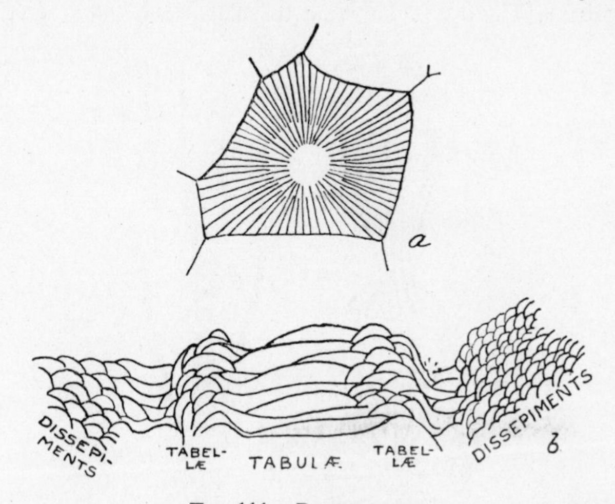

FIG. 111.—PALAEOSMILIA

a, *P. regia* (Phillips), surface view of one calice and parts of its neighbours (× 2); *b*, *P. murchisoni* Edwards and Haime, portion of a vertical section (× $\frac{3}{2}$; after Edwards and Haime).

occasional connecting tubes (**phaceloid**); in others (*L. maccoyanum*) they are tightly packed together and become polygonal (mostly hexagonal) in section, the whole corallum being described as **massive** or **cerioid** (compare Figs. 111*a*, 113*f*).

The essential feature of the genus is the laterally compressed columella. In Great Britain its species are confined to, and very abundant in, the upper stage (Viséan) of the Carboniferous Limestone.

3. **Palaeosmilia regia** (until recently generally called *Cyathophyllum regium*) is a compound massive coral found in one of the highest zones (D_2) of the Carboniferous Limestone (Fig. 111*a*). As

with massive species of *Lithostrotion*, each corallite is bounded by a distinct wall, so that the corallum is **cerioid** (as distinguished from the astraeoid type, in which the walls are indefinite, and where the septa usually alternate in adjacent corallites and the **thamnastraeoid** type, in which the walls are non-existent and the septa are confluent with those of the neighbouring corallite). The septa are far more numerous than those of *Lithostrotion*, and the minor septa are more than half as long as the major: consequently the dissepimentarium is relatively much wider, and the tabularium much narrower. There is no columella. The tabulae are very flat domes, and break up marginally into more convex **tabellae.** The dissepiments are numerous, regular, small and globose. There is a key-hole fossula at the outer part of the

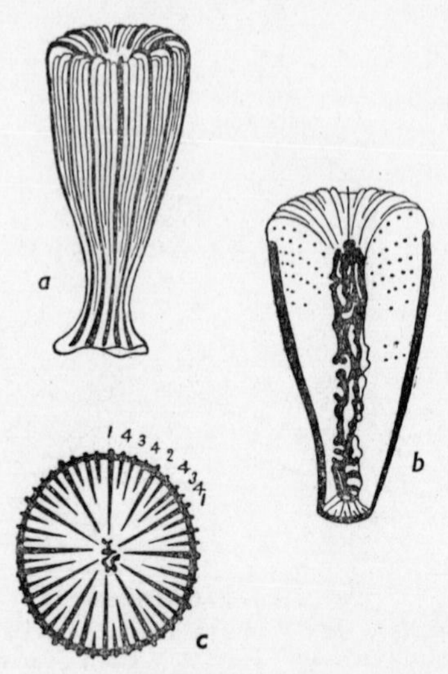

FIG. 112.—PARASMILIA CENTRALIS (Mantell)

a, Young specimen (× $\frac{3}{2}$); *b*, Vertical section (× $\frac{3}{2}$); trabecular columella in centre, black; dots represent granulations on septa; thick marginal lines represent wall in section; edges of half the septa seen at the top and (broken section) at the bottom; *c*, Calice (× 2). Numbers denote septa of the four cycles (After Edwards and Haime).

tabularium. Individual corallites may be as large as 50 mm. diameter or as small as 15 mm.

Another species, *P. murchisoni* (Fig. 111*b*), also found in the Carboniferous Limestone, like "*Zaphrentis*," is a **simple** or **solitary**: it appears earlier than *P. regia*.

4. **Parasmilia centralis** (Fig. 112) is a simple coral found in the White Chalk of England. It is attached to molluscan or echinoid

shells by a spreading base, from which rises a short cylindrical peduncular portion, soon expanding into a conical form which may attain a length of 25 mm. and a diameter of 12 mm. If further growth takes place, the form becomes cylindrical; a total length of 75 mm. or more may be attained. The calice is nearly circular, and shows a prominent columella (which in a section is seen to be spongy in texture), and forty-eight septa showing almost perfect radial symmetry).

Instead of being in two series, major and minor, there are four series (**cycles**) of unequal length. The first cycle consists of six long septa, reaching the columella; the second, of six, slightly shorter, alternating with the first six; the third, of twelve, decidedly shorter, alternating with those of both first and second cycles; and the fourth of twenty-four, still shorter, and alternating with all the others. This arrangement differs altogether from that of "*Zaphrentis*" and *Lithostrotion*, and as it is generally characteristic of the Mesozoic and later corals classed as Scleractinia in contrast to the Palaeozoic Rugosa.

A vertical section shows the lumen to have very few tabulae and dissepiments; the surfaces of the septa bear little tubercles. At the edge of the calice the septa project and extend across the wall into continuity with a series of vertical ridges (**costae**) on its outer surface. These costae show by their varying strength an arrangement in the same cycles as the septa, except that the first and second cycles are indistinguishable.

The columella is not a solid pillar (as in *Lithostrotion*) but an irregular network of trabeculae. The septa have a similar structure, and in some hexacorals (not *Parasmilia*) the ends of the trabeculae project so as to give a **dentate** edge to each septum.

The classification of Corals is still in an unsettled condition. That here adopted is a shortened version of the most recently published; it is that given by J. W. WELLS and Dorothy HILL in the 1956 Treatise of Invertebrate Paleontology (See BAYER and others in Bibliography p. 228).

Before setting out this classification, something must be said of the methods by which the simple form, in which every species begins its life, may develop into the varieties of compound coral.

(1) **Fission,** where the usually circular corallite or calice becomes nipped into the shape of the figure **8**, the two halves continuing growth as equal individuals, the process being repeated (Fig. 113*e*).

(2) **Lateral budding** or **gemmation,** differing from fission in that division is into unequal units, the larger being counted as parent and

retaining its individuality in later growth; in some forms the corallites are connected by transverse stolons.

(3) **Axial, calicinal** or **parricidal gemmation,** where the parent ceases further growth, and several new individuals start as buds from within the calice.

(4) **Peripheral gemmation,** where the parent corallite does not necessarily cease growth but the new corallites arise on its outer septate area.

(5) **Intermural gemmation** seen in massive (cerioid) Tabulata or Rugosa, where new corallites appear to project laterally through or spring from the walls of the older ones. This may only be a form of lateral gemmation.

The most familiar structural units in the Coral skeleton have already been explained—**septa, costae, columella, tabulae, tabellae, dissepiments.** Two others must be added: **Pali** are vertical plates radiating from the columella, in line with the septa but not joining them. **Synapticulae** are calcareous rods or bars which, in some Scleractinia, connect the opposing faces of adjacent septa, perforating the intervening mesenteries.

CLASS: **ANTHOZOA** (or Actinozoa)

SUBCLASS: CERIANTIPATHARIA. Miocene–Recent.

ORDER 1. **Antipatharia.**—Compound corals with slender and branching colonies with a horny skeleton; 6, 10, or 12 complete mesenteries in each corallite. Mio.–Rec.

ORDER 2. **Ceriantharia.**—Simple corals without a skeleton; having 2 rings of tentacles. Rec.

SUBCLASS: OCTOCORALLIA (**Alcyonaria**). Exclusively colonial (compound). Mesenteries in multiples of eight. Skeleton in most cases consisting of loose calcareous spicules only, but continuous in the modern *Tubipora* (organ-pipe corals).

Alcyonarian spicules have been found in Cretaceous and Tertiary deposits, and forms allied to *Tubipora* in the same formations.

Graphularia (Eoc., London Clay), a rod-like body, with some resemblance to a belemnite-guard but without the alveolus of the latter, is the axis of an alcyonarian colony. Range of subclass ?Sil.; Cret.–Rec.

SUBCLASS: ZOANTHARIA (**Hexacorallia** in part, with **Rugosa**). Ord.–Rec.

Includes simple and compound forms. Polyps have simple or divided tentacles, never pinnate; cyclically arranged. Mesenteries paired, number varies.

ORDER 1. **Heterocorallia.**—Simple corals. Septa inserted within "bifurcated" outer ends of 4 original septa (cardinal, counter and alars). Tabulae formed of fibrous calcium carbonate. Range Carboniferous only. Examples: *Hexaphyllia* only 6 septa. *Heterophyllia* numerous septa.

2. **Zoanthinaria.** Mostly compound, no skeleton. Unknown as fossils.

3. **Rugosa** (**Tetracoralla**).—Extinct group. Mesenterial arrangement known from septal insertion which, after insertion of first six septa, takes place in 4 quadrants—between cardinal and alar septa and between alar and counter-lateral septa—with tendency to produce bilateral symmetry. Septa typically of 2 orders, major and minor alternating in size. Skeleton formed of fibrous calcite (subsequent recrystallization not unusual). Ord.–Perm. (abundant in Silurian, Devonian and Lower Carboniferous).

Suborder (i) STREPTELASMATINA.—Simple or compound, with a septal stereozone (area of dense skeletal deposits) or dissepimentarium composed typically of small globose interseptal dissepiments. Tabulae typically domed. Ord.–Perm.

Superfamily CYATHAXONIICAE.—Small simple corals with marked longitudinal costation. Narrow septal sterozone without dissepiments. Ord.–Perm. *Cyathaxonia* (Carb.), small; tall columella developed independently of major septa but in contact with them. *Amplexus* (L. Carb.), long, cylindrical; tabulae flat, with cardinal and alar fossula depressions; major septa thin, continuous vertically only at peripheral edges. *Zaphrentites* (L. Carb.), fossula on concave side of corallum; septa may withdraw from fossula so that radial replaces earlier pinnate arrangement. *Zaphrentoides* (L. Carb., Fig. 107), fossula on convex side of corallum.

Superfamily ZAPHRENTICAE.—Simple or compound; septal stereozone with septa bearing axial lobes, or a regular dessepimentarium or one with inosculating dissepiments. Tabulae conical or domed. Ord.–Perm. *Streptelasma* (M. Ord.–M. Sil.), simple; septa long, axial lobes few. *Kodonophyllum* (M.–U. Sil.), simple or fasciculate; wide septal stereozone; with flat or sloping calice surrounding centre, calicinal gemmation. *Acervularia* (Sil.), phaceloid, or cerioid; dissepimentarium with 3 zones, an outer composed of 2 or more series of globose plates, a middle composed of flat plates just outside inner wall and an inner composed of globose plates inside the wall merging into tabulae; third order septa may occur. *Zaphrentis* (Dev.), simple; major septa long, in contact axially. *Heliophyllum* (Dev.), large corallites weakly compound; septa with yard arm-like flanges which may be dilated in the tabularium; fossula marked by narrowing of wide

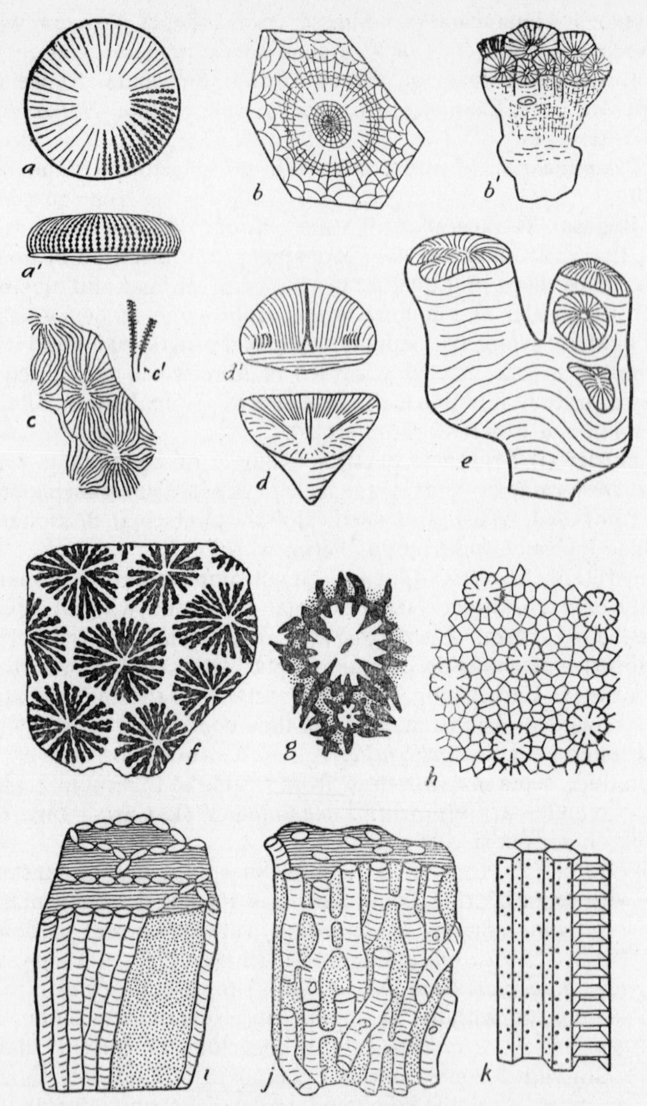

FIG. 113.—FOSSIL CORALS

a, *Palaeocyclus* [*Porpites*] *porpita* (Linné), Wenlock Limestone; plan of calice, detail omitted from three quadrants; *a′*, Side view (× $\frac{3}{2}$). *b*, *Lonsdaleia floriformis* (Fleming), Viséan (Carboniferous Limestone); cross-section of single corallite (× 2); *b′*, Side of young corallum (× $\frac{1}{2}$). *c*, *c′*, *Phillipsastraea pengellyi*; *c*, section through two corallites and parts of several others, showing continuity of septa (× $\frac{1}{2}$); *c′*, two of the septa much enlarged, showing septal dissepimenarium. *Phillipsastraea* (Dev., Fig. 113*c*), compound, massive, thamnastraeoid (no wall between corallites, septa continuous from one corallite to another); septa dilated especially at inner margin of dissepimentarium. *Lithostrotion* (Carb.), phaceloid or cerioid, with columella (already described, Fig. 110). *Orionastraea* (L. Carb.) encrusting, astraeoid or thamnastraeoid; columella weak or absent; septa withdrawn from axis and in some from periphery. *Diphyphyllum* (L. Carb.) like phaceloid *Lithostrotion* but without columella. *Clisiophyllum* (L. Carb.) simple, cylindrical; regular dissepimentarium; axial structure wide with thin septal lamellae abutting on a short median plate. *Aulophyllum* (L. Carb.), like *Clisiophyllum* but axial structure is a well-defined column cuspidate in transverse section, with packed lamellae and tabellae and no median plate. *Dibunophyllum* (L. Carb.), simple; inner parts of minor septa degenerate and inner dissepiments inosculating; axial structure typically consisting of a median plate with 4–8 septal lamellae on either side giving a "spider's web" appearance. *Palaeosmilia* (L. Carb.), solitary or compound already described on p. 219, Fig. 111). The next two genera to be mentioned show Columnariine features. *Caninia* (Carb.–Perm.), simple; in youth long major septa slightly sinuous with lobation in tabularium particularly in cardinal quadrants; septa become amplexoid and less dilated in adult stages when concentric or lonsdaleoid dissepiments may develop peripherally; tabulae flat with downturned margins. *Siphonophyllia* (L. Carb.), large, solitary like *Caninia* but with wide lonsdaleoid dissepimentarium.

Suborder (ii) COLUMNARIINA.—Compound, rarely simple. A single series of impersistent peripheral dissepiments or more commonly a wide septal stereozone may be replaced by long lonsdaleoid dissepiments which replace the major septa. Septa thin in tabularium; none lobed. Tabulae horizontal or flat-topped domes or axially sagging.

FIG. 113.—FOSSIL CORALS (*continued*).

ridges in section. *d*, *Calceola sandalina* (Linné), Middle Devonian (× ½); *d′*, operculum. *e*, *Thecosmilia annularis* (Fleming), Oxfordian (Coral Rag) (× ½). *f*, *Isastrea oblonga* (Fleming), Portlandian; Section of silicified corallum; walls and septa white (× 3). *g*, *Cyathophora* (*Holocystis*) *elegans* (Lonsdale), Aptian (Atherfield Clay); surface of part of corallum; septa and columella, black; wall, dotted (× 6). *h*, *Heliolites porosus* (Goldfuss), Middle Devonian; transverse section (× $\frac{5}{2}$). *i*, *Halysites catenularius* (Linné), Silurian; diagrammatic view of portion of a corallum (× $\frac{3}{2}$). *j*, *Syringopora ramulosa* (Goldfuss), Lower Carboniferous. (Natural size.) *k*, *Favosites gothlandicus* Lamarck, Wenlock Limestone; diagrammatic view of four corallites. The exterior of three is seen, with mural pores, the fourth is split vertically in half, showing tabulae (× 4). *d*, after Goldfuss; *i*, and *k*, original; the rest after Edwards and Haime.

Ord.–Perm. *Ketophyllum* formerly called *Omphyma* (Sil.), simple, top-shaped with root-like outgrowths; flat grouped tabulae and long septa. *Lonsdaleia* (Carb., Fig. 113*b*), fasciculate or cerioid; corallites with lonsdaleoid dissepiments (no septa in peripheral area) predominate, 2 orders of septa; axial structure of septal lamellae, medial plate or columella; conically arranged, shallowly curved tabellae.

Suborder (iii) CYSTIPHYLLINA. Simple or compound. A dissepimentarium of small globose dissepiments typically with septa represented by separate trabeculae resting on upper surface of dissepiments, or a stereozone in which cores of trabeculae are set in a mass of lamellate sclerenchyme. Tabulae flat and complete or sunken in centre and incomplete. Includes the only corals with a lid (operculum) covering the calice. Ord.–Dev. *Palaeocyclus* (Sil., Fig. 113*a*, *a'*), simple; discoidal with central cone of attachment; stereozone of monacanthine septa, no tabulae or dissepiments. *Cystiphyllum* (Sil.), simple; cylindrical; with long septa each represented by trabeculae on top of globose dissepiments and tabellae; dissepimentarium very wide; calice floor sunken in centre. *Goniophyllum* (Sil.), simple; pyramidal; square in transverse section; calice with operculum of 4 triangular plates of dense sclerenchyme; septa thick, dissepiments and tabulae numerous. *Calceola* (Dev., Fig. 113*d*), simple, semiconical, shaped like a Turkish slipper, semielliptical in transverse section; epitheca, thick; semielliptical operculum, composed of dense sclerenchyme.

4. **Actiniaria.** Sea-anemones without skeleton.

5. **Scleractinia (Hexacorallia** in part). Simple or compound, polyps less specialized than Actiniaria. Calcareous external skeleton (composed of fibrous calcite) consisting of radial partitions or septa which are intermesenterial and formed by upward infoldings of basal part of column wall. They differ from rugose corals essentially in the mode of addition of new septa which appear in regular cycles after the first six, in the sextants defined by these. M. Trias.–Rec.

Suborder (i) ASTROCOENIINA. Compound. Corallites small. Septa formed by relatively few (up to 8) simple trabeculae. Trias.–Rec. *Thamnasteria* [*Thamnastraea*] (Trias.–Cret.), massive, ramose or encrusting colonies, septa confluent between centres with row of granulations or discontinuous ridges (thamnastraeoid); columella styliform. *Cyathophora* (*Holocystis*) (L. Cret., Fig. 113*g*), massive, costate corallites; superficially resembles some Rugosa there being four septa longer than the others, tabulae well developed.

Suborder (ii) FUNGIINA. Simple or compound. Septa fenestrate, formed by simple or compound trabeculae united by simple or compound synapticulae, margins dentate or beaded. Trias.–Rec. *Cyclo-*

lites (Cret.–Eoc.), simple, discoidal, septa mostly perforate. *Isastrea* (M. Jur.–Cret., Fig. 113*f*), encrusting to massive cerioid; corallite walls partly absent with septa confluent. *Chomatoseris* [*Anabacia*] (Jur.), simple, discoidal; columella feeble. *Stephanophyllia* (Eoc.–Rec.), simple, discoidal, septa much perforated; columella trabeculate. *Goniopora* [*Litharaea*] (Cret.–Rec.), compound; massive or ramose; septa generally in 3 cycles.

Suborder (iii) FAVIINA. Simple or compound. Septa laminar, formed by simple or compound trabeculae, imperforate, margins dentate, synapticulae rare. Trias.–Rec. *Montlivaltia* (Trias.–Cret.), simple; conical, trochoid or discoid; no columella; covered externally by a perishable calcareous skin (**epitheca**). *Thecosmilia* (Trias.–Cret., Fig. 113*e*), similar to *Montlivaltia* but compound by equal fission.

Suborder (iv) CARYOPHYLLIINA. Simple or compound. Septa laminar with smooth or nearly smooth margins composed of simple trabeculae. Jur.–Rec. *Turbinolia* (Eoc.–Olig.), simple, conical, wall perforated or pitted between prominent costae; with projecting columella. *Parasmilia* (Cret., Fig. 112) already described. *Flabellum* (Eoc.–Rec.), simple; free; compressed, calice elliptical, no columella.

Suborder (v) DENDROPHYLLIINA. Simple or compound. Septa laminar of simple trabeculae but generally secondarily thickened, irregularly porous; synapticulae present. U. Cret.–Rec. *Dendrophyllia* (Eoc.–Rec.), compound, dendroid.

SUBCLASS: TABULATA. M. Ord.–Perm. ?Trias., ?Eoc.

An extinct group characterized by an exclusively compound growth and by a calcareous exoskeleton of slender tubes crossed by many tabulae. The presence of these tabulae and inconspicuousness or absence of septa suggested name of group. They differ from Octocorallia and resemble Rugosa and Scleractinia in having skeletons of fibrous and not spicular calcite arranged in trabeculae in the septa and in sheets in the transverse plates. Unlike Rugosa and Scleractinia they have septa only of one series; in many the number is 12. *Tetradium* (Ord.) massive, slender aseptate corallites, quadripartite axial gemmation. *Heliolites* (Sil.–Dev., Fig. 113*h*), massive, each corallite with 12 spinose septa and complete tabulae, space between corallites occupied by coenenchyme of tubules of which more than 12 bound each corallite. *Favosites* (U. Ord.–M. Dev.), massive, corallites cerioid (prismatic), thin-walled, **mural pores** perforating the walls, tabulae complete and even-spaced; intermural gemmation. *Thamnopora* (Sil.–Perm., especially Dev.), massive and branched, corallites prismatic, thick-walled with tunnel-like mural pores, tabulae thin. *Alveolites* (Sil.–Dev.), massive encrusting or branching, corallites not prismatic, upper wall vaulted, large mural pores at bases of upper walls. *Pleurodictyum* (L. Dev.), discoid with large corallites, thick-walled with tunnel-like mural pores; few or very thin complete tabulae; may grow around a worm-tube. *Michelinia* (U. Dev.–Perm.), like *Pleurodictyum* but with numerous incomplete convex tabulae. *Vaughania* (Carb.), discoid, free, atabulate, walls and floor of dense fibrous tissue. *Halysites* (Ord.–Sil., Fig. 113*i*), phaceloid coralla composed of linear series of cylindrical or oval corallites which interconnect so that transverse section appears as an anastomosing network of chains; walls imperforate, tabulae horizontal or gently arched. *Aulopora* (Dev.), an encrusting network of tubular corallites with trumpet-like apertures; growth by lateral gemmation. *Syringopora* (Sil.–U. Carb., Fig. 113*j*), corallites cylindrical connected by transverse tubes (fasciculate); tabulae close set or coalesced; growth by lateral gemmation.

Short Bibliography

BAYER, F. M. and 9 other authors.—*Treatise on Invertebrate Paleontology, Part F, Coelenterata*, 498 pp. (1956) Kansas Univ. Press.

BUEHLER, E. J.—"The Morphology and Taxonomy of the Halysitidae," *Bull. Peabody Mus. Nat. Hist. Yale Univ.*, 8 (1955).

CARRUTHERS, R. G.—(1) "Primary Septal Plan of Rugosa," *Ann. Mag. Nat. Hist.* (7), xviii (1906); (2) "Revision of some Carboniferous Corals," *Geol. Mag.* (5), v, (1908); (3) "Evolution of *Zaphrentis delanouei*," *Quart. Journ. Geol. Soc.*, lxvi, 523–38 (1910).

DUERDEN, J. E.—"Morphology of the Madreporaria," *Ann. Mag. Nat. Hist.* (7), x, xi, xvii, xviii (1902–6).

DUNCAN, P. M.—"British Fossil Corals," (2nd Series), *Palaeont. Soc.*, 1866–72.

EDWARDS, H. MILNE and HAIME, J.—"British Fossil Corals," *Palaeont. Soc.* (1850–7).

FOWLER, G. H. and BOURNE, G. C.—"Coelentera" in E. R. Lankester's *Treatise on Zoology*, pt. 2 (1900).

HILL, Dorothy.—(1) "British Terminology for Rugose Corals," *Geol. Mag.*, lxxii, 481 (1935); (2) "Carboniferous Rugose Corals of Scotland," *Palaeont. Soc.* (1938–41).

HUDSON, R. G. S.—"The Development and Septal Notation of the Zoantharia Rugosa (Tetracoralla)," *Proc. Yorks. Geol. Soc.*, xxiii, pt. 2, for 1935, 68–78 (1936).

HUDSON, R. G. S.—Various papers on Carboniferous Corals in *Ann. Mag. Nat. Hist.* (1926 on), *Proc. Leeds Phil. Soc.* (1928 on), *Proc. Yorks. Geol. Soc.* (1936 on), *Geol. Mag.* (1943 on).

LANG, W. D.—(1) "Trends in Carboniferous Corals," *Proc. Geol. Assoc.*, xxxiv, 120 (1923); (2) "Further considerations of Trends in Corals," *Proc. Geol. Assoc.*, xlix (1938).

LANG, W. D., SMITH, S. and THOMAS, H. D.—*Index of Palaeozoic Corals* (1940), Brit. Mus. (Nat. Hist.).

LANG, W. D. and SMITH, S.—"Critical Revision of Rugose Corals . . . ," *Quart. Journ. Geol. Soc.*, lxxxiii, 448–91 (1927).

NICHOLSON, H. A.—"British Stromatoporoids," *Palaeont. Soc.* (1886–92).

SMITH, Stanley.—"*Aulina* . . . and *Orionastraea*," *Quart. Journ. Geol. Soc.*, lxxii, 280 (1917).

STUMM, E. C.—"Revision of the families and genera of Devonian Tetracorals," *Mem. Geol. Surv. Amer.*, xl (1949).

THOMAS, H. D.—"Some Aspects of Evolution," *Trans. S.E.U. Sci. Soc.*, xl, 57–79 (1935).

VAUGHAN, A.—"The Carboniferous Limestone Series (Avonian) of the Avon Gorge," *Proc. Bristol Nat. Soc.*, i (1906).

VAUGHAN, A. and other authors.—Various papers on zonal distribution of Carboniferous Corals in *Quart. Journ. Geol. Soc.*, from 1905 onwards.

VAUGHAN, T. W. and WELLS, J. W.—"Revision of the Suborders, Families and Genera of the Scleractinia," *Special Papers Geol. Soc. Amer.*, No. 44 (1943).

X

THE PORIFERA AND PROTOZOA

SUPPOSEDLY more primitive than the Coelenterata there are two grades of animal structure, both of importance in palaeontology—the Porifera or Sponges, and the Protozoa, simplest of all animals.

The **PORIFERA** are so named because they take in their food, not by a single mouth as do all animals from Coelenterates "upwards," but by innumerable pores scattered over the surface of the body. As a consequence of this, and of their fixed mode of life, the form of the body is much more variable than in any phylum yet considered, and is sometimes very indefinite indeed. The skeleton in the more primitive forms is composed of great numbers of separate **spicules** usually of amorphous silica (opal), but of calcite in one class. These spicules form a support to all parts of the body, but are only united by soft tissues, so that on the death of the sponge they usually fall apart and are scattered on the sea-floor. In many cases two kinds of spicules may exist in the same sponge—**body spicules,** elongated and often branched, and **dermal spicules,** flat or rounded. A simple spicule in the fresh state, when seen under the microscope, has the appearance of a fragment of thermometer-tube, being glassy and with a fine central canal. Some of the most characteristic forms of spicule are shown in Figs. 114*a*, *b*, *c*. Loose siliceous spicules of this kind accumulate in such amounts on the sea-floor today as to form a "spicular ooze," as, for instance, at a depth of 120 fathoms off Kerguelen. Similar deposits among the stratified rocks have passed into cherts (or phthanites), the spicules becoming cemented by a secondary deposit of chalcedony into a very hard rock. Such sponge-cherts are found in Britain in the Carboniferous Limestone, at several Jurassic horizons, and particularly in the Lower and Upper Greensands of the Cretaceous.

In the more specialized orders of sponges the spicules are united into a meshwork, and in such cases the form of the body is preserved in the fossil. The substance never persists as unaltered opal; it may simply change into chalcedony, or may be replaced by pyrites, marcasite (oxidizing into limonite), or calcite.

The common bath-sponge belongs to a class having a continuous skeleton of a horny organic substance, but such forms are not preserved as fossils, and palaeontologists are only concerned with two classes, distinguished by the chemical composition of the skeleton.

CLASS: **SILICISPONGIAE.**—Skeleton of silica.

Siliceous sponges attain their greatest abundance in Europe in the Cretaceous system, being the source of the silica of chalk-flints and greensand cherts. The highest zones of Chalk at Flamborough Head contain a rich fauna of siliceous sponges.

ORDER 1: **Monaxonida.**—Spicules one-rayed (Fig. 114*a*). Only known fossil by loose spicules, except one genus *Cliona*, a sponge boring in lamellibranch shells and belemnite-guards, known by its borings, commonly preserved as solid moulds in flint.

2. **Tetraxonida.**—Spicules four-rayed (Fig. 114*b*, *c*).

SUBORDER: **Choristida.**—Spicules not united.

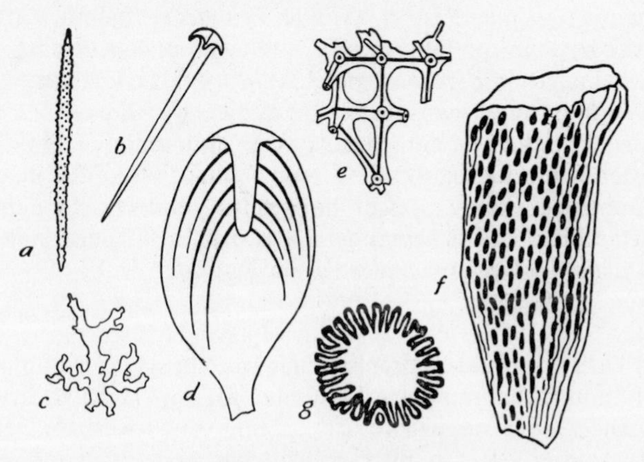

FIG. 114.—PORIFERA

a, Monactinellid (× 15), and *b*, Tetractinellid (× 10) spicules from interior or hollow flint; *c*, Dermal spicule of a Lithistid, Upper Chalk (× 32); *d*, *Siphonia tulipa* Zittel, Cenomanian (Upper Greensand), Warminster (Wilts); vertical median section (stalk broken off) showing vertical canals opening into central cloaca; *e*, Fragment of hexactinellid meshwork, showing central canals (black), Neocomian (× 30); *f*, *g*, *Ventriculites infundibuliformis* (S. P. Woodward), Upper Chalk; *f*, Side view of imperfect specimen; *g*, cross-section. (Both × $\frac{1}{2}$.) Tetractinellid (All after Hinde).

SUBORDER: **Lithistida.**—Spicules united into a continuous skeleton (Fig. 114*c*). Genera: *Siphonia* (Cret., Fig. 114*d*), pear-shaped, stalked; *Doryderma* (Cret.), cylindrical, branched; *Verruculina* (Cret.), cup-shaped or irregular, with short stalk.

3. **Hexactinellida.**—Spicules of six rays, intersecting at right-angles (Fig. 114*e*). In one of the earlier genera *Protospongia* (Cam.), the spicules were very feebly united, as they are found only a few together, the form of the body being unknown. The same is the case with other Palaeozoic genera, but from the Triassic onwards a continuous skeleton is found. Genera: *Ventriculites* (Cret., Fig. 114*f*, *g*), more or less inverted conical, with roots, wall of cone folded; *Plocoscyphia* (Cret.), frilled laminae uniting into a coarse net-work; *Coeloptychium* (Cret.), mushroom-shaped, with folded wall.

CLASS: **CALCISPONGIAE.**—With calcareous skeleton. There are several orders with three-rayed spicules, not forming a continuous skeleton, but little is known of them in the fossil state. Each spicule is, mineralogically, a crystal of calcite.

ORDER: **Pharetrones.**—Spicules united into a fibrous meshwork. Genera: *Tremacystia* (Cret.), clusters of double cylinders, the space between the inner and outer cylinder being crossed by partitions much coarser than the tabulae of a coral; *Peronidella* (Trias.–Cret.), hollow, cylindrical; *Raphidonema* (Cret.), more or less cup-shaped; *Porosphaera*, more or less spherical or conical (U. Cret.).

Calcareous sponges are locally very abundant in the Upper Jurassic of South Germany, the Lower Cretaceous of Faringdon in Berkshire and Upware in Cambridgeshire, and the Upper Cretaceous (Cenomanian) of Westphalia, but are rare in general. Species of *Porosphaera* are of zonal value in the English Chalk.

The **PROTOZOA** are animals composed of a single cell, or a small number of undifferentiated cells. Only two orders leave fossil remains—the Radiolaria, which form a beautiful lattice-work skeleton of silica, and the Foraminifera which have one of calcite, or of agglutinated foreign bodies. The majority of forms in both orders are very small, ranging from 1 mm. diameter downwards, but some of them occur in such abundance at certain horizons as to be quite important rock-forming organisms. A few genera of Foraminifera attain a much greater size, up to a maximum of perhaps 80 mm., and these giant forms are of value as indices of age, though unfortunately their geographical range is limited.

The **Radiolaria** range from possibly Pre-Cambrian to the present day: they are divided into two main suborders—Spumellaria, spherical or discoidal, with the range of the order, and Nassellaria, of various forms, but always with dissimilar ends, dating from the Devonian only. Beds of Radiolarian chert are known in the Ordovician of South Scotland, the Devonian of New South Wales, the Carboniferous of Devon and Cornwall, and the Mesozoic of the Alps. These beds have very generally been taken as analogous to the modern Radiolarian ooze of the greatest depths of the oceans; but in the case of the British Lower Carboniferous cherts their shallow-water character seems now decisively proved. On the other hand, the Miocene Radiolarian earths (not consolidated into chert) of Barbados and other West Indian islands, as well as the Alpine phthanites, are probably of deep-sea origin.

The **Foraminifera** form a shell which is nearly always chambered, the chambers being arranged sometimes in a straight line, sometimes in a zig-zag, but most frequently in some sort of spiral. They were at one time taken for cephalopods, but the resemblance in the shells of the two groups is only homoeomorphy of the most general sort. There are no gas-chambers in a foraminifer, the cell-protoplasm filling all the chambers as well as extending to a greater or less extent outside the shell; this it does through a well-defined aperture (much smaller than the diameter of the chamber) in the last-formed chamber, and (in one group) through numerous perforations of the whole surface. Moreover, the foraminifer does not cut off new chambers by secreting septa—each "septum" formed the terminal part of the outer shell before it was converted into an internal structure by the building of a new chamber in front of it.

The ontogeny reveals in many cases a striking **dimorphism**; in the same species growth may begin by a very small chamber (**microspheric** form) or by a much larger one (**megalospheric** form). The study of living forms has shown that this results from an alternation of generations. The microspheric form is produced by a sexual process, and its ontogeny is complete; the megalospheric is produced asexually, and the early stages are skipped altogether (Fig. 115*b*, *c*). The microspheric form in some species attains a much larger size than the megalospheric; in others there is no difference in size. Both forms occur together in the same strata, the megalospheric forms being more abundant than the microspheric—a natural result of the fact that a number of megalospheric generations intervene between one microspheric and the next.

The Foraminifera have been divided into three groups, based on the structure and composition of the test: (1) Perforata or Hyalina, with test of calcite, in prisms normal to the surface, transparent (at least in the more primitive forms) and with abundant small perforations in addition to (rarely without) a main aperture in the last formed chamber; (2) Imperforata or Porcellanea, with shell less translucent and of a brown colour by transmitted light, of felted calcite fibres and without perforations; (3) Arenacea, in which the test is formed of foreign bodies, such as sand-grains, sponge-spicules, or the small tests of other Foraminifera, held together by an organic secretion. The first two groups are well defined, but the third is probably a rather mixed assemblage of forms which have been led by special circumstances to adopt this curious way of protecting themselves in place of secreting a shell.

The following orders may be recognized, the first seven of them being Perforate:

1. **Lagenidea,** the simplest forms, with thin transparent shell. Genera: *Lagena*, single-chambered, flask-shaped; *Nodosaria*, a

FIG. 115.—DIMORPHISM IN A FORAMINIFER, *Pyrgo bulloides* (d'Orbigny), Recent
a, Exterior showing aperture with T-shaped partition (× 17·5); *b*, Cross-section of microspheric and *c*, of megalospheric form (× 125). Only the early chambers are shown, up to the attainment of the regular alternation of chambers. Chambers numbered in order of development (After Schlumberger).

straight row of chambers (orthocone); *Lenticulina* [*Cristellaria*], spiral (ophiocone to sphaerocone). These all appear to range from Triassic to Recent.

2. **Textularidea,** with a zig-zag alternation of chambers. Sometimes arenaceous. Chief genus: *Textularia* (Jur.–Rec.).

3. **Globigerinidea,** with chambers globose, usually in a somewhat irregular spiral, perforations coarse. This small order should perhaps be united with the next. Chief genus: *Globigerina* (Jur.–Rec.), a most abundant pelagic form at the present day (Fig. 116*a*). Associated with it is the perfectly spherical *Orbulina*, of which most specimens when broken open are found quite hollow, but occasionally a *Globigerina* is found inside: this suggests that *Orbulina* may be the final stage of *Globigerina*, the majority of specimens being megalospheric and omitting the *Globigerina*-stage in ontogeny.

4. **Rotalidea** with chambers coiled in an asymmetric spiral, all the whorls being visible on one face, and only the last whorl on the other. In this order first appears the **supplemental skeleton,** an outer calcareous layer plastered over the surface of the ordinary perforate

wall, penetrating between the two folds of the latter which here form the septa, and sometimes forming a surface pattern (even long spines in some) which quite obscures the arrangement of the chambers. Chief genera: *Discorbis* [*Discorbina*] (Cret.–Rec., Fig. 116*d*); *Planorbulina* (Carb.–Rec.); *Cibicides* [*Truncatulina*] (Carb.–Rec.,

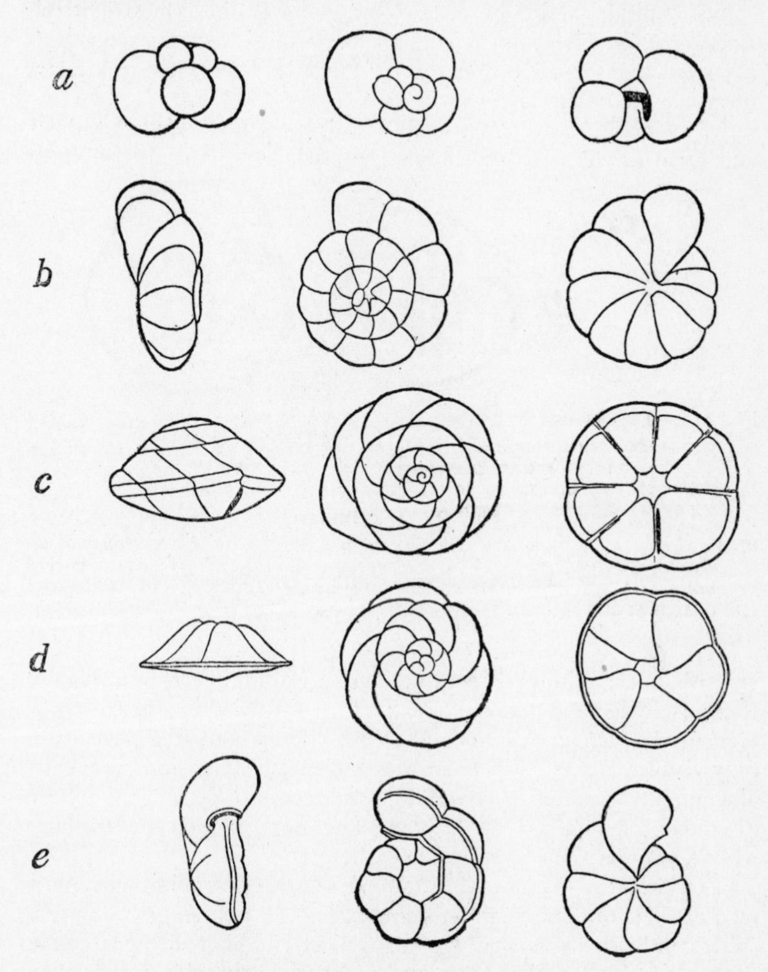

FIG. 116.—GLOBIGERINIDEA AND ROTALIDEA

In each row, the left-hand figure is an edge view, the middle an apical view and the right-hand a basal view. *a*, *Globigerina bulloides* (d'Orbigny), Recent (× 28). *b*, *Rotalia* (*Turbinulina*) *beccarii* (Linné) (× 34). *c*, *Eponides karsteni* (Reuss). Rec. (× 37·5). *d*, *Discorbis elegans* (Hantken), U. Eoc. (× 16). *e*, *Cibicides lobatula* (Walter and Jacob) (× 34).

Fig. 116*e*); *Anomalina* (Cret.–Rec.); *Eponides* [*Pulvinulina*] (Carb.–Rec., Fig. 116*c*). These all have the form described above, but differ in the relative convexity of the two faces, and are without supplemental skeleton; *Rotalia* (Jur.–Rec., Fig. 116*b*), with supplemental skeleton, but recognizably spiral; *Calcarina*, discoidal with peripheral spines, spiral internally (Cret.–Rec., chiefly in Maestrichtian); *Baculogypsina* [*Tinoporus*], similar shape but chambers irregularly arranged, and surface with pattern of tubercles (Mio.–Rec., abundant on modern coral-reefs).

5. **Nummulitidea,** symmetrically spiral, with subdivided chambers, and great development of supplemental skeleton, surface generally

FIG. 117.—NUMMULITES

a, *N. vicaryi* (d'Archiac and Haime); edge view (× ½); *b*, *N. complanatus* (Lamarck), edge view (× ½); *c*, *d*, *N. garansensis* (Joly et Leymerie); *c*, Part of horizontal section (× 8); *d*, Part of vertical section (× 12) (All after d'Archiac and Haime).

smooth, never spinose. Most of the giant Foraminifera, and most of those useful as time-indices come in either this order or the next two. Genera: *Elphidium* [*Polystomella*] (Cret.–Rec.), lenticular, last whorl only visible externally, its chambers and their subdivisions recognizable on the surface; *Nummulites* [*Camerina*] (Upper Cret.–Olig.), discoidal or lenticular (Fig. 117); *Assilina* (Eoc.), similar, but the later whorls only partially overlap the earlier.

6. **Orbitoidea.**—Discoidal, attaining a large size; without supplemental skeleton; spiral growth confined to earliest stages of microspheric individuals; normal growth cyclical, chambers being added in concentric rings, which more or less overlap the sides: a median horizontal section shows the structure in its simplest form, and enables the genera to be distinguished easily. Thus the "equatorial" chamberlets of *Orbitoides* (Upper Cret.), are rounded; those of *Discocyclina* [*Orthophragmina*] (Eoc., Fig. 118) rectangular; of

Lepidocyclina [*Cyclosiphon*] (Olig.–Mio., Fig. 119) more or less hexagonal; while in *Miogypsina* (Mio.) there is a more distinct spiral excentric "nucleus." The generic distinctions given above are those most readily made out in thin sections of limestones, but are not the only ones.

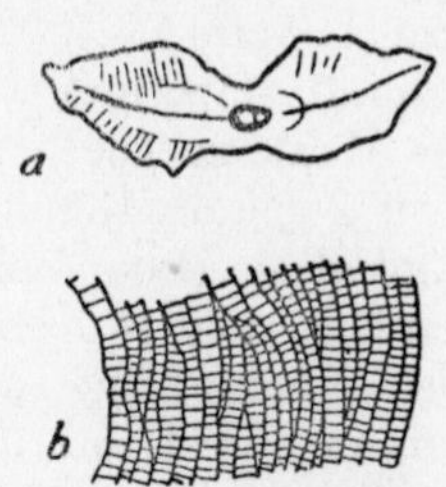

FIG. 118.—DISCOCYCLINA UMBILICATA (Deprat), Eocene, New Caledonia *a*, Vertical section (× ½); *b*, Portion of a horizontal section (× 10; after Deprat).

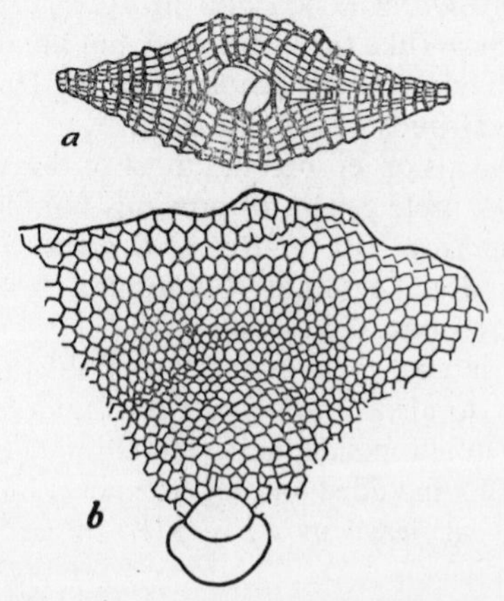

FIG. 119.—LEPIDOCYCLINA TOURNOUERI Douvillé and Lemoine, Miocene *a*, Vertical section (× 13); *b*, Part of horizontal section (× 18). Megalospheric form (After R. Douvillé and Lemoine).

7. **Fusulinidea.**—Developing, from plano-spiral, through globular and subcylindrical to fusiform, being spirally coiled around the long axis. Septa usually thrown into wavy folds, or otherwise complicated to form a distinctive pattern recognizable even in random sections: usually absorbed along a broad median band, so that chambers are connected by a continuous tunnel. Perforations of test very fine, but usually expanding below the surface into conspicuous alveoli.

Confined to Carboniferous and Permian, mainly in Eastern Alps, Asia and U.S.A.

Chief genera: *Millerella*, discoid (uppermost L. Carb.–U. Carb.); *Profusulinella*, subfusiform (U. Carb.); *Fusulinella*, fusiform, i.e., spindle-shaped (U. Carb.); *Fusulina*, elongate fusiform (U. Carb.); *Schwagerina*, subspherical (L. Perm.).

8. **Miliolidea.**—The typical Imperforate Order, the test being formed of a felt of fibres of calcite. The simplest example is *Cornuspira* (Jur.–Rec.), which is a spirally-wound tube, without internal septation. In *Pyrgo* (*Biloculina*] (Eoc.–Rec., Fig. 115) the adult chambers are hemispheroidal, only two being externally seen, the aperture (with a T-shaped septum) being alternately at one end and the other as new chambers are added. In *Triloculina*, *Quinqueloculina* and *Spiroloculina* (all Jur.–Rec.), the same alternation occurs, but with differences in chamber-shape, 3, 5 or all the chambers being externally visible. *Alveolina* (mainly Eoc.) and *Alveolinella* (Mio.–Rec.) are similar to *Fusulina* in shape and coiling, but subdivided internally in a different way. *Orbitolites* (Eoc.) and *Marginopora* (Mio.–Rec.) are discoidal and cyclical (like the Orbitoids), but an abraded surface (or section) shows a characteristic pattern like the "engine-turning" on a watch, or the leaf-buds of a house-leek.

9. **Lituolidea.**—This order includes most of the arenaceous genera, many of which resemble genera of previous families, at least externally. Thus *Ammodiscus* (Carb.–Rec.), *Reophax* (Carb.–Rec.), and *Cyclammina* (Jur.–Rec.) resemble *Cornuspira*, *Nodosaria* and *Lenticulina* respectively; but *Cyclammina* has a labyrinthine internal division of its chambers. *Endothyra* (Dev.–Trias.) resembles some of the Rotalidea, as do also *Orbitolina* (Cret.) and *Dictyoconus* (Eoc.), in which the Rotaline lop-sidedness results in a conical form.

Textularia is often included among Arenacea, but is always perforate and hyaline, at least in early life, though it may become arenaceous later.

The Perforata are usually well preserved as fossils, whether in limestone or clay, or even in sands in some of the latest formations. The Imperforata are more easily dissolved away. A very special mode of preservation (mainly of Perforata) is that of internal moulds in glauconite. The formation of these goes on at the present day on certain areas of the ocean-floor. By some means not yet explained this complex hydrated silicate of iron, aluminium, and potassium is precipitated in the interior of these minute tests, which themselves may afterwards undergo solution. The resulting moulds form "green sands," or on a muddy bottom "green muds." Similar deposits are found at many geological horizons, but most abundantly in the Cambrian and Cretaceous systems; they commonly contain fossils indicating the deposits to have been of marine origin.

As rock-forming organisms the Foraminifera are important. The earliest cases are the *Fusulina* limestones of the Upper Carboniferous, found in the Eastern Alps, Central Russia, and over large areas of Asia, the East Indies, and the Southern United States. The arenaceous *Orbitolina* [*Patellina*] gives its name to a limestone in the Aptian of Switzerland. Certain beds of the White Chalk are made up in great measure of the remains of *Globigerina* and *Textularia*. The Eocene Calacire Grossier of the Paris basin (the Paris building-stone) is full of Miliolidae, and an *Alveolina* limestone occurs in the Hampshire Basin among other regions. Far more important are the Eocene Nummulitic limestones, found throughout the area of the great ocean (Tethys) of that period, i.e., in the Atlas, Pyrenees, Alps, Carpathians, Balkans, Egypt, Persia, the Himalayas, the East Indies and New Caledonia. In many of these limestones *Discocyclina* associated with *Nummulites*. The Oligocene *Lepidocyclina* limestones (commonly known as orbitoidal) have a similar distribution and are found in the West Indies also. Although the accumulation of these giant Foraminifera to a sufficient extent to form masses of limestone is confined to the site of this great tropical and subtropical ocean, the genera are occasionally found beyond its limits. Thus *Nummulites* spread as far as Hampshire, and both it and *Discocyclina* into Bavaria, in the middle part of Eocene times. On the other hand *Nummulites* is almost unknown in America, and completely so in West Africa, south of Senegal, though found in many places in East Africa.

This completes the survey of the Animal Kingdom as known in the fossil state, but for a few minor groups that have been omitted and may briefly be mentioned here.

The **BRYOZOA** (or **POLYZOA**) are a phylum or subphylum of fixed compound animals, having some superficial likeness to the corals, but belonging to a more specialized grade of structure. They were at one time united with the brachiopods under the name Molluscoidea. Their skeleton (**zoarium**) is made up of many more or less tubular individual skeletons (**zooecia**), which are much smaller in size than most corallites, and have no septa. In a few simple cases the zoarium is like a branching thread; commonly it forms a flat lamina, either spreading freely in the water or encrusting other organisms; on occasions it is massive, in which cases it may be difficult to draw the line between this phylum and the Tabulata, especially as structures analogous to tabulae may be present. Certain Palaeozoic forms have been referred to the one phylum by some investigators, and to the other by others. The Bryozoa are almost exclusively marine, and range from Ordovician to Recent times. Occasionally they occur in such abundance as to form bryozoal deposits, for instance, certain Silurian and Lower Carboniferous limestones, and the Coralline Crag (Upper Pliocene) of East Anglia.

The **ANNELIDA** (segmented worms) are mostly soft-bodied and rarely preserved; but one order, Tubicola, is characterized by a fixed habitat and the secretion of a tube. These tubes may be nearly straight, regularly curved, spiral or irregularly twisted. The curved tubes might be confused with *Dentalium* (p. 89), the irregular for an uncoiled, and the spiral for an ordinary gastropod, and these are cases where the proper placing of such a fossil is still doubtful; but in general there is a roughness and want of regularity about annelid tubes that prevents confusion between them and molluscan shells.

The two most familiar genera are *Serpula* (Sil.–Rec., but mainly Jur.–Rec.), and *Microconchus* [*Spirorbis*] (Ord.–Carb.). The former is straight or sin uous, sometimes spiral, with rough surface, usually attached to shells, often an inch or more in length. The latter is smaller, of few spiral coils, smooth, attached to shells or sea-weed: it has been mistaken for *Planorbis*. These organisms are generally marine but *Microconchus* appears to tolerate non-marine surroundings in the Upper Carboniferous of Western Europe, in particular in the so-called *Spirorbis* limestones.

The burrows of soft-bodied worms are common in some formations: they may be simple vertical tubes, or U-shaped with two openings, or irregular and more or less horizontal. They may be filled with material differing from that in which they were bored (as where the top layer of the Chalk contains borings filled with glauconitic sand of Eocene age): in such cases they mark a non-sequence in deposition. In other cases, where there was no such break, the infilling material is similar to that around, but makes a break in the lamination or stratification of the latter.

In Palaeozoic strata, small black chitinous lustrous bodies have been referred to as annelid jaws or scolecodonts. These should not be confused with similar-sized translucent resinous small bodies called conodonts which are composed of calcium phosphate and are of unknown affinities.

Short Bibliography

PORIFERA

HINDE, G. J.—(1) "Catalogue of Fossil Sponges," *Brit. Mus.* (1883); (2) "British Fossil Sponges," *Palaeont. Soc.* (1887–1912); (3) "On Beds of Sponge-Remains in the Greensands of the South of England," *Phil. Trans. Roy. Soc.* (1885).

LAUBENFELS, M. W.—"Porifera" in *Treatise on Invertebrate Paleontology, Part E, Archaeocyatha and Porifera* (1955) Kansas Univ. Press.

MINCHIN, E. A.—"Sponges," in E. R. Lankester's *Treatise on Zoology*, pt. 2 (1900).

OAKLEY, K. P.—"Cretaceous Sponges: some biological and geological considerations," *Proc. Geol. Assoc.*, xlviii, 330 (1937).

PROTOZOA

BRADY, H. B.—*Challenger Reports*, vol. ix (Zoology): "Foraminifera" (1884).

CAMPBELL, A. S. and MOORE, R. C.—*Treatise on Invertebrate Paleontology, Part D, Protista* 3 (*chiefly Radiolarians and Tintinnines*) (1955) Kansas Univ. Press.

CARPENTER, W. B., PARKER, W. K. and JONES, T.R.—"Introduction to the Study of the Foraminifera," *Roy. Soc.* (1862).

CUSHMAN, J. A.—(1) *Foraminifera: their Classification and Economic Use*, 4th edn. (1955) Sharon, Mass, U.S.A. [Standard work with extensive bibliography]. (2) "Parallel Evolution in the Foraminifera," *Amer. Journ. Sci.* (Special Daly vol.), cxvii (1945).

DAVIES, A. M.—*Tertiary Faunas* (1935), London (Murby).

DAVIES, L. M.—(1) "The genus *Dictyoconus* and its allies." *Trans. Roy. Soc. Edinburgh*, lvi, 485 (1930). (a) "The genera *Dictyoconoides*, *Lockhartia* and *Rotalia*," *ibid.*, lvii, 397 (1932). [Good examples of modern systematic work.]

GALLOWAY, J. J.—*Manual of Foraminifera* (1933), Bloomington, Indiana, U.S.A.

LISTER, J. J.—"Foraminifera" in E. R. Lankester's *Treatise on Zoology* (1903). Biology of Foraminifera, including dimorphism.

OVEY, C. D.—"Difficulties in establishing relationships in Foraminifera," *Proc. Geol. Assoc.*, xlix, 160 (1938).

SCHLUMBERGER, C.—"Dimorphism of Foraminifera," *Ann. Mag. Nat. Hist.* (5), xi (1883); and papers in *Bull. Soc. Géol. France* (1884–1900).

SHERBORN, C. D.—(1) "Bibliography of Foraminifera recent and fossil, 1565–1888" (1888). (2) "Index to genera and species of Foraminifera," *Smithsonian Misc. Coll.*, vol. xxxvii (1893–6).

THALMANN, H. E.—"Index and Bibliography of new genera and species of Foraminifera for the year 1931"—and subsequent years, *Journ. Paleont.*, vii–xv (1933–41).

TOUTKOWSKI, P.—"Index bibliographique de la littérature sur les Foraminifères vivants et fossiles," *Mém. Soc. Nat. Kiev*, xvi (1898). A continuation of Sherborn.

WOOD, A. and BARNARD, T.—"*Ophthalmidium*: a study in nomenclature, variation and evolution in the Foraminifera," *Quart. Journ. Geol. Soc.*, cii, 77 (1946).

POLYZOA

BASSLER, R. S.—*Treatise on Invertebrate Paleontology, Part G, Bryozoa*, 253 pp. (1953) Kansas Univ. Press.

ANNELIDA

ARKELL, W. J.—"U-shaped burrows in the Corallian beds of Dorset," *Geol. Mag.*, lxxvi, 455 (1935).

BATHER, F. A.—(1) "Upper Cretaceous Terebelloids from England," *Geol. Mag.* (5), viii, 481, 529 (1911); (2) "U-shaped burrows near Blea Wyke," *Proc. Yorks. Geol. Soc.* (n.s.), xx, 185 (1925).

XI
THE PLANT KINGDOM

PLANTS are divided by botanists into four grades, corresponding broadly to an advance from an entirely aquatic to an entirely terrestrial life. From what we have learned from fossil animals we may expect that such an advance would not take place along one line but along several parallel lines, and that there would be occasional reversions to an aquatic life on the part of plants whose ancestors had become adapted to the land. Both these expectations are justified.

1. **Thallophyta.**—This, the most primitive grade, corresponds in a general way to the Protozoa among animals, its members being composed of cells only, though they may form aggregates of much greater size and more differentiation than do the Protozoa. An alternation of sexual and asexual generations (as in the Foraminifera and other Protozoa) exists, and is important for its bearing on the life-history of the higher plants. The Thallophyta include the **Algae** (the great majority of all aquatic plants) and the **Fungi.**

2. **Bryophyta.**—In this grade (mosses and liverworts) the alternation of generations is much more regular than in the Thallophyta, the sexual generation (**gametophyte**) being the larger and more important plant, the asexual (**sporophyte**) being short-lived and always attached to the gametophyte: hence it is commonly taken for a mere fruit of the latter. The male and female reproductive cells can only function in a liquid environment, hence the gametophyte can only live in damp situations; but the asexual **spores** produced by the sporophyte are adapted to resist drought and to be disseminated by the wind. The plant-body of the Bryophyta shows greater differentiation in its cells than the Algae, and the beginning of formation of vessels, but it never attains a truly woody structure and is hence almost unknown in the fossil state.

3. **Pteridophyta.**—In this grade (Ferns, etc.) there is a reversal of the relative importance of sporophyte and gametophyte. The former is now the larger and longer-lived, it is well differentiated into root, stem, and leaf, and has woody and other tissues capable of preservation in the rocks. It can live in dry soils, but must not be far from water. The spores are borne in spore-cases (**sporangia**) usually on the ordinary leaves, sometimes on special leaves (**sporophylls**), and are wind-scattered, germinating into gametophytes in damp places.

The gametophyte (or prothallus) is very small, delicate, as simple in structure as any of the Bryophyta, and short-lived. Sometimes the sporangia and spores are of two kinds, large **megaspores** from which female gametophytes grow, and small **microspores** which produce males. The study of spores found in various geological formations is increasing in intensity and the name palynology has lately come into use.

4. **Spermaphyta** (Seed-Bearing Plants).—In these, the highest plants, the megaspore germinates, forms a female gametophyte, which is fertilized and produces an embryo of a new sporophyte—all without ever being shed from the megasporangium, which, when all these developments have taken place within it, is called the **seed.** The microspores are called **pollen-grains,** and germinate on the megasporangium. Thus the dependence of the gametophyte on an aquatic (or even a damp) environment is entirely avoided, and adaptation to the driest conditions becomes possible.

Algae.—Most of the Algae leave no trace in the rocks except indefinite carbonaceous remains. There are two groups which form exceptions to this rule.

1. The Diatomaceae (diatoms), microscopic forms with a siliceous skeleton, with a very fine network-structure, resembling that of Radiolaria but much finer. These occur in immense numbers in fresh and salt waters, and their remains accumulate to form masses of diatomaceous ooze, in the cold Arctic and Antarctic seas especially. Similar deposits occur in Tertiary and Cretaceous rocks, and are worked for polishing powder and as the inactive basis of dynamite. The deposits of Bilin (Bohemia) and Richmond (Virginia) are well-known examples but a small deposit is worked in Co. Derry, Northern Ireland. Sometimes the deposit becomes a chert. Below the Cretaceous rocks they are unknown.

2. Among the red and green sea-weeds there are a number of families in which calcium carbonate is deposited in the cell-walls. These forms are collectively known as "calcareous algae," and some of them have been shown to play a large part in the building of modern coral-reefs, and they were no less important as rock-builders in the past. Their remains have often been misunderstood and attributed to Foraminifera, Sponges or Hydrozoa. The calcareous green sea-weeds (Chlorophyceae) form in general branching tubular structures, which often break up into segments of characteristic appearance. Such are *Gyroporella* (Perm.–Trias.) and *Diplopora* (Trias.), which play a part in the building of the Alpine dolomites, and *Ovulites* (Eocene), of which the segments, common in the Calcaire Grossier of Paris, are egg-shaped with a central tubular cavity.

The calcareous red sea-weeds, on the other hand, are more often massive or encrusting. Important are *Solenopora* (Cam.–Jur., Fig. 120*a*), which contributes to the substance of many British Palaeozoic limestones, and *Lithothamnion* (Cret.–Rec.), a rock-former in the Miocene of the Vienna basin.

Allied probably to these are the Palaeozoic forms *Girvanella* (Cam.–Carb.), *Garwoodia* (Carb.), *Mitcheldeania* (Carb.–Trias.) and *Ortonella* (Carb.), which form nodular concretionary masses in limestones. In the first, fine unbranched tubes (averaging 0·01 mm. in diameter) are twisted irregularly together; in the second, similar but larger tubes (0·05 to 0·07 mm. diameter) are arranged radially, giving off branches which turn abruptly to the radial direction; in the third the tubes are of much smaller (0·02 mm.) diameter; in the fourth, the tubes (0·01 to 0·03 mm. in diameter) closely pressed against each other bifurcate at angles of from 30° to 45°.

Two non-calcareous Algae are found in the highest Silurian and Lower Devonian: *Pachytheca* and *Parka.* The former is a small spherical body, about 2 or 3 mm. in diameter, composed of radial and branching filaments ; the latter consists of flat discoidal thalli, and as they produced cuticularized spores, it is probable that the genus may show an early stage in adaptation to aerial conditions.

An interesting group, which attains a higher grade than the true Algae, having definite stems and leaves (but no vessels), is the Characeae, of freshwater habitat and secreting a calcareous skeleton. It is best known by its spirally-marked fruits, found in the Purbeck and Oligocene beds of the south of England (Fig. 120*b*).

Pteridophyta (Vascular Cryptogams).—Three distinct lines of descent may be recognized, in all three of which two subgrades are passed through (a lower **homosporous,** with only one kind of spore and sporangium, and a higher **heterosporous,** with mega- and micro-sporangia and spores), while two of them evolve independently into the higher seed-bearing grade. These three lines are the Lycopodiales (represented today by the creeping, moss-like *Lycopodium* and *Selaginella*) the Equisetales (by the marsh-loving "horse-tails") and the Filicales (ferns). The two first, at least, existed in the Upper Devonian and attained their highest development in the late Palaeozoic, and alongside them are found both undoubted ferns and another group, Sphenophyllales, which appears to represent the common stock from which they sprang. The extensive working of coal in the Carboniferous rocks has enabled the flora to be collected very extensively and studied very thoroughly, and some explanation is necessary of the conditions of preservation of these fossils.

The remains of land-plants are of two kinds—**impressions** of

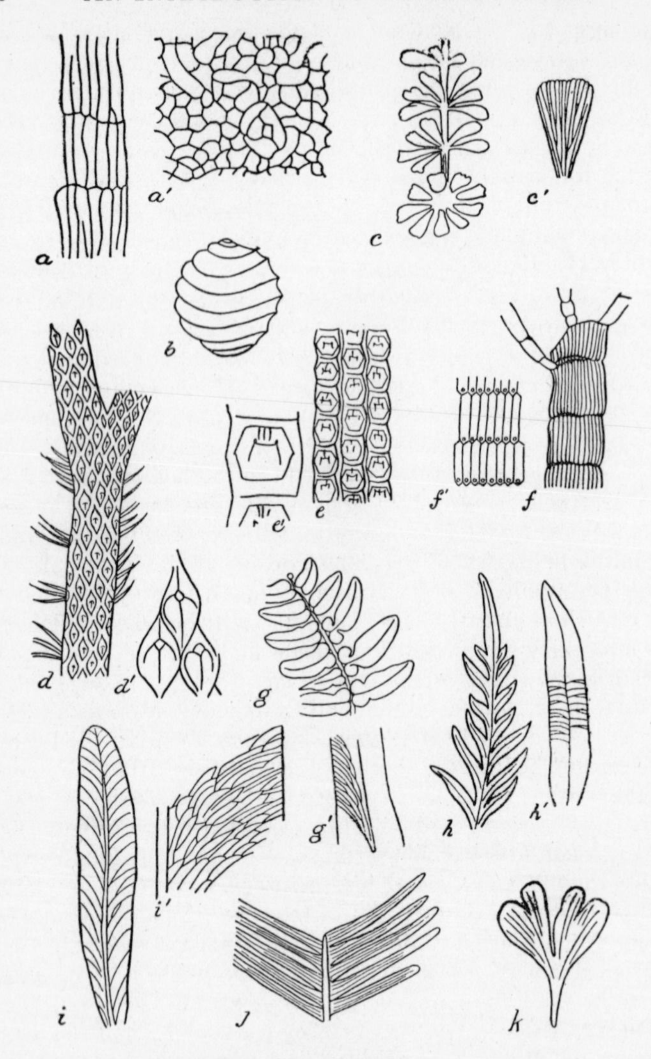

FIG. 120.—FOSSIL PLANTS

a, *Solenopora compacta* (Billings), Ordovician; *a*, Vertical section (× 25), showing part of three concentric layers of cells; *a′*, Horizontal section (× 50; after Brown). *b*, *Chara lyelli* (Forbes), Oligocene; fruit (× 10; after Forbes). *c*, *Sphenophyllum schlotheimi* (Brongniart), Coal Measures; *c*, Stem with whorls of leaves (× $\frac{1}{2}$); *c′*, single leaf (Natural size). *d*, *Lepidodendron sternbergi* (Brongniart), Coal Measures (× $\frac{1}{4}$); *d′*, Leaf-traces of *L. gracile* (Lindley) (× $\frac{1}{2}$). *e*, *Sigillaria tessellata* (Brongniart), Coal Measures; *e*, Part of trunk (× $\frac{1}{2}$); *e′*, Leaf-traces (× $\frac{4}{3}$). *f*, *Calamites mougeoti* (Brongniart), Coal Measures; trunk (× $\frac{1}{3}$); *f′*, *C. decoratus* (Brongniart), Coal Measures, bark

leaves, bark, etc., which show in a carbonaceous film the outward form and surface markings, commonly very perfectly, but showing no internal structure; and **petrifactions,** either in silica or calcite, which preserve internal structure often as perfectly as in a recent plant kept in methylated spirit, but are usually too fragmentary to show the full outward form. Further, the various parts of a complete plant—root, stem, leaves, sporophylls, seeds—are rarely found in organic continuity, and even such a single part as the stem may present very different appearances according as the outer cortex is preserved or not. It is evident therefore that the palaeobotanist has to reconstruct his plants out of very scattered and fragmentary evidence. Great difficulties of nomenclature arise: leaf impressions receive one set of names, stems another, roots another, and even when leaf, stem, and root have been pieced together it may be found that leaves whose forms differ so much that they have been put into distinct genera belong to stems which cannot be distinguished, and again stems which seem justifiably to be divided into two or more genera have indistinguishable roots. A triple nomenclature at least is therefore inevitable, and palaeobotanists speak of "leaf-genera," etc. In the case of the most fully investigated Coal-Measures plant, it is known that at least five generic and five specific names had been given to its separate parts.

Psilophytales.—These are the most primitive Pteridophytes, occurring in the Upper Silurian and Lower and Middle Devonian. They show some relations to Algae, but have definite vascular tissues arranged in a simple **stele** (vascular cylinder of both wood and bast) and sporangia of pteridophyte type. They have no true roots, and the simpler forms (*Psilophyton*, *Rhynia*, *Hornea*) have no leaves. *Psilophyton* is known from impressions in the Upper Silurian and Lower Devonian, but most of the other genera are found in the Rhynie Chert, a silicified peat-bog of Middle Old Red Sandstone age, in Aberdeenshire which represents the oldest Palaeozoic truly land flora known in the world. *Rhynia* was the chief constituent of the peat: it

---

FIG. 120.—FOSSIL PLANTS (*continued*).

(× ½). *g*, *Neuropteris heterophylla* (Brongniart), Coal Measures; part of leaf (× ½); *g′* Venation of *N. blissi* (Lesquereux) (× 3/2; after Kidston). *h*, *Alethopteris lonchitica* (Schlotheim), Coal Measures; leaf (× ½); *h′*, leaflet (Natural size) to show venation. *i*, *Glossopteris browniana* (Brongniart), Permo-Carboniferous, New South Wales; leaf (× ½); *i′*, Venation (× 3/2; after Seward). *j*, *Williamsonia pecten* (Phillips), Lower Estuarine Beds (Aalenian), Scarborough (× ½; after Seward). *k*, *Ginkgo digitata* (Brongniart), Middle Estuarine Beds (Bajocian), Scarborough (× ½; after Seward) (After Brongniart or Lindley, where not otherwise stated).

had submerged rhizomes, from which delicate leafless stems rose in the air, ending in sporangia. *Hornea* differed mainly in the sporangia. In *Asteroxylon* there were definite leaves, and some of the rhizomes seem to have shown a tendency to be transformed into roots.

Sphenophyllales.—Only one genus is known, *Sphenophyllum* (Dev.?, Carb.–L. Perm.), which seems to have been a climbing plant, very much like the modern bedstraws in appearance. The leaves are in whorls round the stem, and are wedge-shaped, narrow at the attachment, widening out and ending abruptly at their broadest (Fig. 120*c*).

Equisetales.—This group attained its highest development in the Carboniferous period, when it was represented by trees which attained a height of fifty feet or more and a thickness of several feet, the latter acquired by a process of secondary thickening like that of modern trees. Like the little *Equisetum* of today, they had a hollow stem, and are most commonly represented by internal moulds (pith-casts), the markings on which are very like those of the exterior of *Equisetum*—vertical flutings, alternating in position at the nodes (where leaf-whorls arose). Chief genera: *Archaeocalamites* (U. Dev.–L. Carb.), *Calamites* (U. Carb., Fig. 120*f*, leaves known as *Annularia*, *Asterophyllites*, *Calamocladus*, reproductive cones as *Calamostachys*, *etc.*), *Schizoneura* (Perm.–Trias.), *Equisetites* (Rhaetic–Wealden).

Lycopodiales.—The living members of this group are all small herbaceous plants, but in the Devonian and Carboniferous periods it was also represented by forest-trees of great size and with secondary thickening. Genera: *Lepidodendron* (Upper Dev.–Carb., Fig. 120*d*), *Sigillaria* (U. Carb., Fig. 120*e*). These two genera are readily distinguished by the leaf-scar pattern on the bark—the former having large rhomboidal scars covering the whole surface, the latter having small scars of varying shape in vertical rows: in both, the full pattern is only shown on the exterior of the bark, its internal surface or the decorticated stem showing a ghost of the same pattern. The roots of both are alike, branching dichotomously (*Stigmaria*); the reproductive cones of *Lepidodendron* are known as *Lepidostrobus*, those of *Sigillaria* as *Sigillariostrobus*. In some species of *Lepidodendron* the megasporangium seems to have become a true seed, the microspores germinating on it and fertilization taking place inside it.

Filicales.—The modern ferns are divided into three groups, (1) Eufilicinae, which includes all the familiar ferns: they are described as **homosporous** (producing one kind of spores only), and **leptosporangiate** (the sporangia developing from a single cell); (2) Hydropteridae, a small group of water-ferns, which are leptosporangiate but **heterosporous** (with mega- and micro-sporangia); (3) Marattiales, a very small group which are **eusporangiate** (sporangia developed from a group of cells) and homosporous.

What may be a very primitive member of the Eufilicinae (*Gosslingia*) has recently been described from the Lower Old Red Sandstone of Brecon. Fossil Eufilicinae, related to the recent *Osmunda*, are known from Upper Permian onwards, and representatives of other families from the Upper Triassic onwards. In the various estuarine beds of the Jurassic of Yorkshire, for instance, there are found fern-leaves closely similar to those of ferns now found only in the Malayan region (*Dictyophyllum*, *Hausmannia*). An abundant genus in the Wealden is *Tempskya*. Hydropteridae are known from the Eocene onwards.

The Marattiales, however, are very well represented, from the Carboniferous onwards. Some at least of the species placed in the leaf-genus *Pecopteris* (Carb.–Jur.) seem to belong to the tree-fern stem *Psaronius* (U. Carb.–L. Perm.).

But the great majority of what have been called ferns in the Carboniferous flora have proved to belong to a higher grade. They may be related to the Marattiales as the Hydropteridae are to the Eufilicinae, but they had advanced much beyond merely being heterosporous, and form the first division of the Spermaphyta.

SPERMAPHYTA.—The seed plants again fall into two subgrades, according to whether the seed is borne on the surface of a sporophyll (**Gymnosperms**) or in a closed ovary formed by the folding over of one or more sporophylls (**Angiosperms**). The Gymnosperms include the following five groups:

1. **Pteridospermeae.**—This extinct group is intermediate between the eusporangiate ferns (Marattiales) and the cycads. It is confined to the Upper Palaeozoic, and best known from the fern-like leaves of the Coal Measures. By piecing together evidence from the microscopical structure of petrifactions with that from leaf-impressions, palaeobotanists have been able to establish its true position. Common leaf-genera are *Alethopteris* and *Neuropteris* (Fig. 120*g*, *h*), corresponding to the stem genus *Medullosa* and the seed *Trigonocarpum*; *Sphenopteris*, to the stem *Lyginodendron* and the seed *Lagenostoma*. Other leaves probably belonging here are *Glossopteris* (Permo-Carb.–Rhaetic, Fig. 120*i*) with its rhizome *Vertebraria*, and the allied *Gangamopteris* (Permo-Carb.). The two seeds just mentioned and a third, *Cordaicarpus*, represent three distinct types of Coal-Measures seeds.

2. **Cycadophyta.**—The modern cycads are a group of plants with palm-like foliage, but much more primitive in their sporophylls than

the palms (which are angiosperms). They are confined to tropical regions, but extend all round the world. Their Mesozoic allies had no such restriction in latitude, being common in England for instance, from the Rhaetic, Jurassic, and Lower Cretaceous. The Jurassic period has been termed the "age of cycads."

Their remains are abundant in the Jurassic estuarine beds of the North of England, in some of the Purbeck Beds, where they occur in the position of growth in the "fossil forest" of Dorset cliffs, and in the Wealden Beds. But the most famous locality for them is the Black Hills of Dakota, where wonderfully preserved specimens of Upper Jurassic age show the details of the floral organs, some of which attained a far higher grade of structure than the modern cycads. This indicates a near relationship to the Angiosperms (typical flowering plants with seeds enclosed in a fruit). Leaf-genera: *Pterophyllum* (Trias.), *Cycadeoidea*, *Williamsonia* (Fig. 120*j*), *Ptilophyllum* (all Jur.), *Otozamites* (Trias.–L. Cret.), *Nilssonia* (Jur.–L. Cret.), *Zamites* (L. Cret.).

3. **Cordaitales.**—The genus *Cordaites* (Dev.–Rhaetic) was a forest tree resembling in habit the Australian Kauri pine, to which it may have been not very distantly related. Its leaves were long, narrow, thick and parallel veined; its seeds (*Cordianthus*) were borne on a cone.

4. **Ginkgoales.**—The maiden-hair tree, *Ginkgo*, with its bifid wedge-shaped leaves (Fig. 120*k*), is now confined to parts of China and Japan; but closely allied forms had a world-wide distribution in Jurassic times, ranging from Greenland to Tasmania. Its time-range in Europe is Rhaetic to Eocene.

5. **Coniferales.**—These include the familiar firs and pines, the yew and monkey-puzzle (*Araucaria*). This last is now native only in South America and Australasia, but it also had an almost world-wide distribution in the Jurassic and Lower Cretaceous. Allied forms are *Walchia* (Perm., rare in U. Carb.) and *Voltzia* (Perm.–Trias.). Coniferous wood is a common object in many Jurassic and Cretaceous rocks, both marine and freshwater.

Angiospermae.—A primitive group, the Caytoniales, has been found in the Middle Jurassic, but at the beginning of Upper Cretaceous times there was such a sudden appearance of an extensive angiosperm flora in Europe, that it is obvious that such a cryptogenetic flora must have migrated from some other part of the world, as yet unknown. Its fossil remains are found, not in England, which was then covered by a fairly deep sea, but in regions like Saxony, which were close to a shore-line and where sandy deposits occur in place of chalk. From this time onwards, the angiosperms are the dominant members of the flora everywhere. Their natural orders and genera are far too numerous to be mentioned in detail here.

Fossil plants are valuable indices of the physical and climatic conditions under which they lived, but as indices of age their value is usually general rather than precise. Thus the floral stages claimed to exist in the Carboniferous are determined more by associations than by zonal species, and even so are less trustworthy than the goniatite- or freshwater mussel-zones. Yet in particular cases an evolutionary sequence of species may date a series of strata, as in the case of the seeds of the freshwater plant *Stratiotes*, from Upper Eocene to Recent (see CHANDLER (1) in Bibliography).

A special contribution to very detailed stratigraphy has been made in recent years by the study of the **pollen-grains** preserved in the late Tertiary and Pleistocene deposits, in Sweden and Britain. These give evidence of successive waves of forest with different genera of trees predominating, and have proved of great value in correlating deposits varying greatly in character from place to place.

The possible existence of marine life-provinces in past times has been referred to in Chapter V. Fossil plants have provided the most unquestionable example of terrestrial life-provinces. The land floras of Devonian and Lower Carboniferous times are practically the same all over the world, but in the Upper Carboniverous there were two sharply contrasted floras—a northern flora, including *Lepidodendron*, *Sigillaria*, *Calamites*, and many pteridosperms, and a southern flora, without any large trees, but chiefly characterized by *Glossopteris* and *Gangamopteris*. The former is found over most of the Northern Hemisphere, both in the Old and New Worlds, the latter is principally found in the Southern Hemisphere and India. So entirely different are the two, that for a long time many geologists would not admit them to be contemporaneous, but held the *Glossopteris* flora as Mesozoic, it having in fact nearer affinities to the European Mesozoic than to the European Carboniferous flora. The discovery of marine strata having an Upper Carboniferous or Lower Permian fauna, alternating with beds containing the *Glossopteris* flora, at length gave conclusive proof of the Permo-Carboniferous age of the latter. It appeared, therefore, that there were two continental masses, a northern on which the descendants of a portion of the Lower Carboniferous flora continued their evolution, and a southern (called Gondwanaland) on which that flora had been exterminated (possibly by glacial conditions of which there is good evidence), and a new and much less rich flora developed. Later discoveries have shown that there was some geographical overlap of the two floras, *Sigillaria* having been found in South Africa, and *Glossopteris* in N.E. Russia (along with Therapsid reptiles and freshwater mussels of South African genera); but the exact nature of this overlap remains to be explained. By the Rhaetic epoch a uniform flora had once again established itself over the land-surfaces of the world.

Short Bibliography

ARBER, E. A. Newell.—(1) "The Glossopteris Flora," *Brit. Mus. Cat.* (1905). (2) Several papers on Coal Measures Plants in *Phil. Trans. Royal Soc.* (B) (1904–16). (3) *The Natural History of Coal* (1911) Cambridge University Press.

CHANDLER, Marjorie E. J.—(1) "Geological History of the genus *Stratiotes*," *Quart. Journ. Geol. Soc.*, lxxix (1923); (2) "Upper Eocene Flora of Hordle," *Palaeont. Soc.* (1925–6).

CROOKALL, R.—(1) *Coal Measure Plants* (1929), London. (2) "Fossil Plants of the Carboniferous Rocks of Great Britain [Second Section]", *Mem. Geol. Surv. Gt. Britain: Palaeont.*, iv, pt. 1 (1955), pt. 2 (1959).

GARDNER, J. S. and ETTINGHAUSEN, BARON.—"Eocene Flora," *Palaeont. Soc.* (1879–86).

GARWOOD, E. J.—"Calcareous Algae" (Pres. Address Sect. C. Brit. Assoc., 1913), *Geol. Mag.*, l, 440, 490, 545 (1913).

GODWIN, H.—"Pollen-Analysis and Quaternary Geology," *Proc. Geol. Assoc.*, lii, 328 (1941).

HICKLING, G.—"Contribution to the Micro-Petrology of Coal," *Trans. Inst. Mining Engineers*, liii (1917).

KIDSTON, R.—(1) Many papers on Coal Measures Plants in *Trans. Roy. Soc. Edinburgh* (1883–1925). (2) "Fossil Plants of the Carboniferous Rocks of Great Britain," *Mem. Geol. Surv. Gt. Britain: Palaeont.*, ii, pts. 1–6 (1923–5) see also Crookall (2) above.

KIDSTON, R. and LANG, W. H.—Papers on Devonian Plants in *Trans. Roy. Soc. Edinburgh* (1917–24).

LANG, W. H. and COOKSON, Ida C.—"On a Flora . . . associated with *Monograptus* . . . from Victoria, Australia," *Phil. Trans. Roy. Soc.* (B), ccxxiv, 421 (1935).

OLIVER, F. W. and SCOTT, D. H.—"On the Structure of the Palaeozoic Seed, *Lagenostoma ovoides*," *Phil. Trans. Roy. Soc.* (B), cxcvii, 193 (1904).

REID, Eleanor M. and CHANDLER, Marjorie E. J.—"The London Clay Flora," *Brit. Mus.* (1933).

SCOTT, D. H.—(1) *Studies in Fossil Botany* (3rd edn., 1923). (Microscopic Structure of Coal Measures Plants.) (2) Presidential Addresses, *Royal Micr. Soc.* (1905–7). (3) *Extinct Plants and Problems of Evolution* (1924).

SEWARD, A. C.—(1) *Fossil Plants*, vols. i–iv, Cambridge Biological Series (1898–1919). (2) "The Jurassic Flora," *Brit. Mus. Cat.*, vol. i (1900), ii (1904).

STOPES, Marie C.—(1) *Ancient Plants* (1910). (2) "Studies in the Composition of Coal," *Proc. Roy. Soc.* (B), xc (1919).

STOPES, Marie C. and WATSON, D. M. S.—"On the present distribution and origin of . . . Coal-balls," *Phil. Trans. Roy. Soc.* (B), cc, 167 (1909).

THOMAS, H. Hamshaw.—"The Caytoniales: a new group of Angiospermous Plants from the Jurassic Rocks of Yorkshire," *Phil. Trans. Roy. Soc.* (B), ccxiii, 299 (1925).

WIELAND, G. R.—*American Fossil Cycads*, Carnegie Institution of Washington (1906).

WOOD, Alan.—"Lower Carboniferous Calcareous Algae," *Proc. Geol. Assoc.*, lii, 216 (1941).

XII

THE COLLECTION AND PRESERVATION OF FOSSILS

If a fossil be defined as any trace of an organism preserved as part of a rock, we must distinguish between various conditions of preservation of fossils.

1. The preservation of the whole body of an animal or plant unchanged. It is doubtful if any fossil really satisfies this description. The most often quoted case is that of insects in amber—a fossil resin which enveloped them as it trickled down the bark, preserved them by its antiseptic properties, and became hardened by loss of its volatile constituents. It is only the exoskeleton of the insect, however, which is preserved, not the internal organs. If the amber is treated with a normal organic solvent, strangely enough nothing remains insoluble so presumably the chitin has been changed chemically.

The frozen mammoths and woolly rhinoceroses of Siberia are certainly preserved whole, but not as part of a sediment. As they are extinct species, however, they may pass as fossils.

2. While the soft parts decay, the hard parts—mainly of mineral substance, but sometimes organic—may be preserved without any chemical change. This is a very common case among animals, if we interpret the term "chemical change" rather broadly. Some amount of chemical change is almost inevitable—the filling-up of minute cavities with secondary mineral matter, as in echinoderm skeletons and vertebrate bones, is rarely escaped. Among plants, where the hard parts are organic, some carbonization nearly always takes place; but in the paper-coal of Tula, in Russia, the cuticles of the plants are said to be absolutely unchanged. The same may be the case with some Graptolites, Trilobites and Eurypterids.

3. The hard parts may be preserved, but with more or less complete chemical change. If the original skeleton was of calcite this may be replaced by silica, dolomite, chalybite, limonite, calcium phosphate, pyrite, fluorite or selenite. Aragonite may be replaced by any of the same, or may undergo a paramorphic change into calcite. The extent to which these changes may alter the minute structure of the skeleton varies very much: when any recrystallization takes place the minute structure is usually destroyed, wholly or in part. Amorphous

silica (opal) may preserve the finest details, but if it passes into chalcedony it will commonly produce a structure due to imperfect crystallization in rings (**beekite** structure), which will not only destroy minute structure but even obscure the finer details of outward form, as in the Lower Carboniferous corals of Tournai, Belgium. When calcite is replaced by selenite (gypsum), few fossils but belemnites can survive the change in recognizable shape; but this chemical change differs from most others, as it is usually a very recent result of weathering near the surface, while most others took place soon after the entombing of the fossil.

Originally siliceous skeletons may be altered to calcite, pyrite or marcasite, the last in turn passing into limonite under conditions of weathering. Sponges in the chalk may sometimes be found, partly as silica, partly limonite.

Hard organic tissues generally undergo some degree of carbonization, but they may be replaced by pyrite (as with graptolites, though here in many cases, pyrite only forms an infilling or internal mould), or by calcite or silica (as in plant petrifactions).

Phosphatic skeletons are generally stable, and beyond having their minute cavities filled with secondary phosphate undergo little change; very rarely they may be changed to vivianite (a blue iron phosphate). In the Forest of Dean, red sandstones of early Carboniferous age, however, contain empty moulds of fish teeth and spines.

4. An organism may be represented by moulds of its skeleton—external or internal. External moulds are composed of the matrix of the rock, that is to say the sediment in which the organism was buried: they are produced when a fossil is extracted, unless the rock is too friable, but they may be found naturally produced through the removal of the fossil in solution as in the case of the fish remains mentioned in the previous paragraph. Infillings of the interior (internal moulds) may also consist of the matrix, when that had free access to the interior; when it had no access, they consist of some material deposited from solution, or partly the one and partly the other, when the sediment could only penetrate into part of the interior. Very rarely does a hollow skeleton remain really empty, though infilling may often be incomplete. In limestones, the material deposited from solution will be calcite, or chalcedony (as in the chalk-flints); in clays it is usually pyrite, or marcasite (the two forms of FeS_2). In cases where deposition on the sea-floor was extremely slow, calcium phosphate is a usual infilling, or for minute cavities glauconite: these two substances being deposited before burial is complete.

In any of these cases it is only by subsequent removal of the shell itself that the infilling appears as an actual internal mould. Such removal may be by solution on the sea-bottom or within the rock, or it may be artificially produced, either accidentally during the extraction, or deliberately for the purpose of investigating internal structure. When naturally produced, the external and internal moulds together are often described as "hollow casts": by pouring melted wax or gutta-percha, or the various compositions used by dentists in the preparation of dentures, into the cavity, a solid cast of the fossil can be obtained. An impression for temporary use may be obtained by using plasticene, slightly wetted to ensure removal: should a perfect cast of an object with re-entrant angles be needed, a liquid which sets into an elastic substance like rubber should be employed as the casting medium. Polyvinyl choride has gained use for this purpose in recent years especially and it has the advantage of speed in execution as described by RIXON and MEADE (see Bibliography, p. 261).

5. There are other traces of living organisms in rocks that may rank as fossils, though even less definite than moulds. Such are footprints and other impressions of moving animals in sand or mud (preserved by being covered by a lamina of a different kind of sediment), burrows in sand or mud (preserved by being filled with slightly different material) and borings in hard rocks or in fossils. Among boring organisms are minute algae, sponges (*Cliona*), some echinoids, many lamellibranchs (*Lithophaga*, *Pholas*, *Teredo*). Sometimes when a shell or coral is bored into, the material filling the boring may resist subsequent solution better than the shell or coral itself, and a solid mould of the boring may result.

All these various ways of preservation must be borne in mind by the intelligent collector when at work on cliffs and quarries. While the spoil-heap of a quarry will often provide the choicest specimens, he should never neglect to search for fossils *in situ* in the undisturbed rock. Only by so doing can he know whether all the fossils belong to one fauna, or whether there are several, and through what thickness a single fauna persists; only so can he see whether *Lingula* or lamellibranchs, for instance, are in the vertical position of living burrowers or with separated valves are in the horizontal position of drifted shells; **derived fossils** may be found, whose original home was in some older bed and whose worn and rounded surface, and perhaps different means of preservation or content of matrix distinguishes them from the contemporaneous fossils with which they may be mixed.

The collector would be well advised not to attempt to clean or select his fossils on the spot where he obtains them, except when their weight is serious, or when they are too fragile for transport and must at once be treated with some preservative. Many a good specimen has been spoiled by the attempt to clean it on the spot. Even broken specimens may often prove to show special features not seen in perfect examples. The writer [**A.M.D.**] once picked up seventy-three specimens of *Chonetes laguessianus* from the Carboniferous shales of Fourstones (Northumberland). They all looked alike at the time, but on cleaning and sorting them at leisure afterwards, fifty-eight were found to be complete specimens but with the dorsal valve crushed in; fourteen were ventral valves, and one was a dorsal valve showing both external and internal features. That one dorsal valve would probably have been lost had any attempt to clean or select been made on the spot.

Care must be taken to pack up specimens so that they do not get broken in transit home: heavy fossils should not be placed on top of fragile small ones, for instance. Fossils collected from wet clay or in wet mudstone must be allowed to dry out slowly. Each specimen must be so labelled that there will be no mistake as to the collecting locality and the particular bed from which they came. At the time it may seem quite safe to trust to memory, but a week afterwards an unlabelled specimen will prove to be an enigma to the conscientious collector.

The extracting and cleaning of fossils—the removal of fossil from matrix or matrix from fossil—is an art in itself. In extracting large specimens from hard rocks, such as tough limestones, the latter must be subjected to great stress—the blow of a sledge-hammer may make a fossil jump out quite clean. A screw-movement stone-breaking machine may be effectively used in the same way. At other times when the fossils are fragile and firmly cemented to the matrix, their violent extraction may be almost hopeless, but heating the rock and cooling it by plunging it into cold water may have the required effect. In the extraction of small fossils, such as foraminifera and ostracods, from sands and friable rocks, the rock is first broken up and then passed through a series of sieves: the great majority of fossils will be found between a mesh of 32 to the inch and one of 64 to the inch. In the case of stiff clays, another method must be used: the clays are broken up into small pieces, about the size of a pea or bean, and heated on a metal plate over a flame until thoroughly dry, when they are dropped while still hot into cold water in a circular dish. The clay quickly disintegrates into mud, and by keeping up a swirling of the water by constant movement of the dish, decanting the muddy water and adding fresh over and over again, all the argillaceous material is at length removed and a concentrate left, consisting of sand-grains and small fossils. This is dried and then spread out, a little at a time, on black paper, examined with a lens, and fossils picked out with a moistened sable brush, or in the case of heavier ones with fine forceps. Siftings of friable rocks are picked over in the same way. Shale or mudstone may be converted almost to its constituent mud by heating fragments for a short time in a solution of hydrogen peroxide.

Chalk may be scrubbed under water with a stiff brush, and a concentrate obtained: the chalk from the interior of large fossils (such as echinoids) is often the most fruitful, because the small shells have been more protected from destructive movements than those in the ooze on the sea-bottom.

The methods of removing matrix from a fossil depend greatly on their respective toughness. When the fossils are strong and the matrix friable, washing with or without the help of a brush (a tooth-brush is very convenient) may be enough; but some fossils (many lamellibranchs) will break into pieces if wetted, and must be cleaned dry with the help of some tool. A shoemaker's awl mounted in a handle is very useful for tough matrix; in other cases a mounted gramophone needle, pin, or steel pen, or even a common pocket-knife may be serviceable. A motor-driven dental mallet or vibro-tool are becoming increasingly used for this purpose. The golden rule about all such tools is that they should never be used to scrape the matrix away from the fossil. The aim should always be to put such a strain on the matrix by means of a tool applied at right-angles to the plane of the fossil in such a way that the matrix breaks away without the tool touching the fossil. The application of this rule to particular cases may be varied.

As an example, the case of a graptolite, part of which is exposed on the lamination surface of a shale, partly buried under other laminae. The natural first idea is to push a knife-blade between the laminae and separate them, but the probable result of this will be that the laminae break away anywhere except where the graptolite lies, and the hidden part may be broken before it is exposed. If, however, the knife-blade is pressed vertically downwards along a line parallel to the hidden part but a quarter-inch or more away, the strain on the concealing laminae of shale will very likely cause them to split off just where wanted.

Brushes, the bristles though stiff being softer than most fossils, can be applied directly to the surface of a fossil. The dental drill used by dentists for the filing of teeth can be adapted to the cleaning of fossils: it enables a rapidly rotating brush to be applied to any point on the surface of the fossil, with more concentrated and rapid effect than that of a hand-brush. Wire brushes can even be used to clean pyritic fossils embedded in hard slates.

Where mechanical means fail, chemical methods may be used. The simplest of these is ordinary weathering, which in many cases disintegrates matrix without injuring fossils (though in other cases the reverse is the case). Very weak acids may effect the same result more rapidly. In general, acids must be resorted to with great care, since most fossils are calcareous and at once attacked by them. Where it is desired to expose more fully a calcareous fossil partly embedded in tough limestone, the exposed surface may be varnished to protect it and acid poured on: careful watch must be kept so that as soon as any more is exposed this may be washed and varnished.

Dilute hydrochloric or acetic acids have yielded excellent results in isolating silicified trilobites, or graptolites or conodonts preserved in limestone. Acetic or formic acid has been used for freeing phosphatic fossils including bones, and chitinous fossils such as insects from matrix.

Graptolites and plants preserved in shale can be isolated by the "transfer method," invented independently by HOLM and WALTON (see Bibliography). The essentials are: (1) Cement the specimen, downwards, by canada balsam, on to a glass slide; (2) cut or grind away as much of the shale as can safely be done with tools; (3) wet the shale surface and dip the whole in melted wax, which may easily be separated from the wet shale and enough left to protect the glass from chemical action; (4) place the whole in an etching bath of hydrofluoric acid, which rots the shale without injuring the organic material; (5) after washing, protect the plant or graptolite by a coverglass and balsam. For necessary precautions and other details the papers quoted in the Bibliography must be studied.

Soft fossils such as graptolites and some plants, thus extracted, may be treated like recent soft organisms, and a continuous series of sections cut by a microtome. From such series large-scale models of the fossil can be constructed. For ordinary hard fossils the same result can only be obtained by a laborious process of grinding, photographs being taken at regular intervals in the grinding. (See SOLLAS in Bibliography.) The objection to this method—that it destroys the fossil under examination—can partly be avoided in certain cases by the use of acid and some cellulose solution (such as Durofix). As first tried on Coal-balls (WALTON, in Bibliography) the process was this: (1) Cut or grind to a flat surface in the required direction for sections; (2) immerse in an etching solution of hydrochloric acid; (3) after washing and drying, pour the cellulose solution over the etched surface; (4) when dry, the cellulose film can be peeled off, bringing the plant-section with it. The process is repeated, to give a complete sequence of sections. When the remains are preserved in silica, hydrofluoric acid is used.

It was later found that the method could be applied to calcareous fossils, such as corals. Owing to the slight difference of resistance to acid in the different structures, a cellulose film peeled off an etched surface of a coral shows the pattern of that surface, and serial sections can be obtained with less trouble than is involved by repeated photographing. Moreover, any number of films can be peeled off each successive etched surface. Nevertheless, this method does actually destroy the coral, but in the case of coal-balls, the plant still exists in the pulled off sections.

Many fossils, though stable in their long home, are physically or chemically unstable as soon as extracted, and need special treatment to save them from disintegration or the risk of it. If they are simply fragile they must be impregnated with some binding material. The most usual one is gelatine, applied as a hot, strong solution with a soft brush: it penetrates the shell and as it hardens it binds the parts together and prevents decay. It has the disadvantage of requiring rapid treatment lest it harden before penetrating. Another material is a solution of collodion in amyl acetate: this has the advantage that it hardens much less quickly by evaporation. It can be brushed over the shell at intervals of a day or more until no more is absorbed, or small specimens may be dropped into the solution and left there indefinitely. Care must be taken not to inhale the amyl acetate.

The chief case of chemical disintegration is that of fossils preserved in marcasite; this mineral is an unstable sulphide of iron which absorbs moisture and oxygen, and gives a crystalline efflorescence of ferrous sulphate. As soon as such an efflorescence is noticed, it must be brushed away and the fossil soaked in ammonia or hot, but not boiling, potash solution to neutralize any free sulphuric acid. It is then washed thoroughly in distilled water to remove the alkali, and thoroughly dried in an oven or on a hot plate, after which it is painted all over with shellac varnish (or if not too large soaked in the varnish). In this way further access of oxygen and moisture is prevented.

The various methods so far described give some idea of the more frequent and general forms of palaeontological technique. Other special methods can only be mentioned, such as the delicate technique employed for exposing, tracing, and reproducing the suture-lines of ammonoids (of which several examples are shown in the Frontispiece). The accurate measurements of fossils by means of sliding callipers, and in the case of microscopic fossils either by an eye-piece or stage micrometer, is commonly necessary in the identification of species and for statistical or developmental studies.

Short Bibliography

BATHER, F. A.—Nathorst's Collodion method, *Geol. Mag.* (5), iv (October, 1907).

BULMAN, O. M. B.—(1) "Dendroid Graptolites," *Palaeont. Soc.*, Introduction, pp. 1–6 (1927). (2) "Recent developments and trends in Palaeontology," *Adv. Science*, no. 62. (1959).

BUCKMAN, S. S.—"A method of removing tests from fossils," *Amer. Journ. Sci.* (4), xxxii, 163 (1911).

BUTLER, A. J.—"Use of Collodion films in Palaeontology," *Nature*, xxxv, 510 (30th March, 1935).

DAVIES, J. H. and TRUEMAN, A. E.—"Revision of the non-marine Lamellibranchs of the Coal Measures . . . ," *Quart. Journ. Geol. Soc.*, lxxxiii, 210 (1927). Methods of studying measurable variations.

FISCHER, A.—"Rubber casts and molds of fossils," *Journ. Paleont.*, xiii, 621 (1939).

GLAESSNER, M. F.—*Principles of Micropalaeontology* (1945) Melbourne University Press.

RIXON, A. E. and MEADE, M. J.—"Casting Techniques," *Mus. Journ.*, lvi, 9 (1956).

SOLLAS, W. J.—"A Method for the investigation of Fossils by Serial Sections," *Phil. Trans. Roy. Soc.* (B), cxcvi, 259 (1904); and *Geol. Mag.* (4), x, 361 (1903).

WALTON, J.—*Annals of Botany*, xxxvii, 379 (1928). Describes the "transfer method."

WALTON, J.—"A Method of preparing sections of Fossil Plants contained in Coal-Balls, etc.," *Nature*, cxxii, 571 (15th March, 1930).

XIII

THE RULES OF NOMENCLATURE

THE subject of the nomenclature, or naming of fossils, is productive of irritation and of sarcastic utterances among some field-geologists, to whom the time spent by many palaeontologists on the accurate determination of the names of fossils appears wasteful. The attitude of mind of these critics is expressed by a rather hackneyed quotation from Shakespeare, of which it is sufficient to say that if Juliet had ordered a rose from her gardener by the name of garlic, the latter would not have smelled as sweet as she desired. The purpose of giving names to things is that persons may be able to speak to one another about them without danger of misunderstanding.

In proportion as we wish to discriminate more and more exactly between things, we require more and more exact names for them. There are purposes for which roses and lilies and buttercups may be spoken of without discrimination as "flowers"; there are others for which they require distinct names; there are yet others for which the need is felt of finer discrimination, by means of qualifying adjectives as "red rose" and "white rose"; while a modern gardener requires separate names for a great number of different kinds in each of these categories.

The naming of fossils is only a part of the naming of animals and plants in general. Leaving plants aside for the moment, the Animal Kingdom is divided into a number of **phyla,** each of which may be divided successively into **classes, orders, families, genera,** and finally into **species.** To define a species is difficult enough among living animals, where the test of interbreeding is available; among fossils there is only one really scientific way of delimiting species, and that is by determining whether variation is continuous or discontinuous. To do this a large number of individuals from precisely the same bedding plane should be taken, note made of those measurable characters in which they show variation, and a series of accurate measurements taken. The results for each character are plotted as a graph, the horizontal coordinates being the measurements and the vertical coordinates the number of individuals giving each particular measurement. If the plotted curve is a simple one with a single maximum, that is evidence for treating all the individuals as one species: but if it shows two or more maxima with definite minima between, then there is ground for regarding them as two or more species. Confirmation should be sought in the curves for the other measurable characters.

This method is only practicable when large numbers of individuals from the same bed are available, and when their variations are expressible in simple numerical form. In the absence of such conditions no better definition of a species than this can be suggested for the purposes of palaeontology: "A species is a collection of individuals so nearly alike that they may conveniently be denoted by the same name". This definition leaves the decision to the judgment of every palaeontologist, as to whether there is sufficient likeness, and judgments will inevitably differ: as such differences constantly occur, this definition recognizes the facts of the case.

A **genus** may be defined either as "a collection of species having certain conspicuous features in common" (the morphological definition) or as "a collection of species believed to be derived from the same immediate ancestral stock" (the genetic definition). The difference between these two conceptions has been illustrated under the Graptolites (Dichograptidae p. 175).

As a rule, each of the categories named above consists of several of the next category below, but it may consist of only one—e.g., a genus may include only one species, and a family only one genus.

If the categories named are not enough for satisfactory classification in any particular case, intermediate grades may be intercalated, denoted by the prefix **sub-**; thus an order may be divided into several suborders, each containing several families. Still further grades may sometimes be necessary: they are denoted by the prefix **super-**; thus a suborder may be divided into superfamilies, and these into families. The term **variety** often formerly used in place of **subspecies** is no longer employed in that sense by modern workers.

The system of nomenclature for species now universally adopted is that of the Swedish naturalist Linné (latinized as Linnaeus) and consists in denoting every species by a double name: hence it is called the **Linnaean** or **binominal** system. Thus "*Dalmanites vulgaris*" is the **scientific name** or **binomen** of a certain species of Trilobite, and is made up of the **generic name** "*Dalmanites*" (the name of the genus to which this species belongs), and the **specific name** "*vulgaris*" which distinguishes this species from others of the same genus.

A generic name is always a single word, either a Latin noun or a word treated as such. In the early days of the Linnaean nomenclature Latin or latinized names of animals were plentifully at hand, but with the increase in the number of genera new kinds of names had to be sought. Names of mythological characters, such as *Astarte* and *Actaeon*, were appropriated, but have long been exhausted. Then came more or less descriptive compound words of Latin or Greek derivation, such as *Micraster* (little star), and *Macrodon* (big tooth): Many genera were named in compliment to individuals, as *Murchisonia* and *Lonsdaleia*, or from the locality where they were first discovered, as *Amberleya*, *Bohemilla*. As the number of generic names increased enormously there came a natural tendency to help the memory by compounding similar names for genera in the same class; thus the great majority of crinoid genera have names ending in *-crinus*, those of many corals end in *-phyllum* or *-phyllia*, of many starfish in *-aster*, or cephalopods in *-ceras*, of brachiopods in *-thyris* and graptolites as *-graptus*. But there is no copyright in such endings and they must not be trusted as a necessary indication of affinities. Thus *Cryptocrinus* is not a crinoid but a cystid; *Holocystis*, not a cystid but a coral; *Megalograptus* is not a graptolite but a Eurypterid; *Camptoceras* and *Platyceras* are gastropods. Most names ending in *-obolus* denote brachiopods, but *Trochobolus* is a sponge. The attempt to combine the name of a man or a place with one of these terminations has led to such uncouth compounds as *Agassizocrinus* and *Quenstedtoceras*.

If a subgeneric name is needed it is written in parentheses after the generic name; thus *Protolenus* (*Latoucheia*) *latouchei* denotes that the species belongs to the subgenus *Latoucheia* of the genus *Protolenus*. Square brackets are used for a different purpose: a name enclosed in them is one that has been rejected, but is perhaps better known than the correct name which precedes it. This helps recognition of changed names.

The best specific name is a simple Latin adjective, either descriptive of some feature of the species (as in *Chama squamosa*, *Neohibolites* [*Belemnites*] *minimus*) or indicating the locality where it was first found (as *Crania parisiensis*, *Amphilichas hibernicus*, or the less pleasing *oystermouthensis* and *czenstochovensis*). The adjective must always agree in gender with the generic name, and must be altered if necessary with a change in the latter. Instead of an adjective a noun may be used "in apposition" according to Latin usage, when it keeps its own gender, as *Condylopyge rex*, *Rhynchonella vespertilio*. Thirdly, it may be a noun in the genitive case, usually (but not always) a personal name, as *Lingulella davisi*, *Asteroceras smithi*.

In view of the mistakes that may arise from over-hasty identifications of a known species at a new locality or horizon, modern palaeontologists frequently use qualifying or cautionary terms. Their lists frequently contain such records as *Hildoceras* cf. *bifrons* (Bruguière) or *Productus* aff. *productus* (Martin). The first (confer or compare with) means "a species of *Hildoceras* very much ilke *bifrons* but probably not exactly identical with it"; and the latter (affinities with) "a species of *Productus* which is certainly not identical with *P. productus*, but appears to be genetically related to it." The term cf. is sometimes used when dealing with imperfectly preserved material the specific identification of which is uncertain.

If a subspecies needs to be distinguished, its subspecific name should be written immediately after the specific name, as *Clonograptus tenellus callavei*; but some older authors wrote *Clonograptus tenellus* var. *callavei*. When a species can be traced, with gradual changes, through a number of zones, each zonal form is called a **mutation***, and may be denoted by the symbol of its zone (e.g. *Productus concinnus* J. Sowerby, mut. D_2) or by an adjective (using mut. instead of subsp. or var.). Mutations which precede in time the typical species are termed descending or praemutations; those which follow it in time are ascending or postmutations.

For over a century the Linnean nomenclature was adopted by naturalists in general, but without any code of rules to settle the claims of rival names. In 1841, the British Association for the Advancement of Science appointed a committee to draft a code of rules, and this produced the Stricklandian code (from the name of the Secretary of the Committee, H. E. Strickland). More recently, a permanent International Committee has been set up by the International Zoological Congress, and this acts as a legislative body (drafting the rules), a court of law (deciding how the rules apply in particular cases) and a court of equity (authorizing the suspension of a rule in cases where it would cause unjustifiable disturbance of accepted names). Some of the decisions on difficult cases involve as much subtlety of argument as can be met with in a Chancery law-suit.

As to the mode of writing or printing specific names, the following rules should be strictly adhered to: (1) A generic name should always bear a capital initial letter; (2) a specific name should always bear a small initial letter†; (3) if a generic or specific name comes in the course of an ordinary sentence it should be printed in different type from the rest of the sentence (normally, in a sentence of roman type it would be in italics). An example will show the importance of adhering to these rules: If in a geological work we come across an allusion to "the *Planorbis* Beds," we should know that it referred to beds in which there were abundant gastropods belonging to the freshwater

* This use of the term mutation, proposed by Waagen in 1869, has priority over the general use of the term by geneticists (de Vries, 1901). Picturing evolution as a staircase, Waagenian mutations are the horizontal treads, de Vriesian mutations are the vertical risers.

† In Botany, however, specific names formed from personal names till very recently have been written with a capital initial letter.

genus *Planorbis*; but when we read of "the *planorbis* Beds," we infer marine beds characterized by the ammonite *Psiloceras planorbis*.

When several species of a genus are referred to in the same paragraph, or where the same species is repeatedly referred to, it is permissible to use only the initial letter of the generic name, to save repetition.

If no author had ever, by accident or otherwise, broken the rules of nomenclature, the generic and specific names would suffice to define a species absolutely. It has been found necessary, however, to avoid confusion, to write after the name of a species the name of its author —i.e., of the palaeontologist who first applied that name to it; and in systematic descriptions it is commonplace to add the date of his giving the name, thus: *Terebratula intermedia* J. Sowerby, 1812.

There is no rule as to the type in which the author's name should be printed, but it is common practice to print it in type different from that used for the specific name or the rest of the sentence in which the name occurs (though the latter is not the case in the present chapter).

The need for further rules arises because authors writing at different times have occasionally (*a*) given the same name to different genera, or to different species of the same genus, either through a mistake in identification or in ignorance that the name was already in use; (*b*) given different names to the same genus or species; (*c*) divided up an existing genus into several genera, or (*d*) united into one, genera previously considered to be distinct.

The most general rule in all nomenclature is the **Law of Priority,** according to which, other things being equal, the earliest name given to any genus or species must stand: all later names are **synonyms,** and as such are rejected. To this rule there are certain necessary qualifications.

In order to have any status in nomenclature and to be available for use, a name must fulfil certain basic conditions. It must be published, i.e., printed and circulated so as to be available for the public, in the case of animal names—after 1757,* and in accordance with the Linnean system of binominal nomenclature (see p. 263). It must not be a name given to a monstrosity or to a hybrid animal as such, or to an imaginary animal or to a form of lower than subspecific rank. It must have been accompanied by either a description (diagnosis) or a figure (or both) which would make it possible to identify the species to which the name was applied. In the case of a new genus the type-species must be stated.

* Valid animal (fossil and living) names date from the publication of the 10th edition of Linneaus' *Systema Naturae* (1758); plant names and valid from various dates on or after 1 May 1753 the date of Linnaeus' *Species Plantarium* ed. 1.

Once a name qualifies as an **available name** it becomes subject to the basic rule of all nomenclature, the Law of Priority, according to which, other things being equal, the oldest available name for a given family, genus or species is the valid name and must be used for it provided it has not already been used in a different sense.

If a generic name has already been applied to another genus, or if a specific name has already been applied to a different species in the same genus, the name in its new application is a **junior homonym.** Thus the name *Avalonia* was given by Walcott in 1889 to a genus of trilobites; the same name was applied by Seeley in 1898 to a reptile. Hence *Avalonia* Seeley, 1898, is a homonym of *Avalonia* Walcott, 1889. Again James Sowerby in 1820 gave the name *Spirifer pinguis* to a Carboniferous fossil; Zieten in 1832 gave the same name to a Jurassic fossil. Hence *S. pinguis* Zieten, 1832, is a homonym of *S. pinguis* J. Sowerby, 1820. When a homonym has to be quoted, it is advisable to make its honomymity clear by putting it thus:

Avalonia Seeley, 1898, *non* Walcott, 1889.
Spirifer pinguis Zieten, 1832, *non* J. Sowerby, 1820.

Generic homonyms must be rejected as soon as discovered, and replaced by the next available synonym if there is one, or by a new name if no synonym exists. Thus Emmrich in 1844 gave the name *Dalmannia* (in honour of the Swedish palaeontologist Dalman) to a genus of trilobites which he separated from *Phacops*; but *Dalmannia* had been previously applied to an insect, therefore Barrande in 1852 substituted the name *Dalmanites*. Sometimes an author is less fortunate. Thus Goldfuss in 1839 gave another trilobite-genus the name *Brontes*; finding this preoccupied he changed it in 1843, to *Bronteus*, but in the meantime de Koninck had given the same genus the name *Goldius*, and although palaeontologists had been accustomed to the name *Bronteus*, *Goldius* was used for a time until it was found that the name *Scutellum* Pusch 1833 had priority over all other names.

Specific homonyms are of two kinds. If two different species named in the same genus are given identical names, they are **primary homonyms,** and the junior name must be replaced by the next available synonym, if one exists, or by a new name if there is no synonym. Thus, *Spirifer pinguis* J. Sowerby, 1820, and *Spirifer pinguis* Zieten, 1832, are primary homonyms, and the fact that the former is now referred to *Brachythyris* and the latter to *Spiriferina* does not alter the fact that the junior name must be replaced. In this case, *Spirifer tumidus* von Buch, 1836, has been adopted in place of Zieten's name, and the species in question should be cited as "*Spiriferina tumida* (von Buch, 1836)."

If two different species with identical names come to be referred to the same genus through the transfer of one or both of them from the genus in which it was first described, they are **secondary homonyms,** and the later published of the two specific names must be replaced. This is true, however, only for those who agree with the proposition that the two species are congeneric.

As an example of a secondary homonym, we may take the following case: J. Sowerby, in 1818, named a lamellibranch *Cardita lirata*, and in 1819 named another *Lutraria lirata*. In 1826 his son J. de Carle Sowerby, named a new genus *Pholadomya*, to which he transferred both the above species, each of them thus becoming *Pholadomya lirata* (J. Sowerby). The 1818 species having the prior claim of the two of this name, the 1819 species became a homonym, and consequently a new name was given to it, and it is now *Pholadomya fidicula* J. de C. Sowerby, of which name *Lutraria lirata* J. Sowerby is a synonym.

A wise precaution to avoid such possibilities of confusion is never to give the same specific name to species in closely similar genera, nor to name a type species *typica*.

It must be clearly understood that the term **homonym** implies the same name for two things, while **synonym** implies two names for the same thing. In the case of generic synonyms a distinction may be drawn between **objective synonyms,** where two authors have founded two genera on the same species as type species (see later, p. 273), and **subjective synonyms,** where the two genera have different type species but are judged to be a single genus. In the latter case there is a possibility that future progress may necessitate the revival of the synonym as a generic name. In other words, an objective synonym is suppressed, a subjective synonym is dormant.

It was at one time a custom for the generic name of fossils to bear the termination *-ites*. Thus a fossil *Nautilus* was called *Nautilites*, not because it was thought to belong to a different genus from the recent *Nautilus*, but merely to emphasize its fossil state. This plan in name-making has long been abandoned, but it has led to some confusion in the case of genera entirely extinct. A name ending in *-ites* is available provided that it was clearly intended to denote a new genus, but if it was only a modification of an earlier name intended to indicate the fossil nature of certain individuals it is not valid, although it renders invalid any junior generic homonym.

One more exception to the law of priority must be noted. It has been found impracticable to insist on its covering both Animal and Plant Kingdoms simultaneously, so botanists and zoologists have agreed, while each accepting it within his own kingdom, to let the nomenclature of the two be quite independent. Thus *Zeilleria* is the name both of a Carboniferous pteridosperm leaf, and of a Jurassic brachiopod. No practical confusion is likely to arise between the two. Unfortunately, before this understanding between zoologists and botanists had been reached, some unnecessary substitutions had been made. Thus Bronn in 1828 named a fossil lamellibranch *Posidonia*, but finding that this name had been given by König to a plant, he later replaced it by *Posidonomya*, a name which has long been in use, but *Posidonia* is now restored to the lamellibranch.

These are the only valid exceptions to the law of priority. There are one or two other exceptions that have been proposed but are now quite rightly rejected. In the first place a name must not be rejected because it is inappropriate: to allow this would cause endless confusion, since appropriateness is partly a matter of individual judgment, and if one palaeontologist thinks the appearance of a certain brachiopod sufficiently suggestive of a coin to justify him calling it *Obolus*, while a later author, failing to see the likeness, is allowed to substitute another name, who is to settle which name is correct? After all, a name is a name, not a description, and in treating it as such we are following the common-sense customs of ordinary speech. Some houses possess kitchen "coppers" made of galvanized iron, drawing-room fire-"irons" made of brass, and candle-"sticks" made of china; men named Short or Little may be taller than others named Long, yet we neither expect them to change their names nor allow them to do so except with much expense and trouble; and no registrar of births would think of objecting to such a name as Thomas Thomas on the ground of tautology, or to Violet Green as a contradiction in terms. We need not be shocked at finding that a crustacean with an enormous number of legs was called *Apus* (footless), or that *Agoniatites* is not "devoid of an angle" as its name would imply.

Again, just as the change of manufacture from one material to another has led to such anomalies as brass irons, so the transfer of species from one genus to another has produced curious results. Sometimes a species has been thought to have only a superficial resemblance to a genus to which it has afterwards been found to belong, and the specific name indicating that resemblance has had to be kept: thus we get *Axinus axiniformis* (Phillips), *Dendrophyllia dendrophylloides* (Lonsdale), etc. Or, again, a species has been named from its resemblance to some object, such as a sword or an acorn, and the Latin name of that object has later been applied to a new genus into which the species has been transferred: thus we get *Ensis ensiformis* (S. V. Wood) and *Balanus balanoides* (Ranz). Yet again, a character unusual in its genus is taken for the trivial name of a species which is afterwards transferred to another genus in which that character is normal: thus a spiny shell from the Chalk was taken by J. Sowerby to belong to the genus *Plagiostoma*, and as the only spiny species in that genus it was appropriately called *Plagiostoma spinosum*, but later it was transferred to the spiny genus *Spondylus*, but the name *Spondylus spinosus*, though meaningless as a distinction from other species, serves suitably as a name.

The original Stricklandian code of 1841 proposed that a name might be changed when it implies a false proposition which is likely to propagate important error. For instance, Hyatt named an ammonite *Mantelliceras indianense*, a name apparently implying that it was found in the State of Indiana, whereas it actually came from India. It might seem reasonable in this case to alter the specific name to *indiense* or *indicum*, were it not that such a change might lead to the idea that two distinct species were in question. Really serious cases of this sort must be so few that they can easily be kept in mind, and once we depart from extreme cases we find individual judgment introducing uncertainties; consequently later zoologists have wisely dropped this proposal and agreed that once a name has been published it may not be altered, even by the author himself, except in correction of an evident slip of the pen or misprint. The current laws, however direct that hyphens in generic and specific names must be eliminated and that letters bearing diacritic marks must be replaced with letters in latinized form, e.g., *ölandicus* becomes *oelandicus*.

The Stricklandian code also proposed that where the name given to a species was taken as the generic name for a new genus in which that species was included, the trivial name must be altered, because they regarded the exact duplication of generic and specific names as uncouth. For example, Martin, in 1809, named a certain brachiopod *Anomites productus*. J. Sowerby, in 1812, took it as the type of a new genus, which he called *Productus*, and changed Martin's species into *Productus martini* (quite rightly according to the Stricklandian code). Later zoologists have rejected this exception to the law of priority, and the fossil which had long been known as *Productus martini* must be called *P. productus* (Martin). Strickland's objection to such names as inelegant has little validity when we remember that Scottish lairds are proud to call themselves "of that ilk." *Parkinsonia parkinsoni* is no worse than "Kinloch of Kinloch."

As knowledge increases and becomes more exact, a later author may transfer a species from its original genus to another. Thus the specific name *Cheirurus insignis* Beyrich, 1845, is a synonym of the name *Paradoxides bimucronatus* Murchison, 1839. The law of priority here settles that the earlier specific name must stand; but as it was recognized that *Paradoxides bimucronatus* Murchison was not a true *Paradoxides*, the species was transferred to Beyrich's new genus *Cheirurus*; but the specific name having priority over Beyrich's name *insignis*, stands as *Cheirurus bimucronatus* (Murchison), the author's name being placed in parentheses as an indication that an alteration in the generic name has been made. An alternative way of expressing this is *Cheirurus bimucronatus* Murchison sp.

A later author may divide a genus into several genera. The original generic name must then be retained for one of these divisions, and, if possible, for that division which was regarded by the author of the original genus as most typical. With a view to this possibility the founder of a genus should always name one species as most typical: such a species is called the **type-species** of the genus (the term **genotype** which has a different meaning in genetics should not be used in a nomenclatural context). The absence of a specified type-species from the original definition of a genus has tended to be a source of confusion when the genus has been subdivided later.

The subdivision of a genus often takes place in two stages: one author divides it into subgenera, and a later author raises these to the rank of genera. To provide for this possibility it is a rule that when a genus is divided into subgenera, one of these (that including the type species) must bear the same name as the genus itself. To distinguish the subgeneric name from the generic name, the former may be written with the addition of the abbreviation *s. str.* (*sensu stricto*, in the restricted sense). Thus Hall and Clarke divided the old genus *Orthis* of Dalman into a number of subgenera, to all of which except one new names were given, such as *Schizophoria*, etc., but the one which included Dalman's type-species was called *Orthis s. str.* Under this scheme the specific names of some well-known species were:

Orthis (*Schizophoria*) *resupinata* (Martin);
Orthis (*Orthis*) *callactis* Dalman.

Later authors have raised these subgenera to the rank of genera, so that the same species now stand as:

Schizophoria resupinata (Martin);
Orthis callactis Dalman.

The printing of the author's name with or without parentheses in the different cases should be noted carefully.

A later author may unite several previously separate genera into one. This is a much rarer case than the last, as the dominant tendency of advancing classification is to increase rather than to diminish the number of divisions of all grades. In this case the application of the law of priority leads to the rule that out of the names of the several genera which are being united the earliest name shall be retained as the name of the united genus.

The names of categories higher than genera cannot be subject to the same rigid law of priority as generic and specific names.

For families the rule is that the name of the family is formed by combining the name of the type-genus with the patronymic Latin termination -idae. Thus the family of which *Olenus* is the most typical genus is called Olenidae. If for reasons of priority it is found necessary to change the name of the type-genus, the family name must not be changed accordingly; but if such change has been made and has been generally accepted, the rejected family name is not now to be revived. The purpose of this rule is to prevent changes of family names for technical causes and to retain static family names widely used outside specialist circles.

In the case of superfamilies the name is similarly formed by combining the root of the name of the type family with the termination *-acea*. Thus the superfamily Phacopacea, typified by the family Phacopidae and the genus *Phacops*, includes three families, named Dalmanitidae, Phacopidae, and Pterygometopidae, typified by the respective genera *Dalmanites*, *Phacops*, and *Pterygometopus*. In the case of subfamilies, the suffix is -inae and the Phacopidae is divided into Phacopinae, Bouleinae and Phacopidellinae. It will be noted that the genus *Phacops* is typical both of the superfamily, the family and of one of the subfamilies.

As there have been many cases where opinions have differed as to the particular form which was denoted by a certain name, and as neither printed descriptions nor figures can always be interpreted with certainty, it is important that the actual specimens which an author had before him when founding a species should be preserved for reference in case of doubt. These specimens are called type-specimens or types, and they are now preserved with the greatest care by all museums which possess any, being usually marked with some special label—e.g., in the British Museum types are indicated by a small circular green label.

Types are of several kinds.

If the author of a new species defines that species with special reference to one particular individual, that individual is called the **holotype** of the species.

If, while specifying a single type of holotype, he also refers to other specimens in his original description, or figures them along with the holotype, these additional specimens are called **paratypes.**

If, however, he uses several specimens in his original description without specifying one of them as the holotype, these specimens are called **syntypes** (or co-types).

If the syntypes of a species are subsequently discovered to belong to two or more distinct species, the original author, or any later author, may select one of them as the type of the original species: this is called a **lectotype.** The remaining syntypes, if not identical with other species already named, may become holotypes of new species at the same time.

If the holotype, or lectotype, or all syntypes of a species are lost or destroyed, and it becomes necessary to determine the exact meaning of a specific name, a **neotype** may be designated, under certain stringent conditions, to replace the lost type-specimen.

Types of a much lower degree of value are the following:

Any specimen of a species coming from the same locality and zone as its holotype or syntypes is a **topotype.** Topotypes can be indefinitely increased in number by collecting, whereas there can never be more than one holotype or lectotype, and never more than a few paratypes or syntypes.

A topotype recognized by the original author as belonging to his species acquires additional value thereby, and is called a **metatype.**

A specimen of a species described or figured by the original or any later author at some subsequent date to the original establishment of the species is a **plesiotype.**

What a type-specimen is to a species, a **type-species** is to a genus: that is, it is a more certain means of settling the application and extent of the generic name than any verbal definition.

Type-species may also be classified in this way:

1. *Types by original designation*, when in establishing a new genus the author definitely names the type-species.

2. *Types by subsequent designation*, when the author fails to do so, and a later investigator selects a lectotype. (Once this has been done, any subsequent selection is void.)

3. *Types by monotypy.*—When a genus was originally established for one species only, however many species may later be included in it.

4. *Types by tautonymy.*—When the name of a species is used as the name of the new genus, as when *Anomites productus* was taken as one of the species of the new genus *Productus* (see p. 270). Two further methods of determining type-species have been used but must be rejected:

5. *Types by elimination.*—An author, Linnaeus for instance, may have named a number of species under a newly-named genus, without designating any one as the type. Later authors may have chosen

one after another of these as type of some new genus, leaving only one finally in its original genus. This has sometimes been accepted as the type-species of the Linnaean genus, but it is now recognized that this is not necessarily the case: one of the others may have a stronger claim and its new generic name may become a synonym.

6. *Types by list-priority.*—When a genus is established and a number of species quoted as belonging to it, it has been sometimes held that the first on the list must be taken as the type. Thus Linnaeus established the genus *Arca* with many species, one of them, popularly known as the "Noah's Ark shell," being named by him *Arca noae*. But this does not stand first on his list, that place being given to *Arca tortuosa*, a very exceptional form certainly not congeneric with the rest of the species. It is unlikely that Linnaeus should have thought of this, rather than *Arca noae*, as typical of the genus.

The collector of fossils who wishes to name his specimens accurately will need to take much trouble in hunting through works in which fossils are described and figured. These works are of several categories.

First, we have monumental works, mostly of rather early date, which set out to describe and figure all the fossils (or all of some group of fossils) found in some one country or State up to the time of publication.

Secondly, there are numerous smaller works, sometimes separately published, sometimes forming parts of scientific journals, in which only some small groups of fossils, or the fossils from some one locality or district, are described.

Thirdly, there are various monographs, such as those published by the Palaeontographical Society of London, which aim at a complete description (with revision of all previous publications on the subject), either of some great group as the Brachiopoda, for all geological systems, or for some one system (as the Cretaceous Lamellibranchs or the Carboniferous Trilobites), or, more rarely, the whole fauna of some one system or formation. Such monographs, if properly prepared, are of enormous value to students, as they sum up all that is known up to the date of their publication, and only very special investigators need go behind them to earlier works. These three categories are fairly distinct, but it is not always possible sharply to separate them.

Fourthly, there are catalogues and indexes which, without themselves describing any new genera or species, enable students to find with the least labour those works where they are described.

Some of these monographs, etc., have already been noted in the short bibliographies appended to each chapter. It remains to list a few of the works of more general kind. Fuller bibliographies of particular groups of fossils will be found in some of these.

General Bibliography

I.—RULES OF NOMENCLATURE

INTERNATIONAL COMMISSION ON ZOOLOGICAL NOMENCLATURE.—*International Code of Zoological Nomenclature* (with official list of accepted generic names). A new edition was published in 1961 by the Commission, c/o British Museum (Natural History), London, S.W.7. The Commission also publishes a series of *Opinions and Declarations* on matters of special difficulty which have been referred to them.

LANJOUW, J. and others.—*International Code of Botanical Nomenclature* (1956) Utrecht.

SCHENCK, E. T. and MCMASTERS, J. H.—*Procedure in Taxonomy* (1935) Stanford, Calif. and London.

II.—INDEXES

SHERBORN, C. D.—*Index Animalium.* A complete index to all animal genera and species, recent and fossil. Vol. 1 (1902) covers names published from 1758 to 1800. The subsequent nine volumes (1922–39) extend the index to 1850.

NEAVE, S. A.—(1) *Nomenclator Zoologicus*, 1758–1935, 4 vols. (1939-40), Zool. Soc. London. (2) *Nomenclator Zoologicus*, 1936–45, vol. v (1950), *ibid.* These deal with genera only, on the same lines as Sherborn, but continuing to a much more recent date. It occupies much less space than Sherborn.

ETHERIDGE, R.—*Fossils of the British Islands, Vol. I, Palaeozoic* (1888). Species described up to 1886. No further volumes published.

MORRIS, J.—*British Fossils*, 2nd edn. (1854). Records all British species up to date of publication, with references. Nomenclature out of date.

ZOOLOGICAL RECORD.—Annual publication by the Zoological Society of London, including lists of new genera and species (Recent and Fossil) of animals published during the year.

PALAEONTOLOGIA UNIVERSALIS.—A series of photographic reproductions of type-specimens and their original figures and descriptions (chiefly of species published before 1865). Ceased publication after 1914; restarted in a new series in 1955.

FOSSILIUM CATALOGUS, formerly edited by F. Frech, now by W. Quenstedt (Berlin, 1913–).—A series of lists of species compiled by specialists, with geological and geographical range, synonymy, and references, covering considerable portions of the animal and plant kingdoms, but with large gaps.

III.—LARGE GENERAL TEXTBOOKS

MOORE, R. C. (Editor).—*Treatise on Invertebrate Paleontology.* Published in volumes dealing with separate groups (1953–) sponsored by the Paleontological and the Palaeontographical Society, Kansas Univ. Press.

MOORE, R. C., LALICKER, C. G., and FISCHER, A. G.—*Invertebrate Fossils* (1952) New York and London (McGraw Hill).

PIVETEAU, J.—*Traité de Paléontologie* Paris (Masson).

SHIMER, H. W. and SHROCK, R. R.—*Index Fossils of North America* (1944).

ROMER, A. S.—*Vertebrate Paleontology* (1945) 2nd edn. Chicago.

ZITTEL, K. A.—(1) *Handbuch der Paläontologie*, 4 vols. (1876–93), also French translation by Ch. Barrois (1883–94). The most complete textbook, but suffering from the lapse of time since its publication. (2) *Grundzüge der Paläontologie* (1895, new edn. 1910–11), also English translation by C. Eastman. *Textbook of Paleontology*, Vol. I: *Invertebrata* (1900, 2nd edn. 1913), Vol. II: *Vertebrata* (Fishes to Birds) (1932), Vol. III: *Mammalia* (1925). (3) *History of Geology and Palaeontology* translated by Maria M. Ogilvie-Gordon (*Contemporary Science Series*, 1901); the chapters on Palaeontology and Stratigraphical Geology are indispensable to any student wishing to understand the history of research and the work of the great pioneers.

IV.—EARLY MONOGRAPHS OF EXTENSIVE NATURE

BARRANDE, J.—*Système Silurien du Centre de la Bohème* (1852–99). A series of very detailed and magnificently illustrated monographs on the fossils of the Lower Palaeozoic rocks of Bohemia. Barrande's own work covers the Trilobites, Mollusca, and Brachiopods; and after his death other authors added volumes on Graptolites, Corals, Bryozoa, and Cystids.

DESHAYES, P. G.—*Description des Coquilles Fossiles des Environs de Paris*, 2 vols. and Atlas (1824–37): Cainozoic Mollusca.

D'ORBIGNY, A.—*Paléontologie Française* (1840–55, with later volumes by other authors). An attempt, never completed, to figure and describe all the fossils of France. It covers Jurassic and Cretaceous Corals, Echinoderms, Bryozoa, Brachiopoda, Mollusca and Plants, also Eocene Echinoids.

GOLDFUSS, A. and MÜNSTER, G.—*Petrefacta Germaniae*, 2 vols. (1826–40). An equally ambitious design for Germany, limited in execution to Sponges, Corals, Echinoderms, and Mollusca.

HALL, J.—*Palaeontology of New York* (1848–98). A series of elaborate monographs on the Palaeozoic Fossils of New York State.

SOWERBY, J., continued by SOWERBY, J. DE C.—*Mineral Conchology of Great Britain* (1812–46). Published in monthly parts, with interruptions; its object was to figure British fossil shells, i.e., Mollusca as then understood (including Foraminifera, Cirripedia, Annelida, Brachiopoda, and, by mistake, one Coral). In all about 1,250 species were described, and figured on 648 plates, engraved on metal and coloured by hand with great faithfulness to the natural appearance. There is no system in the sequence of species, and, as British stratigraphy underwent its early development during the publication, more details are given of the geological age in the later than in the earlier parts. Unfortunately many of the fossils described were derived from the Glacial Drift.

The above lists might be extended indefinitely, but their object is to give the student a first general idea of the sort of material available for the purpose of identifying the fossils in his collections, and for extending his knowledge in any department that may specially interest him.

V.—ENGLISH TEXTBOOKS, ETC.

HAWKINS, H. L.—*Invertebrate Palaeontology* (1920) London (Methuen).

NEAVERSON, E.—*Stratigraphical Palaeontology. Study of the Ancient Life Provinces*, 2nd edn. (1955) Oxford Univ. Press.

SWINNERTON, H. H.—*Outlines of Palaeontology*, 3rd edn. (1947) London (Arnold).

WOODS, H.—*Palaeontology: Invertebrate*, 8th edn. (1946) Cambridge Univ. Press.

The various Museum Guides of the British Museum (Natural History) give useful summaries. A new series of illustrated guides started with British Cenozoic Fossils (1959) is specially useful.

VI.—WORKS OF GENERAL INTEREST

HOLMES, A.—*The Age of the Earth* (1927), London (Benn) and a larger work, with the same title (Harper, 1913).

SIMPSON, G. G.—*Tempo and Mode in Evolution* (1944), Columbia Univ. Press (N.Y.).

THOMPSON, d'Arcy W.—*On Growth and Form*, new edn. (1942) Cambridge.

Much in these works has a bearing, direct or indirect, on Palaeontology in general.

VII.—PERIODICALS IN ENGLISH

Palaeontology.—Started in 1957 and published in volumes of which at present four parts a year are issued by the Palaeontological Association.

Journal of Paleontology.—Started in 1927 and published in volumes, six parts a year by the Paleontological Society of America and the Society of Economic Paleontologists and Mineralogists.

The annual volumes of the Palaeontographical Society are mentioned on p. 274.

APPENDIX

DIVISIONS OF GEOLOGICAL TIME

THESE are tabulated here for reference. For the basis of the minor divisions works on Stratigraphy must be consulted: the palaeontological characters of the major divisions are given on pp. 283–9, and the recognized palaeontological zones on pp. 290–304. The meaning of the rock- and time-divisions is given on p. 13.

Group Era	System Period	Series Epoch	Stage Age	Local Stratigraphical Terms, chiefly British
CAINOZOIC		Holocene	—	Recent Alluvium, etc.
		Pleistocene	Monasterian Tyrrhenian Sicilian Calabrian	Raised Beach Corton Beds Early glacial deposit Red Crag
	NEOGENE	Pliocene	Astian Plaisancian Diestian	Coralline Crag Lenham Beds
	25 million years*	Miocene	Pontian Sarmatian Vindobonian Burdigalian Aquitanian	*Congeria* Beds, etc., of Caspian facies in S.E. Europe Swiss Mollasse; Faluns of Touraine, etc.
	PALEOGENE [NUMMULITIC]	Oligocene	Stampian Tongrian	Fluvio-marine of Isle of Wight
		Eocene	Bartonian Auversian Lutetian Ypresian	Barton Clay U. Bracklesham L. Bracklesham Calcaire Grossier (Paris) London Clay
	70±2 m. years	Paleocene	Landenian Montian	Thanet Sand, etc. —
MESOZOIC	CRETACEOUS	Upper Cretaceous	Danian Maestrichtian Senonian Turonian Cenomanian	— — Upper Chalk Middle Chalk Lower Chalk, etc.
	135±5 m. years	Lower Cretaceous	Albian Aptian Neocomian	Gault and Lower Greensand Wealden
	JURASSIC	Upper Jurassic	Purbeckian Portlandian Kimmeridgian Oxfordian Callovian	Freshwater Purbeck Beds Portland Stone and Sand Kimmeridge Clay Corallian, Oxford Clay, Kellaways Rock and Clay

* Approximate age of the start of the period, based on maximum thicknesses and control points fixed by lead ratios, following A. Holmes, "A Revised Geological Time Scale," *Trans. Edinb. Geol. Soc.*, xvii, 254 (1960).

Group Era	System Period	Series Epoch	Stage Age	Local Stratigraphical Terms chiefly British
MESOZOIC	JURASSIC	Middle Jurassic	Bathonian	Great Oolite Series
			Bajocian	Inferior Oolite
		Lower Jurassic (Liassic)	Toarcian	Upper Lias
			Pliensbachian Sinemurian Hettangian	Middle and Lower Lias
	180±5 m. years			
	TRIASSIC	Upper	Rhaetian	Rhaetic Beds
			Norian Carnian	Keuper (non-marine)
		Middle	Landinian	Muschelkalk of Germany, etc.
		Lower	Anisian Scythian	Bunter (non-marine)
	225±5 m. years			
NEWER PALAEOZOIC	PERMIAN	Upper	Tatarian	—
			Kazanian	Magnesian Lmst.
			Kungurian	—
			Artinskian	—
	270 ± 5 m. years	Lower	Sakmarian*	—
	CARBONIFEROUS	Upper (Pennsylvanian)	Stephanian	Keele Beds
			Westphalian	British Coal Ms.
			Namurian	Millstone Grit
		Lower (Mississippian)	Viséan Tournaisian	Carboniferous Limestone
	350±10 m. years			
	DEVONIAN	Upper	Famennian Frasnian	Upper Old Red Sandstone
		Middle	Givetian Eifelian	Middle O.R.S.
		Lower	Emsian Siegenian Gedinnian	Lower O.R.S. including Dittonian and Downtonian*
	400±10 m. years			
OLDER PALAEOZOIC	SILURIAN	Ludlow	— / Salopian	Upper Ludlow Aymestry Lmst. Lower Ludlow
		Wenlock	Salopian	Wenlock Lmst., W. Sh., Woolhope Lmst.
	440±10 m. years	Llandovery		—
	ORDOVICIAN	Ashgill Caradoc		Bala Beds
		Llandeilo		
		Llanvirn Arenig		Skiddaw Slates
	500±15 m. years			
	CAMBRIAN	Tremadoc	—	Shineton Shales
		Upper	Dolgelly Ffestiniog Maentwrog	Lingula Flags
		Middle	Menevian Solva	Upper Comley Sandstone
	600±20 m. years	Lower	—	Lower Comley Sandstone

* The trend of recent opinion is to place the Sakmarian in the Permian and the Downtonian in the Devonian; in Britain, the Tremadoc was placed in the Cambrian but writers in other countries classify it with the Ordovician.

STRATIGRAPHICAL PALAEONTOLOGY

I.—OLDER PALAEOZOIC FOSSIL ASSEMBLAGES

GENERAL FEATURES

PRESENCE in abundance of **Graptolites, Brachiopoda,** and **Trilobites**; in fair abundance of CORALS, CRINOIDS, and CYSTIDS, NAUTILOIDEA (mainly straight or slightly coiled). Rarity of Lamellibranchs and Gastropods. Complete or almost complete **absence** of Echinoids, Ammonoidea, Vertebrata, and Plants.

SPECIAL BIOLOGICAL FEATURES OF THE SEPARATE SYSTEMS

1. **CAMBRIAN.**—**Absence** of Graptolithina (except in uppermost beds), Corals, Crinoids, Lamellibranchs, and Cephalopods (except in upper beds).

BRACHIOPODS are mainly **horny Inarticulata**; the **Orthoids** are the only other common forms. **Trilobites** are rarely isopygous; they serve as zone-fossils. The index-species in North-West Europe of the Upper Cambrian are mainly of the family Olenidae, those of the Middle Cambrian, Paradoxididae and Agnostidae, and the Lower Cambrian is characterized by genera of the Olenellacea.

The only known plants are calcareous algae.

2. **ORDOVICIAN.**—Abundance of **branched Graptolites,** the most complexly branched being the earliest. The Graptolites serve as zone-fossils (see p. 303), but are almost confined to the black-shale facies. The Lower Ordovician is characterized by Dichograptids, the Middle and Upper by Leptograptids, Dicranograptids and Diplograptids.

CORALS: appear in the Middle Ordovician; Tabulata more abundant than Rugosa.

BRACHIOPODS: **Orthoids** and **Strophomenoids** abundant; horny Inarticulata reduced in numbers; Rhynchonelloids and Spire-bearers appear, but are rare.

CEPHALOPODS: Nautiloids only.

TRILOBITES of many superfamilies abundant; most distinctive are the **Trinucleacea, Cyclopygacea** and **Asaphacea.** Trilobites and Brachiopods may serve as zone-fossils in beds of "shelly" facies.

VERTEBRATA: First trace of Agnathous Fishes in N. America.

3. **SILURIAN.**—*Monograptus* is the predominant genus, and its species (chiefly) serve as zone-fossils (p. 302). Diplograptids survive in Valentian.

Corals and **Crinoids** very abundant in the limestones.

BRACHIOPODS: Large **Pentameroids** first appear in Britain and are abundant; Ordovician forms continue; **Rhynchonelloids** and **spire-bearers** (*Atrypa*, *Meristina*, *Delthyris*) common.

TRILOBITES as in Ordovician except for **absence** of Agnostacea, Shumardiidae, Trinucleidae and Asaphacea. Encrinuridae and Illaenidae do not survive beyond this system.

VERTEBRATA (primitive **Fishes,** represented by skin-teeth and fin-spines) first appear in Britain near the top of the system.

The first primitive LAND-PLANTS occur near the top of the System associated in some beds with marine animals (best known from Victoria, Australia).

II.—NEWER PALAEOZOIC FOSSIL ASSEMBLAGES

GENERAL FEATURES

Fishes and **land-plants** appear and become abundant.

Corals, Crinoids, and **Brachiopods** continue abundantly. Among the latter, **Loop-bearers** (Terebratuloids) first appear, and **spire-bearers** are very abundant.

ECHINOIDS and BLASTOIDS, previously very rare, become less rare, especially Blastoids, which, however, die out in this era.

LAMELLIBRANCHS and GASTROPODS increase in numbers.

AMMONOIDEA (**Goniatites**) first appear and become abundant.

NAUTILOIDEA continue to be common, and fully-coiled forms become more frequent.

TRILOBITES and CYSTIDS become **rarer.**

Graptolites are **extinct** but Dendroids last until Carboniferous.

SPECIAL FEATURES OF THE SEPARATE SYSTEMS

4. **DEVONIAN.**—BRACHIOPODS: of horny Inarticulata, only *Lingula* and Discinids (*Orbiculoidea*) survive (lasting to Recent times). Terebratuloids not yet common. Most other Silurian families survive. **Spire-bearers** abound.

CORALS: Tabulata less dominant than Rugosa in and after Middle Devonian.

Goniatites are common locally (especially in Upper Devonian) with very simple suture-lines; they serve as zone-fossils.

TRILOBITES: much reduced in numbers, but the same families as in the Silurian **except** Illaenids and Encrinurids.

FISHES abundant, particularly in Old Red Sandstone facies, especially **Ostracoderms** and **Crossopterygii**; **Dipnoi** appear. First Amphibia in Upper Devonian of Greenland.

LAND-PLANTS (mainly PSILOPHYTALES and PTERIDOSPERMS).

5. **CARBONIFEROUS.**—**Blastoids** commoner than in any other system.

CORALS: Rugose corals continue to dominate Tabulates in Lower Carboniferous.

BRACHIOPODS: **Productacea, Spiriferacea,** and Rhynchonelloids most abundant. Terebratuloids fairly abundant.

CEPHALOPODA: **Goniatites** with suture-lines generally a little more complex than in Devonian; used as zone-fossils. Tightly coiled NAUTILOIDS and Orthocones.

TRILOBITES rare, of only three families (see p. 142).

VERTEBRATA: Most groups of fishes survive from Devonian: **fish-teeth** and **spines** common in some beds. OSTRACODERMS are **extinct.** AMPHIBIA, and, in Upper Carboniferous, the first primitive reptiles.

PLANTS abundant in the coal-bearing beds. In the northern hemisphere (except India) the fern-like leaves of **Pteridosperms,** and stems of *Lepidodendron*, *Sigillaria* and *Calamites* are common. In India and the southern hemisphere, there is a totally distinct flora (Permo-Carboniferous), with the pteridosperm *Glossopteris*.

The zoning of the Lower Carboniferous is based on shallow-water Corals and Brachiopods, and the zones may be of restricted geographical extent. The Upper Carboniferous (Coal Measures) are classified, but scarcely zoned, by the flora, and zoned by freshwater mussels: their marine equivalents by Goniatites and by Fusulinidea, as are the marine Permian.

6. **PERMIAN.**—The true marine fauna is only known in a few restricted (tropical and subtropical) areas, constituting the Tethyan marine region. It is very similar to the Carboniferous fauna, except that the **Ammonoidea** have far more complex suture-lines, frequently with many-pointed lobes and rounded saddles.

In other areas an impoverished (inland sea) fauna is found, consisting mainly of BRACHIOPODS and LAMELLIBRANCHS, and a terrestrial flora which is an impoverished Carboniferous flora.

The land-fauna includes primitive reptiles, especially those of mammalian affinity.

III.—MESOZOIC FOSSIL ASSEMBLAGES

General Features

Corals are less abundant and of entirely new types.

Cystids and Blastoids are **extinct,** Crinoids become less and less common; **Echinoids** are abundant, and include **Irregular** as well as **Regular** forms.

Brachiopods, mainly Terebratuloids and Rhynchonelloids, are common.

Lamellibranchs are abundant, and **Gastropods** are common.

Ammonoidea are very abundant, except towards the end of the era; their suture-lines are extremely complex as a rule. They serve as zone-fossils (pp. 292–5).

Trilobites are quite **extinct.**

Vertebrata: **Fishes** (Elasmobranchs, Chrondrostei and Holostei) and **Reptiles,** both terrestrial and marine, are abundant. Mammalia are exceedingly rare and small.

Land-Plants: principally **Cycads,** also Conifers and Ferns.

Special Features of the Separate Systems

7. **TRIASSIC.**—As with the Permian, though to a much less extent, the true marine fauna is restricted in area, being preserved principally in the Alps and other great mountain chains.

Echinoids are rare and of very special types.

Brachiopoda: spire-bearing forms are still numerous, as well as Terebratuloids and Rhynchonelloids, but nearly all other Palaeozoic superfamilies are extinct.

Cephalopoda: **Ammonoidea** with ceratitic suture-lines are very abundant, but there are others with more complex (ammonitic) suture-lines.

Primitive Belemnoidea also occur.

Vertebrata: **Reptiles**, especially terrestrial forms, are abundant. Some marine reptiles. Mammalia very doubtful, but mammal-like reptiles continue from Permian.

Zone-Fossils: Ammonoids (and Lamellibranchs).

8. **JURASSIC.**—**Echinoids** are abundant, especially in certain beds. *Clypeus* is confined to this system, *Hemicidaris* nearly so. *Acrosalenia*, *Holectypus*, *Pygaster*, *Nucleolites*, and *Collyrites* are Cretaceous also.

The last spire-bearing Brachiopods occur in the Lower Jurassic.

Lamellibranchs are abundant, especially *Trigonia* and **Oysters.**

Ammonoidea with highly complex suture-lines (**Ammonites**) and typical **Belemnites** are abundant.

Large marine and terrestrial **Reptiles** are common, flying forms (Pterodactyls) rare.

In Upper Jurassic, small primitive mammals, rare; reptilian **birds,** very rare.

LAND-PLANTS (CYCADS, etc.) are common in estuarine deposits.

The **Lower** Jurassic is characterized by the presence of the brachiopod *Spiriferina*, the **absence** of Irregular Echinoids, and the dominance of **keeled** forms among the Ammonites (though in particular zones this dominance may be reversed). The **Upper** Jurassic is characterized by the absence of *Spiriferina*, the presence of **Irregular Echinoids,** the dominance (reversed in particular zones) of **unkeeled** Ammonites and the frequency of **lappets** on the apertural margin of Ammonites (whether keeled or unkeeled).

ZONE-FOSSILS.—Ammonites or, in their absence, Brachiopods (pp. 11–14), or occasionally (and locally) Lamellibranchs.

9. **CRETACEOUS.**—**Echinoids** of several new families appear, among which the Spatangoida at once become abundant.

In the Alpine-Mediterranean province, numerous very peculiarly-shaped inequivalve Lamellibranchs (Rudistes: *Monopleura*, *Caprotina*, etc.) occur in the Lower Cretaceous, and are followed in the Upper by the **Hippurites,** whose habit of growth is more like that of rugose Corals than Lamellibranchs.

The last British *Trigonia* is found in the Cenomanian (early Upper Cretaceous).

Ammonoidea and **Belemnoidea** become less abundant and finally die out. More or less uncoiled **Ammonoids** are common.

In certain regions (especially North Africa, Syria, South America, and the southern U.S.A.) there are Cretaceous Ammonoids with suture-lines resembling those of the Triassic *Ceratites* (**pseudo-ceratites**).

In the Upper Cretaceous there is a sudden appearance of new families of Teleostean **Fishes** taking over dominance from Holosteans and land-plants (**Dicotyledons**) of much more modern types than those of the Jurassic and Lower Cretaceous.

IV.—CAINOZOIC FOSSIL ASSEMBLAGES

GENERAL FEATURES

Foraminifera are abundant and include forms of much greater size than in most earlier systems. These, where they occur (mainly in tropical and sub-tropical regions), are the best zone-fossils.

Corals are not abundant except in warm latitudes.

Crinoids and Brachiopods are rare.

Echinoids are common, especially in warm latitudes.

Three orders of Irregularia which appeared in the Cretaceous become dominant: Clypeastroida, Cassiduloida and Spatangoida. The same is the case with some Regular families, e.g., Echinidae.

VERTEBRATA: **Fishes** are of modern types; many Mesozoic types of Reptiles are extinct, only those surviving which last to Recent times (crocodiles, turtles, lizards, snakes). **Mammals** are abundant, mainly in freshwater deposits, and serve as zone-fossils.

PLANTS are of modern types, but many now confined to warm climates (such as palms) are found fossils as far north as the Arctic regions.

SPECIAL FEATURES OF THE SEPARATE SERIES

10. **EOCENE** (including **Paleocene**).—The foraminifer *Nummulites* is very abundant, especially in the Middle Eocene. *Discocyclina* is confined to this system, except for highest Cretaceous.

The Echinoids *Conoclypeus* and *Nucleolites* survive from the Cretaceous, but not beyond the Eocene.

The same is the case with **Pycnodonts** (Holostean fishes).

The British Molluscan fauna of the Eocene resembles that now living in the Indian Ocean.

The **Mammalia** of the Paleocene and Lower Eocene are all small and primitive, with forty-four low-crowned teeth, five-toed limbs and small brains. In the Middle and Upper Eocene they rapidly become larger and more specialized. In the Paleocene (and to a lesser extent in the Lower Eocene) of North America, Mesozoic families survive.

11. **OLIGOCENE.**—The foraminifera *Nummulites* and allied genera continue in some abundance. *Lepidocyclina* replaces *Discocyclina*.

The Irregular Echinoid *Clypeaster* (rare in uppermost Eocene) becomes common.

Mammalia.—In Europe and America new families appear, including the first (hornless) rhinoceros. The most characteristic family are the ANTHRACOTHERES (bunoselenodonts related to the hippopotamus). In Africa a distinct fauna was developing, including the early ancestors of the elephant.

12. **MIOCENE.**—*Nummulites* is extinct. *Lepidocyclina* persists, and is accompanied by *Miogypsina*, both dying out in Upper Miocene.

The Spatangid *Micraster*, which has survived from the Cretaceous, now appears for the last time, alongside *Schizaster* and others which continue to Recent times.

Many new genera of Gastropods suddenly appear, or if previously represented by very rare examples, become suddenly abundant (*Nassa*, *Columbella*, *Thais* [*Purpura*], *Oliva*).

Mammalia include the first antlered deer, first horned rhinoceros, first ape. Ancestors of elephants suddenly appear in Europe.

13. **PLIOCENE.**—During this epoch there is in the northern hemisphere a gradual southward migration of the marine fauna, Arctic species appearing in Britain and, later, in the Mediterranean area. The relative horizons of any normal marine Pliocene beds in the same latitude can be determined by the percentage of their species which are (*a*) extinct, (*b*) surviving in more northern or (*c*) in more southern latitudes.

Mammalia attain their greatest size. *Mastodon* is characteristic.

The fourteen following tables give the most widely recognized palaeontological zones, from Miocene down to Cambrian.

1. FORAMINIFER ZONES OF THE CAINOZOIC AND CRETACEOUS

Series	Stages	Nummulites	Lepidocyclina Old World	Lepidocyclina America	Other large Foraminifera
MIOCENE	VINDOBONIAN	—	—	—	*Archaias malabaricus*
	BURDIGALIAN	—	*sumatrensis*	*canellei*	*Operculina niasi* *Miogypsina* *Cycloclypeus*
	UPPER AQUITANIAN	—	*tournoueri* *marginata*	—	*Miogypsina*
	MIDDLE AQUITANIAN	—	*dilatata*	*chaperi*	—
	LOWER AQUITANIAN	—	*formosa*	—	—
OLIGOCINE	STAMPIAN	*intermedius, vascus*	*formosa* *raulini*	—	—
	TONGRIAN	*intermedius* *vascus, bouillei*	—	*mantelli*	*Orbitolites martini* *Operculina complanata*
			DISCOCYCLINA		
EOCENE	BARTONIAN	*bouillei, chavannesi,* *fabianii, yawensis*	*pratti* *radians*		—
	AUVERSIAN	*striatus, variolarius*	*nummulitica*		—
	UPPER LUTETIAN	*gizehensis* *laevigatus* *aturicus*	*sella, discus* *omphalus* *stellata*		*Orbitolites complanatus* *Heterostegina reticulata* *Alveolina bosci*

EOCENE	Lower Lutetian	*atacicus, irregularis*	*archiaci*	*Assilina spira*
	Ypresian	*planulatus*	*archiaci*	*Assilina granulosa* *Assilina exponens*
PALEO-CENE	Landenian	*globulus, nuttalli*	—	—
	Montian	—	—	—
UPPER CRETACEOUS			Orbitoides	Series leading to Orbitolites
	Danian	*deserti, fraasi, mengaudi*	—	—
	Maestrichtian	—	*gensacica, socialis, apiculata, media, Omphalocyclus*	*Fallotia* *Maeandropsina*
	Senonian	—	*media, mammillata*	*Broeckina, Praesorites*
	Turonian	—	—	—
		Orbitolina		
	Upper Cenomanian	*concava*	—	—
	Lower Cenomanian	*plana*	—	—
LOWER CRETA-CEOUS	Albian	*subconcava*	—	—
	Aptian	*lenticularis*	—	—
	Barremian	*discoidea, conoidea*	—	—

2. ZONES OF THE CRETACEOUS

Stages	Zones	English Formations
DANIAN		not represented
MAESTRICHTIAN	*Ostrea lunata*[1] *Belemnitella mucronata*[2] (in part)	Norfolk Chalk
SENONIAN	*Belemnitella mucronata*[2] (in part) *Actinocamax quadratus*[2] *Offaster pilula*[3] *Marsupites testudinarius*[3] *Micraster coranguinum*[3] *Micraster cortestudinarium*[3]	Upper Chalk
TURONIAN	*Holaster planus*[3] *Terebratulina lata*[4] *Inoceramus labiatus*[1] (or *Orbirhynchia cuvieri*[4])	Chalk Rock Middle Chalk
CENOMANIAN	*Holaster subglobosus*[3] *Schloenbachia varians*	Lower Chalk
ALBIAN	*Stoliczkaia dispar* *Mortoniceras inflatum* *Euhoplites lautus* *Hoplites dentatus* *Douvilleiceras mammillatum* *Leymeriella tardefurcata*	Upper Greensand and Gault
APTIAN	*Diadochoceras nodosocostatum* *Parahoplites nutfieldensis* *Cheloniceras martini* *Deshayesites deshayesi*	Lower Greensand
BARREMIAN[5]	*Costidiscus recticostatus* *Heteroceras astierianum* *Crioceratites emericianus*	Snettisham Clay
HAUTERIVIAN[5]	*Pseudothurmannia angulicosta* *Subsaynella sayni* *Crioceratites duvali* *Acanthodiscus radiatus*	Tealby Limestone and Clay Claxby Ironstone
VALANGINIAN[5]	*Kilianella roubaudiana* [unnamed gap here] *Platylenticeras heteropleurum*	Hundleby Clay
BERRIASIAN[5]	*Thurmanniceras boissieri*	Spilsby Sandstone

[1] Lamellibranchs. [2] Belemnites. [3] Echinoderms. [4] Brachiopods. The remainder are Ammonites.
[5] Substages of Neocomian.

Note.—For a revision of the zones of the Aptian see R. Casey, *Palaeontology*, iii, pt. 4, p. 497 (1961).

3. ZONES OF THE UPPER JURASSIC

Stages	Zones	English Formations
PURBECKIAN	[No Ammonites]	Purbeck Beds[1]
PORTLANDIAN	*Titanites giganteus* *Glaucolithites gorei* *Zaraiskites albani*[2]	Portland Beds
UPPER KIMMERIDGIAN	*Pavlovia pallasioides* *Pavlovia rotunda* *Pectinatites pectinatus*	Kimmeridge Clay
MIDDLE KIMMERIDGIAN	*Subplanites wheatleyensis* *Subplanites spp.* *Gravesia gigas* *Gravesia gravesiana*	
LOWER KIMMERIDGIAN	*Aulacostephanus pseudomutabilis* *Rasenia mutabilis* *Rasenia cymodoce* *Pictonia baylei*	
OXFORDIAN	*Ringsteadia pseudocordata* *Decipia decipiens* *Perisphinctes cautisnigrae* *Perisphinctes plicatilis*	Corallian Beds
	Cardioceras cordatum— *Quenstedtoceras mariae*	Oxford Clay
CALLOVIAN	*Quenstedtoceras lamberti* *Peltoceras athleta* *Erymnoceras coronatum* *Kosmoceras jason*	
	Sigaloceras calloviense *Proplanulites koenigi*	Kellaways Beds
	Macrocephalites macrocephalus	Upper Cornbrash

[1] Zoning by ostracods (*Cypridea punctata*, *Metacypris forbesii*, *Cypris purbeckensis*, in descending order).

[2] Perhaps represents the zone of *Zaraiskites scythicus* (Pomerania–Russia).

4. ZONES OF THE MIDDLE JURASSIC

Stages	Zones	English Formations
	Clydoniceras discus	Lower Cornbrash
BATHONIAN	*Oppelia aspdioides* *Tulites subcontractus* *Gracilosphinctes prograciIis* *Oppelia fallax* *Zigzagiceras zigzag*	Great Oolite Series with Fuller's Earth
UPPER BAJOCIAN (VESULIAN)	*Parkinsonia parkinsoni* *Garantiana garantiana* *Strenoceras subfurcatum*	Upper Inferior Oolite
MIDDLE BAJOCIAN (BAJOCIAN, *s.str.*)	*Stephanoceras humphriesianum* *Otoites sauzei* *Sonninia sowerbyi*	Middle Inferior Oolite
LOWER BAJOCIAN (AALENIAN)	*Ludwigia murchisonae* *Leioceras opalinum* *Tmetoceras scissum*	Lower Inferior Oolite

5. ZONES OF THE LOWER JURASSIC

Stages	Zones	British Formations
UPPER TOARCIAN (YEOVILIAN)	*Lytoceras jurense*[1]	mainly transition sands
LOWER TOARCIAN (WHITBIAN)	*Hildoceras bifrons* *Harpoceras falcifer* *Dactylioceras tenuicostatum*	Upper Lias
UPPER PLIENSBACHIAN (DOMERIAN)	*Pleuroceras spinatum* *Amaltheus margaritatus*	Middle Lias
LOWER PLIENSBACHIAN (CARIXIAN)	*Prodactylioceras davoei* *Tragophylloceras ibex* *Uptonia jamesoni*	Lower Lias
SINEMURIAN	*Echioceras raricostatum* *Oxynoticeras oxynotum* *Asteroceras obtusum* *Caenisites turneri* *Arnioceras semicostatum* *Arietites bucklandi*	Lower Lias
HETTANGIAN	*Schlotheimia angulata* *Psiloceras planorbis* *Ostrea liassica*[2], *Pleuromya tatei*[2] (pre-*planorbis* Beds)	Lower Lias

[1] With subzones: *moorei*, *levesquei*, *dispansum*, *struckmanni*, *striatulum*, *variabilis*. For subzones in the remainder of the Lias, see SPATH, L. F., *Geol. Mag.*, lxxix, 264 (1942).

[2] Lamellibranchs.

6. ZONES OF THE TRIASSIC

Stages	Zones	Famous Localities
RHAETIC	*Pteria contorta* (Lamellibranch)	Aust and Garden Cliffs (Gloucestershire)
NORIAN	*Sirenites argonauta* *Pinacoceras metternichi* *Cyrtopleurites bicrenatus* *Cladiscites ruber* *Sagenites giebeli*	Austrian Alps, Himalayas, Timor
	Discophyllites patens	Alaska
CARNIAN	*Tropites subbullatus* *Carnites floridus*	Austria, Sicily, Himalayas
	Trachyceras aonoides *Trachyceras aon*	St. Cassian (Tirol)
LADINIAN	*Protrachyceras archelaus* *Protrachyceras reitzi*	Austrian Alps
ANISIAN	*Paraceratites trinodosus* *Paraceratites binodosus*	Austrian Alps, Kashmir, Nevada
	Nicomedites osmani	Austria, Himalayas
	Neopopanoceras haugi	California
UPPER SCYTHIAN	*Prohungarites similis* *Columbites parisianus* *Tirolites cassianus* *Anasibirites multiformis* *Meekoceras gracilitatis*	Widely scattered over N. hemisphere from Spitsbergen to Timor and Nevada to N.W. India
LOWER SCYTHIAN	*Flemingites flemingianus* *Koninckites volutus* *Xenodiscoides fallax* *Prionolobus rotundatus* *Protychites rosenkrantzi*	Salt Range (Punjab)
	Vishnuites decipiens *Ophiceras commune*	Himalayas
	Otoceras woodwardi	Julfa (Armenia)

In Germany, Lorraine, etc., the Muschelkalk (the only marine division of the Triassic) is partly Ladinian, partly Anisian. The overlying Keuper and underlying Bunter (both found in England) cannot be dated with precision, owing to absence of significant fossils. The zones listed above are based on the work of L. F. Spath and B. Kummel (*Treatise on Invertebrate Paleontology*, *Part L Mollusca* 4 *Cephalopoda Ammonoidea* (1957), p. L124.

ZONING OF NEWER PALAEOZOIC

These are difficult zoning problems, still only partially solved, because of the varied facies with their different classes of significant fossils and the existing uncertainties of correlation. Deep-water foraminifera, pelagic goniatites, shallow-water brachiopods and corals, freshwater mussels and land-plants are the guiding fossils chiefly used, and it is only by the occasional interdigitation of two of these four facies that they can be correlated satisfactorily.

7. FORAMINIFER AGES OF CARBONIFEROUS AND PERMIAN

	Stages		Ages	Possible equivalent Ammonite Ages
	U.S.S.R.	U.S.A.		
PERMIAN	Tartarian	Ochoan	—	*Cyclolobus*
	Kazanian	Guadalupian	*Polydiexodina*	*Timorites*
	Kungurian Artinskian	Leonardian	*Parafusulina*	*Waagenoceras* *Perrinites*
	Sakmarian	Wolfcampian	*Schwagerina*	*Properrinites*
CARBONIFEROUS	Orenburgian	Virgilian	*Pseudofusulina*	*Uddenites*
	Gzelian	Missourian	*Triticites*	*Prouddenites*
	Moscovian	Desmoinian	*Fusulina* & *Fusulinella*	*Eothalassoceras*
		Atokan	*Profusulinella*	*Wellerites* *Owenoceras*
	Bashkirian	Morrowan	*Millerella* & *Pseudostaffella*	*Paralegoceras* *Gastrioceras* *Reticuloceras*
	"Namurian"	Chesterian	*Millerella*	*Eumorphoceras*
	Viséan	Chesterian Meramecian		*Goniatites* *Beyrichoceras*

These ages represent broad divisions, distinguished by genera rather than species. The foraminiferal index fossils all belong to the family Fusulinidae, the range of which covers most of the two systems. Early members of the "family" such as *Eostaffella* and *Parastaffella* occur in the Viséan but one is confined to that stage. This foraminiferal facies is found mainly in Eastern Europe, Asia and North America. The Sakmarian is not universally considered as Permian. The Soviet term Bashkirian includes Namurian C with Westphalian A and B of the Western European nomenclature (shown in Table 8). The Moscovian corresponds to Westphalian C and D. Some stratigraphers prefer to retain the *Schwagerina* Beds in the Carboniferous. In the U.S.A. the term Mississippian is used as a system name for beds equivalent in age to the Lower Carboniferous of Western Europe and in addition the *Eumorphoceras* Beds considered to be of Chesterian age. The Pennsylvanian System includes strata corresponding to the remainder of the Upper Carboniferous of Europe.

Formations	Goniatite Ages	Principal Marine Bands	Non-marine Lamellibranch zones
"STEPHANIAN"			*Anthraconaia prolifera*
WESTPHALIAN D			*Anthraconauta tenuis*
WESTPHALIAN C	[1]	Top or Cwm Gorse	*Anthraconauta phillipsii*
[1]		Mansfield or Cefn Coed	Upper Z. of *similis-pulchra*
WESTPHALIAN B	*Anthracoceras* (A)		Lower Z. of *Anthracosia similis* and *Anthraconaia pulchra* Upper part of *Anthraconaia modiolaris*
	[1]	Clay Cross or Amman	
WESTPHALIAN A	*Gastrioceras* (G)	Halifax Hard or Bullion Mine *Gastrioceras subcrenatum*	Lower part of *Anthraconaia modiolaris* *Carbonicola communis*[3] *Anthraconaia lenisulcata*
NAMURIAN C			
NAMURIAN B	Upper *Reticuloceras* (R_2)		
	Lower *Reticuloceras* (R_1)		
NAMURIAN A	*Homoceras* (H)		
	Upper *Eumorphoceras* (E_2)		
	Lower *Eumorphoceras* (E_1)		
overlying Upper Viséan	overlying Upper *Posidonia* Age (P_2)		

[1] The double lines represent stratigraphical boundaries based on marine bands.
[2] The boundary lines of the Coal Measures follow the usage commended by Stubblefield, C. J., and Trotter, F. M., "Divisions of the Coal Measures on Geological Survey Maps of England and Wales," *Bull. Geol. Surv. Gt. Brit.*, no. 13, 1–5 (1957).
[3] Formerly known as *C. ovalis* Zone.

Formations	Typical Plants	Some British strata	
"STEPHANIAN"	*Odontopteris* *Callipteridium*	Upper Coal Measures, Keele Beds	
WESTPHALIAN D	*Alethopteris grandis* *Neuropteris ovata* *N. macrophylla*	Upper Coal Measures[4]	
WESTPHALIAN C	*Mariopteris latifolia* *Sphenophyllum majus*		
WESTPHALIAN B	*Alethopteris decurrens* *Neuropteris obliqua*	Middle Coal Measures[2, 5]	
WESTPHALIAN A	*Mariopteris acuta* *Neuropteris schlehani* *Sphenopteris hoeninghausi*	Lower Coal Measures[2, 6]	
NAMURIAN C	*M. acuta* *N. schlehani*	Millstone Grit	(?) upper part of Roslin Sandstone
NAMURIAN B		including	
NAMURIAN A	*Lyginopteris stangeri* *Sphenopteris adiantoides*	Upper Bowland Shale	lower part of Roslin Sandstone Upper Limestone Group (Scotland) Limestone Coal Group (Scotland)
		overlying Lower Bowland Shale	overlying Lower Lmst. Group (Scotland)

[4] Equivalent to Morganian; the earlier beds to Staffordian and the later to Radstockian. [5] Include beds formerly known as Yorkian.
[6] Include some beds formerly known as Yorkian and others known as Pre-Yorkian or Lanarkian. [7] The Westphalian A and B together constitute the Ammanian which is a division underlying the Morganian.

9. ZONES OF LOWER CARBONIFEROUS OR CARBONIFEROUS LIMESTONE SERIES

Stages	Coral–Brachiopod Zones	Goniatite Ages
VISÉAN	*Dibunophyllum* (D)	Upper *Posidonia* (P_2) or *Neoglyphioceras spirale* Lower *Posidonia* (P_1) Upper *Beyrichoceras* (B_2)
	Seminula (S_2)	*Beyrichoceras hodderense* (B_1)
	Upper *Caninia* (C_2S_1)	*Pericyclus* (Pe)
TOURNAISIAN	Lower *Caninia* ($C_1\gamma$)	
	Zaphrentis (Z)	(?) *Protocanites*
	Cleistopora (K)	(?) *Gattendorfia*

Notes on the names of zones.—*Seminula*, the name fossil is now called *Composita*; Vaughan's *Seminula* Zone originally included S_1 strata. *Zaphrentis* should more appropriately be called Zaphrentoid. *Cleistopora*, the name fossil is now called *Vaughania*. *Posidonia* is, of course, a Lamellibranch genus.

In the south-west and north-west of England, A. Vaughan (*Quart. Journ. Geol. Soc.*, lxi, 1905, 181–305) and E. J. Garwood (*ibid.*, lxviii for 1912, 1913, 449–586), respectively erected zonal schemes based on sequences of brachiopod and coral assemblages. Both these groups of fossils are essentially sedentary organisms requiring particular sedimentary conditions for their existence. It has been claimed that the boundaries between some of Vaughan's zones at different localities represent dissimilar positions in geological time. A discussion on the difficulties of applying these zonal schemes has lately been published (T. N. George, *Proc. Yorks. Geol. Soc.*, xxxi, 1958, 227–318).

Vaughan's *Dibunophyllum* Zone was originally divided into two subzones—D_1 and D_2. The upper subzone D_2, is notably more diversified in its coral fauna including some forms not recorded from D_1, such as the genus *Orionastraea*, and *Palaeosmilia regia*; this genus *Orionastraea* appears to have a limited range but many species of other genera seem to survive through great thicknesses of strata in the north of England.

Goniatites are plentiful only in upper Viséan strata in a limited region of north-central England; only rarely are they found in strata containing corals and brachiopods.

10. DEVONIAN AMMONOID AGES

Series	Stages	Ages
UPPER DEVONIAN	Famennian	VI *Wocklumeria* V *Orthoclymenia* IV } *Platyclymenia* III } II *Cheiloceras*
	Frasnian	I *Manticoceras*
MIDDLE DEVONIAN	Givetian	*Maenioceras*
	Eifelian	*Anarcestes*
LOWER DEVONIAN	Emsian (Coblenzian)	} *Mimosphinctes*
	Siegenian	}
	Gedinnian	No ammonoids known

Notes.—The divisions called ages in this table, like those based on goniatites and ammonites in Tables 8 and 9, represent the time characterized by particular genera. The rocks deposited during these time-units have been divided into true Oppelian or species zones in most of the stages. The generic indices of the three highest Famennian ages are not true Goniatites or Ammonites, but Clymeniids.

11. OLD RED SANDSTONE FAUNAS AND FLORAS

FAUNA (mainly Fishes)

UPPER: *Holoptychius,*[1] *Bothriolepis.*[2]
MIDDLE: *Pterichthys,*[2] *Coccosteus,*[2] *Osteolepis.*[1]
LOWER: *Pteraspis,*[3] *Cephalaspis,*[3] *Pterygotus.*[4]

FLORA

UPPER: *Cyclostigma,*[5] *Archaeopteris,*[6] *Pseudobornia.*[7]
MIDDLE: *Psygmophyllum,*[11] *Sphenopteridium,*[6] *Psilophyton,*[8] *Asteroxylon,*[8] *Rhynia,*[8] *Hornea.*[8]
LOWER: *Psilophyton,*[8] *Arthrostigma,*[8] *Gosslingia,*[9] *Parka,*[10] *Pachytheca.*[10]

[1] Crossopterygians
[2] Placoderms
[3] Ostracoderms
[4] Eurypterid
[5] Lycopodiales
[6] Pteridosperms
[7] Sphenophyllales
[8] Psilophytales
[9] Eufilicinae (?)
[10] Algae
[11] Doubtful position

12. SILURIAN GRAPTOLITE ZONES

After Lapworth, Elles, and Wood

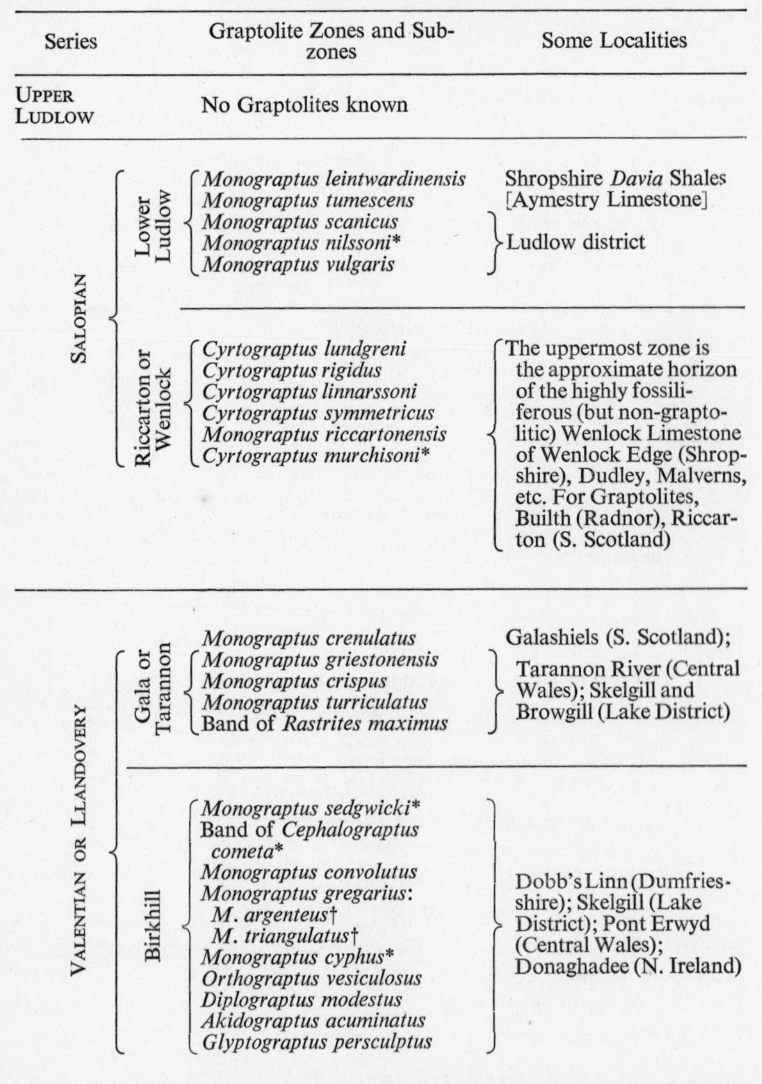

Series		Graptolite Zones and Subzones	Some Localities
UPPER LUDLOW		No Graptolites known	
SALOPIAN	Lower Ludlow	*Monograptus leintwardinensis*	Shropshire *Davia* Shales
		Monograptus tumescens	[Aymestry Limestone]
		Monograptus scanicus	Ludlow district
		*Monograptus nilssoni**	
		Monograptus vulgaris	
	Riccarton or Wenlock	*Cyrtograptus lundgreni*	The uppermost zone is the approximate horizon of the highly fossiliferous (but non-graptolitic) Wenlock Limestone of Wenlock Edge (Shropshire), Dudley, Malverns, etc. For Graptolites, Builth (Radnor), Riccarton (S. Scotland)
		Cyrtograptus rigidus	
		Cyrtograptus linnarssoni	
		Cyrtograptus symmetricus	
		Monograptus riccartonensis	
		*Cyrtograptus murchisoni**	
VALENTIAN OR LLANDOVERY	Gala or Tarannon	*Monograptus crenulatus*	Galashiels (S. Scotland);
		Monograptus griestonensis	Tarannon River (Central Wales); Skelgill and Browgill (Lake District)
		Monograptus crispus	
		Monograptus turriculatus	
		Band of *Rastrites maximus*	
	Birkhill	*Monograptus sedgwicki**	Dobb's Linn (Dumfriesshire); Skelgill (Lake District); Pont Erwyd (Central Wales); Donaghadee (N. Ireland)
		Band of *Cephalograptus cometa**	
		Monograptus convolutus	
		Monograptus gregarius:	
		M. argenteus†	
		M. triangulatus†	
		*Monograptus cyphus**	
		Orthograptus vesiculosus	
		Diplograptus modestus	
		Akidograptus acuminatus	
		Glyptograptus persculptus	

* Zones also recognized in Scandinavia and (nearly all) in Brittany. The remarks made on the Ordovician zones apply here also. † Subzones.

13. ORDOVICIAN AND TREMADOC GRAPTOLITE ZONES

After Lapworth, Elles and Wood

Series		Zones	Faunas	
ASHGILL	Hartfell	*Dicellograptus anceps* *Dicellograptus complanatus**†	Diplograptid Fauna	Orthograptid-Dicellograptid Subfauna
CARADOC	Hartfell	*Pleurograptus linearis**† *Dicranograptus clingani** *Climacograptus wilsoni*	Diplograptid Fauna	
	Glenkiln	*Climacograptus peltifer* and *Diplograptus multidens* *Nemagraptus gracilis**†	Diplograptid Fauna	Nemagraptid-Dicellograptid Subfauna
LLANDEILO		*Glyptograptus teretiusculus*	Diplograptid Fauna	Glyptograptid-Dicellograptid Subfauna
LLANVIRN		*Didymograptus murchisoni* *Didymograptus bifidus*†	Diplograptid Fauna	
ARENIG		*Didymograptus hirundo* *Didymograptus extensus*	Dichograptid Fauna	
TREMADOC (Transition to Cambrian)		*Clonograptus*† *Dictyonema flabelliforme**†	Anisograptid Fauna	

Notes.—Zones marked * occur in the same sequence in Scandinavia as in Britain, and those marked † in North America. The identity of other zones may be masked by difference of name; or greater or lesser thickness of sediment or abundance of fossils may lead to a larger or smaller number of zones being recognized; or zones may be wanting through non-sequence. Faunas are as defined by Bulman, O. M. B., "The Sequence of Graptolite Faunas," *Palaeontology*, i, 170–1 (1958).

14. CAMBRIAN ZONES

Series	Stages	Zones	Some Localities
TREMADOC[1] (Transition to Ordovician)		[*Angelina sedgwicki* Beds]	Merioneth only
		Shumardia pusilla	Shineton Brook (Shropshire); Penmorfa (Merioneth)
		[Brachiopod Beds]	Shropshire only
		Clonograptus tenellus	Shropshire
		Dictyonema flabelliforme	Shropshire; Pedwardine (Herefordshire); Ogof-ddu (Merioneth) where it includes *Niobe* Beds
UPPER CAMBRIAN	Dolgelly	*Peltura*, *Sphaeropthalmus* and *Ctenopyge*	Malverns; Ogof-ddu; Shropshire
		Leptoplastus and *Eurycare*	
		Parabolina spinulosa and *Orusia lenticularis*	Ogof-ddu; Shropshire
	Ffestiniog		Harlech dome only
	Maentwrog	*Olenus* and *Homagnostus obesus*	Dolgelly; Warwickshire
		Agnostus pisiformis	Not recognized in Britain
MIDDLE CAMBRIAN[2]	Menevian	*Paradoxides forchhammeri*	Shropshire; St. Tudwal's (Merioneth)
		P. davidis	Porth-y-rhaw (Pembrokeshire)
		P. hicksi	Pembrokeshire; Shropshire; Warwickshire
	Solva	*P. aurora*[3]	Warwickshire; Pembrokeshire
		Conocoryphe solvensis[3]	Pembrokeshire only
		P. harknessi[3]	Pembrokeshire only
LOWER CAMBRIAN		*Protolenus*	Shropshire; St. Tudwal's (Merioneth)
		Callavia or *Olenellus*	Shropshire Scotland

[1] When the Ordovician System was established by Lapworth in 1879 he excluded the Tremadoc which he left in the Cambrian. Most geologists in Britain retain the Tremadoc in the Cambrian.

[2] In Scandinavia the Middle Cambrian is now divided into (3) *Paradoxides forchhammeri*, (2) *P. paradoxissimus* (formerly called *P. tessini*) and (1) *P. oelandicus* stages, each with one or more zones. The *P. davidis* and *P. hicksi* zones are correlated with the *P. paradoxissimus* Stage. Stage names, however, are usually given a prefix based on a geographical rather than a fossil name.

[3] These zones are known only in South Wales except that the *P. aurora* fauna is known in Warwickshire where it is underlain by the *P. oelandicus* zone; this latter zone has also been recognized in Shropshire.

INDEX